Practical Troubleshooting with the Advanced Video Analyzer

Practical Troubleshooting with the Advanced Video Analyzer

Robert L. Goodman

TAB Books
Division of McGraw-Hill, Inc.
Blue Ridge Summit, PA 17294-0850

FIRST EDITION
FIRST PRINTING

Library of Congress Cataloging-in-Publication Data

Goodman, Robert L.
Practical troubleshooting with the advanced video analyzer / by Robert L. Goodman.
p. cm.
"The SENCORE VA62A Video Analyzer and accessory units are the test instruments used"—Introd.
Includes index.
ISBN 0-8306-4353-2 (pbk.)
1. Television—Repairing—Handbooks, manuals, etc.
2. Videocassette recorders—Repairing—Handbooks, manuals, etc.
3. Signal generators—Handbooks, manuals, etc. I. Title.
TK6642.G653 1993
621.388'87—dc20 92-43107
CIP

Acquisitions editor: Roland S. Phelps
Editorial team: Steve Bolt, Executive Editor
Sally Anne Glover, Editor
Production team: Katherine G. Brown, Director
Ollie Harmon, Typesetting
Susan E. Hansford, Typesetting
Brenda S. Wilhide, Layout
Lorie L. White, Proofreading
Ruth Gunnett, Computer Artist
Janice Stottlemyer, Computer Artist
Jodi L. Tyler, Indexer
Design team: Jaclyn J. Boone, Designer
Brian Allison, Associate Designer
Cover design: Holberg Design, York, Pa.
Cover photography: SENCORE

EL1
4358

Contents

Acknowledgments *xii*

Introduction *xiii*

1 Operating features of the Sencore VA62A Video Analyzer *1*

Video patterns *3*
IF generator *5*
Sound IF generators *5*
Drive signals *5*
Ringing test *6*
Video analyzer operation *6*
Crosshatch pattern *7*
Chroma bar sweep *8*
Multiburst bar sweep *9*
Interlace adder switch *10*
VIR adder switch *10*
Selecting channels *12*
IF troubleshooting generator *14*
Understanding functional analyzing *16*
Signal substitution *19*
Divide-and-conquer technique *19*
Confirming the symptoms first *23*
Finding the real-world problem *23*

2 TV and VCR tuner servicing: VIRS signals *28*

RF-IF generator section *28*
Standard TV channels *28*
Standard cable channels *29*
PROG cable channels *30*
AFT alignment checks *33*
AFT operations check *35*
Tuner substitution *35*
Use as a tuner subber *36*

MATV and CATV systems check *36*
The VIR test signal *37*
VIR sensor *40*
VHF electronic tuner *43*

3 Troubleshooting video amplifiers *47*
Video amplifier comments *47*
Bar and staircase patterns *48*
Locating faulty video stages *48*
Video amplifier frequency response check *51*
Setting the 3.58-MHz trap *52*
Troubleshooting direct-coupled (dc) video circuits *52*
Video circuitry *54*
Video processing (black-and-white TV) *56*
Poor video problems *59*
Video signal frequency relationship *59*
Using the multiburst bar sweep for IF circuit tests *63*
How the video amplifiers affect the picture *64*
How to test response by looking at the CRT *65*
Isolating problems *66*

4 Chroma circuit alignment and troubleshooting techniques *67*
Special patterns *67*
Multiburst bar sweep *68*
Staircase pattern *69*
IF alignment tips *70*
Trap alignment notes *72*
Chroma bar sweep alignment and checks *72*
Pinpointing chroma bandpass troubles *73*
Chroma troubleshooting *75*
Color killer analysis *77*
Chroma demodulator checks *77*
Color circuit adjustments *79*
Chroma processing circuit (chip type) *79*
Color circuit analysis and alignment *81*
Chroma bandpass alignment *82*
Color killer adjustment *85*
Introduction to comb filters *85*
Basic comb filter operation *86*
Zenith comb filter circuitry *98*

5 Video IF/video detector systems *100*
Video IF alignment *100*
Sweep and marker comparison *100*

Bar sweep alignment procedures *102*
TV alignment comments *103*
Trap comments *103*
Checking alignment without a scope *103*
IF signal injection checks *104*
Stage gain checks *106*
Setting the IF traps *106*
Special notes on traps *107*
Using the programmable IF generator *108*
Setting the bandpass coils *109*
Black-and-white TV alignment *111*
Tips for "rough-in" IF alignment *112*
IF alignment comments *113*
Alignment information *115*
The "spot align" alignment technique *115*
More alignment comments *116*
Troubleshooting a typical IF defect *116*
SAW filter and synchronous detector comments *118*
Aligning synchronous video detectors *120*
SAW filter alignment *121*
Analysis of the synchronous detector *123*
Cable-shifted channels *124*

6 Troubleshooting TV sync and AGC systems *131*
Sync circuit testing *131*
Specified NTSC sync *135*
Sync circuit applications *136*
Analyzing sync defects *137*
Integrated vertical pulse *137*
Horizontal keying pulse applications *137*
Vertical countdown circuitry *140*
Closed-loop system checks *142*
Closed-loop testing procedure *142*
The AGC (automatic gain control) system *143*

7 Analyzing vertical sweep systems and sandcastle circuits *147*
Zenith vertical countdown system *147*
Vertical sweep analysis *150*
Vertical drive test signal *151*
Testing the vertical sweep stages *151*
VA62A control settings for vertical sweep testing *152*
Vertical module checks (Magnavox T-809) *153*
Vertical sweep circuits *155*
Sandcastle circuit troubleshooting (RCA) *159*

Checking out "no picture" problems *159*
Sandcastle generator signals *160*
Isolating picture problems *161*
Keying pulse path *163*
Flyback pulse path *163*
How to locate retrace line problems *163*
Detailed sandcastle troubleshooting *163*
Isolating "no picture" problems *165*

8 Troubleshooting horizontal sweep systems *168*
The horizontal oscillator *168*
Horizontal drive checks *169*
Horizontal output transistor testing *172*
Horizontal output SCR gate *173*
Horizontal output tube grid *173*
Split and "hot" chassis flyback testing *175*
Horizontal sweep and B+ regulator circuit checks *177*
SCR drive circuit operation *179*
Function of each SCR during sweep operation *181*
SCR sweep-circuit troubleshooting *186*
Zenith electronic sweep system *191*
Horizontal sweep *192*
Chopper/circuit operation *196*
The in-phase driver *199*
The pulse width modulator (PWM) *200*
Safety tips *203*
Zenith electronic power system (HV shutdown operation) *204*
HV shutdown operation *204*
The Zenith "Z" chassis *205*
High-voltage shutdown checks *206*
RCA CTC-108 shutdown circuit checks *207*

9 The NT64 NTSC color pattern generator and VC63 VCR test accessory *209*
The NT64 patterns *209*
Use with a vectorscope *210*
Color frequency response *211*
NT64 "NTSC" color patterns *212*
"NTSC" color pattern comments *213*
The NT64 patterns *214*
VC63 VCR test accessory *215*
Special VCR signals *215*
FM luminance signal *216*
Down-converted color *217*
Special signals cause problems *218*

10 Servicing video tape recorders (VCRs) *222*
Dual-trace scope *222*
Frequency counter *223*
Digital voltmeter (dvm) *223*
Transistor-diode tester *223*
Special VCR adjustment tools and test tapes *223*
VCR magnetic recording fundamentals *224*
VCR signal crosstalk *231*
Guard band elimination *231*
Video head alignment *232*
VCR recording system *232*
VCR playback system *233*
VHS color system *236*
VCR control system *237*
The color-under system *238*
Color crosstalk elimination *239*
Comb filter operation *239*
Test tapes *241*
Troubleshooting the VCR IF stages *244*
IF alignment notes *247*
Troubleshooting the VCR record circuits *247*
The chroma recording circuits *250*
Analyzing VCR color record circuits *250*
VCR playback circuits (luminance) *253*
VCR (chroma) playback circuits *253*
Servicing VCR servos *258*
Servo troubleshooting examples *258*
Introduction to the VCR VC63 test accessory *260*
VC63 basic operation *260*
Special VCR test signals *264*
Play and record mode signal circuits *264*
Jitter-frequency component *267*
VCR servo comments *268*
Video heads and head amplifier circuit comments *268*
Confirming defective video heads *269*
Checking out the record head amplifiers *270*
Analyzing the record head amplifier circuits *273*
Analyzing the playback head amplifier circuits *275*
The wrap-up *277*

11 All-format VCR analyzer *278*
Some starting tips *278*
VCR overview *279*
Luminance troubleshooting *280*
Video head troubles *280*

Testing rotary transformers *283*
Testing four-head VCRs *285*
"Trick" VCR heads *287*
Checking video-head preamps *288*
Headswitcher checks *290*
Testing video stages *293*
Testing the RF modulator *294*
Super-VHS, super-Beta, and hi-8 testing *295*
Troubleshooting the VHS-C format *297*
Chroma troubleshooting *298*
Down-converted chroma *298*
Identifying head-related chroma problems *300*
Troubleshooting chroma conversion problems *302*
Audio troubleshooting in VCRs *308*
Identifying defective hi-fi stereo audio heads *309*
Baseband audio troubleshooting *312*
Servo troubleshooting *313*
Understanding the servos locked test *313*
Understanding the capstan speed error test *316*
Understanding the capstan jitter test *317*
Understanding the drum speed error test *318*
Understanding the drum jitter test *318*
Making up a work tape *319*

12 Sencore VG91 universal video generator and TVA92 video analyzer *323*

VG91 Universal Video Generator *323*
Instrument operation *326*
Video patterns *328*
Window circle *329*
The MTS/audio generator *330*
Computer-automated bus operation *330*
Using the VG91 as a listener *330*
Talker mode *332*
Analyzer applications *332*
Performance testing *332*
Testing IF response *335*
Synchronous detector testing *335*
Video stages *336*
Testing luminance frequency response *337*
Audio troubleshooting tips *338*
Some "sound" troubleshooting tips *339*
IC sound systems *339*
Audio performance tests *339*
Stereo separation test *340*

Introduction to MTS theory *340*
The MTS signal *342*
Generating the MTS signal *342*
Demodulating the composite audio signal *343*
The TVA92 TV video analyzer *344*
TVA92 features *346*
Operation of the VG91 with the VA62A Video Analyzer *347*
Understanding the horizontal output device sub and drive *348*
Using the horizontal output sub and drive *348*
Preparation for sub transistor *351*
Understanding the device sub current readings *352*
Sub current comparison *354*
Using the horizontal output device sub and drive *355*
Troubleshooting with the horizontal output load test *356*
When to use the horizontal output load test *356*
Understanding the mA readout *357*
Isolating dc loading problems *357*
Isolating ac loading problems *359*
Understand the μsec readout *360*
Isolating timing problems *361*

Index ***364***

Acknowledgments

THIS BOOK HAS BEEN IN THE WORKS FOR MANY YEARS AND HAS HAD TECHNICAL input from numerous individuals. The book is authorized by the Sencore Company, and information in this book was taken from the *Sencore News* magazine, Sencore service manuals, video tapes, and my own experience in troubleshooting video equipment using the Sencore VA62A Video Analyzer.

I want to thank the following good folks from Sencore for the information they furnished me and for the opportunity to write this book:

Al Bowden
George Gonos
Paul Nies
Bob Eiesland
Larry Schnabel

Many thanks.

Introduction

THIS BOOK WILL HELP YOU KEEP PACE WITH THE RAPID ADVANCES THAT ARE taking place in video equipment technology. The book thoroughly explains the Sencore VA62A Video Analyzer test instrument and accessory units, which are used for video troubleshooting.

Chapter 1 tells you how to operate the VA62A Video Analyzer, how to make the setup adjustments, and how to make the connections for various test procedures.

Chapter 2 covers TV video and tuner servicing information, while chapter 3 is devoted to testing and troubleshooting the video amplifier systems.

Chapter 4 explains chroma circuit alignment, checks, and troubleshooting techniques. Chapter 5 looks at video IF/video detector troubleshooting, while chapter 6 goes into TV sync and AGC system troubleshooting.

Chapter 7 covers ways to analyze vertical sweep circuits, and chapter 8 teaches you how to troubleshoot horizontal sweep systems. Next, chapter 9 provides details on using the NT64 pattern generator and VC63 VCR test accessory, which is used with the VA62A Video Analyzer for VCR and color-TV troubleshooting.

Chapter 10 gives advice on VCR servicing, and chapter 11 discusses the new Sencore VC93 All-Format VCR Analyzer. The book concludes with chapter 12, a sneak preview of the Tech Choice System, which covers the New Sencore VG91 Universal Video Generator and the TVA92 TV Analyzer instruments.

1

Operating features of the Sencore VA62A Video Analyzer

THE VA62A UNIVERSAL VIDEO ANALYZER PROVIDES ALL ANALYZING FEATURES needed to effectively service modern video circuits. Signal generators and special tests combine signal substitution and signal tracing for positive tests results every time. Unlike other video generators, the VA62A is more than a performance tester that works only with good systems. The VA62A is designed to isolate problems to a single stage, leaving the video servicer with only a few components to question. The same procedures apply to TVs, video recorders, or other video systems.

The VA62A is designed to resist obsolescence. The functions built into the basic analyzer are common to the FCC-specified signals and the basic blocks of the TV receiver. The VA62A will troubleshoot any video system. Therefore, the VA62A can easily be adapted to special tests and, as video systems change, the VA62A can be easily updated with accessories to meet the changes.

The NT64 NTSC Pattern Generator is an accessory that can be used with the VA62A Universal Video Analyzer to produce NTSC color test patterns. The VC63 VCR Test Accessory can also be used with the the VA62A for troubleshooting videotape machines.

The VA62A consists of five general parts. (See Fig. 1-1.)

1. The video pattern generator.
2. The microprocessor-controlled RF-IF generator.
3. The drive-signal generator.
4. The digital meter.
5. Special analyzing tests.

All of those are part of a single, self-contained unit.

The digital pattern generator develops signals for use by the other sections. All signals are phase-locked to each other, allowing individual signals to be substituted one at a time while the rest of the circuits respond to a reference input signal. Five conventional video patterns, three special analyzing patterns, plus the ability to externally modulate the generator ensure the correct reference pattern to troubleshoot any video problem. Interlace or VIR can be added to any pattern for applications in special circuits.

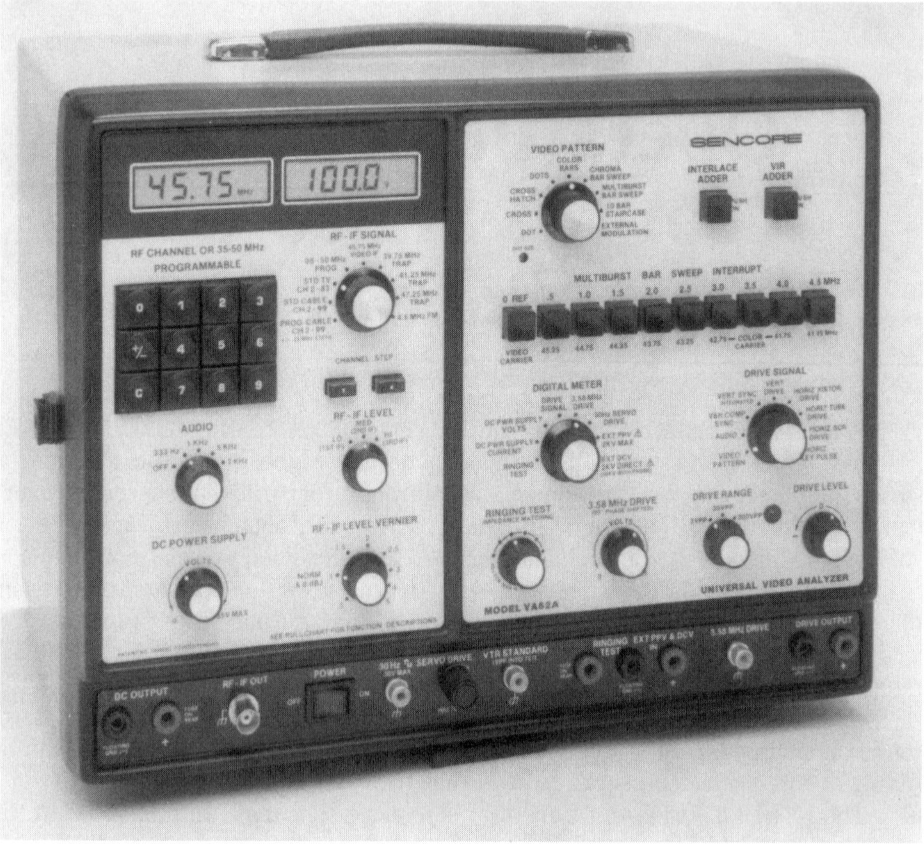

1-1 Front panel controls of the VA62A Video Analyzer.

The RF generator produces 148 channels, covering every possible VHF, UHF, or cable channel. The cable channels can be generated at the designated FCC frequencies or programmed to match any local cable system's carrier offset frequencies for a dynamic tuner test.The RF generator is fully modulated with video information and, to test the audio system, can be modulated with frequency modulated (FM) audio as well. A single microprocessor-compensated attenuator provides flat output across all 148 channels.

The IF generator provides a crystal-controlled, fully modulated carrier or a microprocessor-controlled adjustable frequency generator to troubleshoot or align any IF stage. Special trap-setting signals allow fast and accurate trap alignment without the need of an oscilloscope.

The drive signals provide a known-good signal that can be injected into any tube, transistor, or integrated circuit stage. In addition, special drive signals are available for color or VCR servicing. The drive signals can be injected into any stage without disconnecting components or disabling signals already in the circuit. Special floating grounds prevent damage of interaction due to ground loops. The VA62A is set up for IF tests in Fig. 1-2.

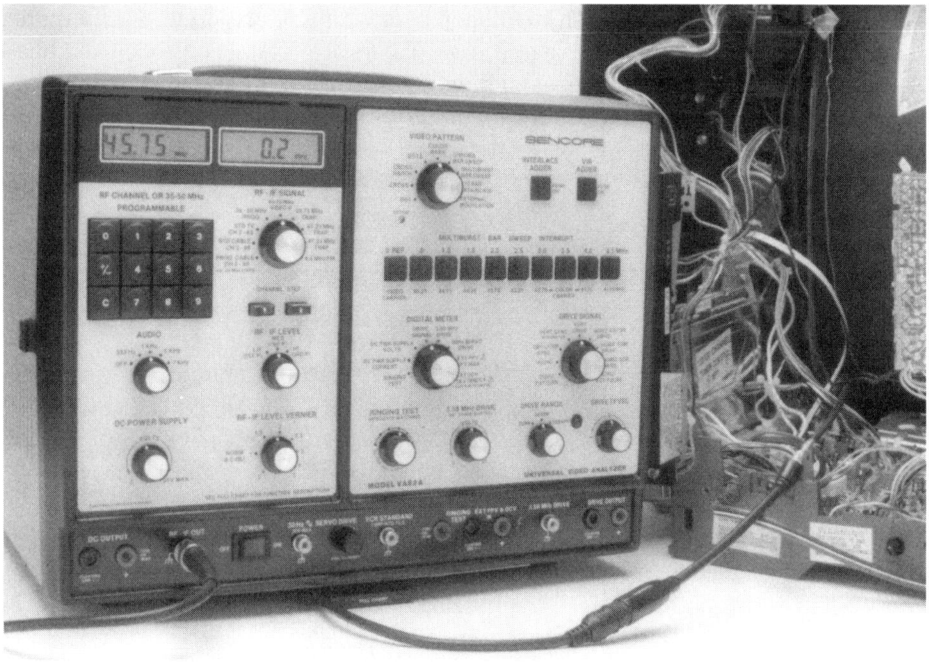

1-2 Sencore VA62A Video Analyzer connected up to IF input for tuner subber check.

The microprocessor-controlled, digital meter provides the ability to monitor any of the drive signals to prevent overdriving circuits and to show signs of loading. The meter also allows fully autoranged external signal measurement of dc and peak-to-peak voltages. High-voltage dc can be measured with optional probes.

The special tests include a highly reliable ringing test of deflection yokes and flyback transformers, a dc power supply for troubleshooting, and special signals to feed directly into a video tape deck. In addition, an accessory jack provides signals to feed optional accessories for special applications.

Video patterns

Digitally generated video patterns can be used to modulate the RF and IF generators, the video pattern position of the drive signal switch, and the VCR standard output.

Dot: Single dot centered between horizontal and vertical blanking intervals.

Cross: Single cross centered between horizontal and vertical blanking intervals.

Crosshatch: 21 vertical lines and 15 horizontal lines.

Dots: 315 white dots.

Color bars: 10 bars, phase-locked, gated-rainbow with true color burst.

Chroma bar sweep: Standard burst and blanking levels during horizontal sync plus three frequency bars for chroma bandpass alignment of the chroma demodulator stages.

Multiburst bar sweep: Ten frequency bars with modulation levels alternating between pure black and 100 percent white. All bars are crystal controlled and phase locked to each other and to horizontal sync. All bars can be switched in or out on the front panel. Note the bar pattern shown in Fig. 1-3.

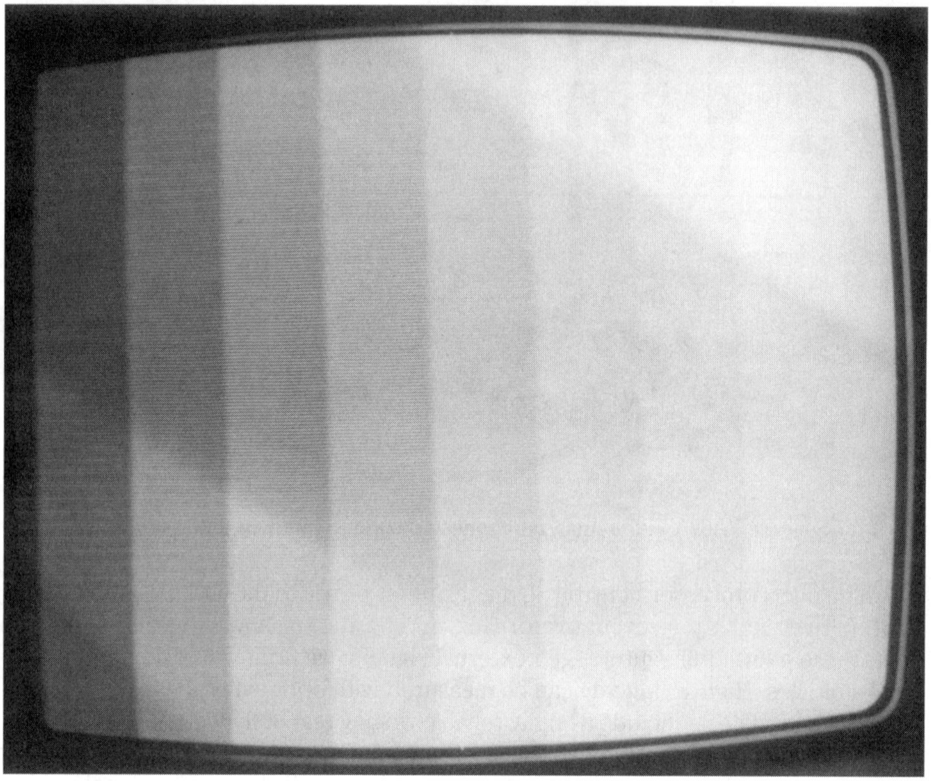

1-3 The 10 chroma bar sweep pattern.

10-bar staircase: Standard EIA 10-level staircase, each within 10 percent tolerance from EIA specifications. Note the 10-bar pattern shown in Fig. 1-4.

External modulation: Maximum 1 VPP input (negative sync); 75-ohm input impedance, frequency range from 0–4.5 MHz.

Interlace adder: All patterns except EXT. MOD. input can be interlaced or non-interlaced with the front panel switch on the VA62A.

VIR adder: A VIR signal can be added to line 19 of all patterns with the front panel switch, except EXT. MOD. RF GENERATOR.

Channels: All standard TV FCC-specified channels. All standard cable TV channels, and programmable cable channels with off-sets.

Channel frequency accuracy: All channels accurate to 400 PPM, within guaranteed temperature range.

Video modulation: Any video pattern selected by the video pattern switch, or with an input from the external modulation jack.

1-4 Multiburst 10-bar test pattern.

Sound modulation: 4.5 MHz ±1 kHz sound carrier ±15 kHz FM deviation, modulates RF or IF. Carrier can be turned off with the audio switch and can be tone modulated by the four frequencies of the audio switch when the switch is on.

IF generator

Video IF: 45.75-MHz, ±10 kHz, crystal-controlled carrier. Also, a 35- to 50-MHz programmable IF provision.

Sound IF generators

Crystal controlled, ±1-kHz FM carrier. Output level: Maximum 50 mV; ±4 dB at 500 mV. Modulation: Tone modulation selected by the audio switch when turned on and unmodulated with the audio switch off. Typical FM deviation: ±15 kHz.

Drive signals

All drive signals, except audio, are phase-locked to the video signal modulating the RF and IF generators. All signals can be monitored with the internal meter to determine the peak-to-peak levels. All signals are continuously variable and have plus or minus polarity. A "0" detent is provided on the drive-signal-level vernier.

Frequency response: Using 10-bar multiburst, standard rolloff of frequencies above 2.5 MHz, comparable to properly operating receiver tuner and IF strip.

Signals available: Video pattern, audio, vertical and horizontal composite sync, integrated vertical sync, vertical drive, horizontal transistor drive, horizontal tube drive, horizontal SCR drive, and horizontal keying pulse.

3.58 MHz drive: Output of 30 VPP to less than 1 VPP.

30 Hz servo drive: Output of 0 to 30 VPP, phase-locked to vertical sync.

Ringing test

Dynamic test of coil Q is determined by counting the number of ringing cycles before reaching a preset damping point. The preset damping point is set at 25 percent of the excitation pulse. The number of cycles is displayed on the digital display and continually updated.

Other VA62A features

- dc power supply current.
- dc power supply volts.
- External PPV.
- External dcV.
- Protection: 2000 V (dc + peak ac), across the inputs.
- dc power supply voltages.
- VCR standard output of 1 VPP.

Video analyzer operation

Let's now look at various features of the VA62A and how to operate them. This will include general operation and setup, operating the signal generators, and some special tests and features.

Isolated outputs

Three of the VA62A outputs are isolated from the VA62A chassis and from each other. This isolation allows connection of the VA62A common connection to some point other than circuit ground. Each floating output must be connected to the circuit through its own ground connection—you can't use one ground connection for all VA62A outputs. You can connect all of the common leads to the same test point in the circuit if all the signals are to be referenced to the same circuit potential.

Video patterns

The *video pattern generator* is the heart of the VA62A. The pattern selected by the video pattern switch is fed to the RF-IF generator, the drive signal generator, and VCR standard output jack. All signals are automatically phase locked to each other because they come from a common source. This allows individual signals to be injected into a video system while a reference signal is fed to its input. The system returns to normal (or near normal) operation when the substitute signal is injected at the correct point because the phase-locked substitute signal duplicates the signal normally found in the circuit.

The first six patterns are used for work in the color circuits. These patterns include the standard convergence patterns and two special patterns (color bars and chroma bar sweep) to test the color circuits. The next two patterns (multiburst bar sweep and the 10-bar staircase) are for troubleshooting the luminance section of a color system or a black-and-white system. The external modulation position allows any video pattern generator that supplies the correct amplitude and polarity of signal to feed the VA62A outputs.

Dot pattern

The *dot pattern* provides a single dot that's electronically centered in the raster. This dot pattern is useful for setting the static convergence of a color TV receiver.

Cross pattern

The *cross pattern* provides a single horizontal line and single vertical line that cross in the exact center of the raster pattern. This can be used to set the picture at the proper center point. The horizontal line of the single cross pattern should be perfectly straight across the center of the screen.

Crosshatch pattern

The *crosshatch pattern* is used to set the linearity adjustments necessary for proper reproduction of a transmitted picture. These adjustments include vertical height, vertical linearity, and pincushion. This pattern has 21 vertical and 15 horizontal lines.

The crosshatch pattern is used for the entire convergence procedure on "one-gun" CRTs that have no electronic convergence controls. Convergence of these CRTs is performed by physical movement of the deflection yoke up and down or side to side, and adjustment of permanent "4-pole" or "6-pole" magnets.

The *dots pattern* provides a standard dynamic convergence pattern for most receivers. The makeup of the dots pattern is identical to the crosshatch pattern, except that the crossover points are shown without the interconnecting lines. When the dots are all pure white, without any color fringing, the convergence adjustments have been set properly.

Both the dots and crosshatch patterns should be examined carefully for any ringing or ghosts. If either of these conditions exist, the IF alignment and the video amplifier system should be checked out.

Dot size control

The *dot size control* adjusts the width of the dots and the vertical lines produced in the crosshatch pattern. Any frequency restrictions in the IF strip or the video amplifiers might cause the dots or vertical lines to disappear. If this is the case, the defective stages must be repaired before the pattern can be used effectively.

Color bar pattern

The *color bar pattern* is similar to the industry-standard 10-bar gated-rainbow pattern used for troubleshooting and alignment of most color TV receivers. The VA62A color bars pattern, however, has three improvements compared to the pattern produced by conventional color bar generators. These improvements expand the uses of this pattern without affecting conventional applications.

The first improvement is that the color information is phase-locked to the horizontal sync information. This allows the color bars pattern to be used in video systems using comb filters, which require the proper phase relationship between chroma and sync to separate color from luminance. Conventional color bar generators have a random phase relationship between sync and chroma.

The second improvement is that the VA62A provides a true color burst that's at the correct amplitude and position on the blanking pedestal and agrees with NTSC specifications. A conventional color bar generator simply inserts an extra color bar in the position normally occupied by the blanking and color burst. The true burst signal is needed for correct operation of the chroma circuits in some systems.

Finally, the VA62A pattern produces a larger amplitude color signal than conventional generators. The amplitude matches the cyan bar of an EIA color pattern (sometimes called an NTSC color pattern). Many procedures reference the EIA cyan bar amplitude. The VA62A color bars pattern can be used directly in place of the EIA color bar pattern in these applications.

Chroma bar sweep

The *chroma bar sweep* pattern is used for troubleshooting or alignment of chroma circuits in TV receivers or for setting the electronic alignment for record and playback of VCR machines.

The chroma sweep bar consists of three frequency bars set at 3.0 MHz, 3.5 MHz, and 4.0 MHz. The center bar is a 75 percent saturated cyan bar, meeting NTSC requirements. The left and right edges of the pattern are 100 percent white. (See the chroma bar sweep pattern shown in Fig. 1-5.) This pattern arrangement gives the white reference plus the cyan level reference needed for accurate VCR alignment. The peak white reference is crucial in VHS machines that use an average AGC detection system instead of a keyed AGC system.

The chroma bar sweep bars are controlled by the multiburst bar sweep interrupt buttons marked 3.0, 3,5, and 4.0 MHz. The label below the buttons shows that the 3.5 button controls the color carrier. The two adjacent buttons control the left and right bars respectively.

All three bars of this pattern are phased as chroma signals. This is important when working with a comb filter that separates chroma signals from composite video using phase rather than frequency. All three bars of the chroma bar sweep should pass through the chroma signal path if the comb filter is properly adjusted and working

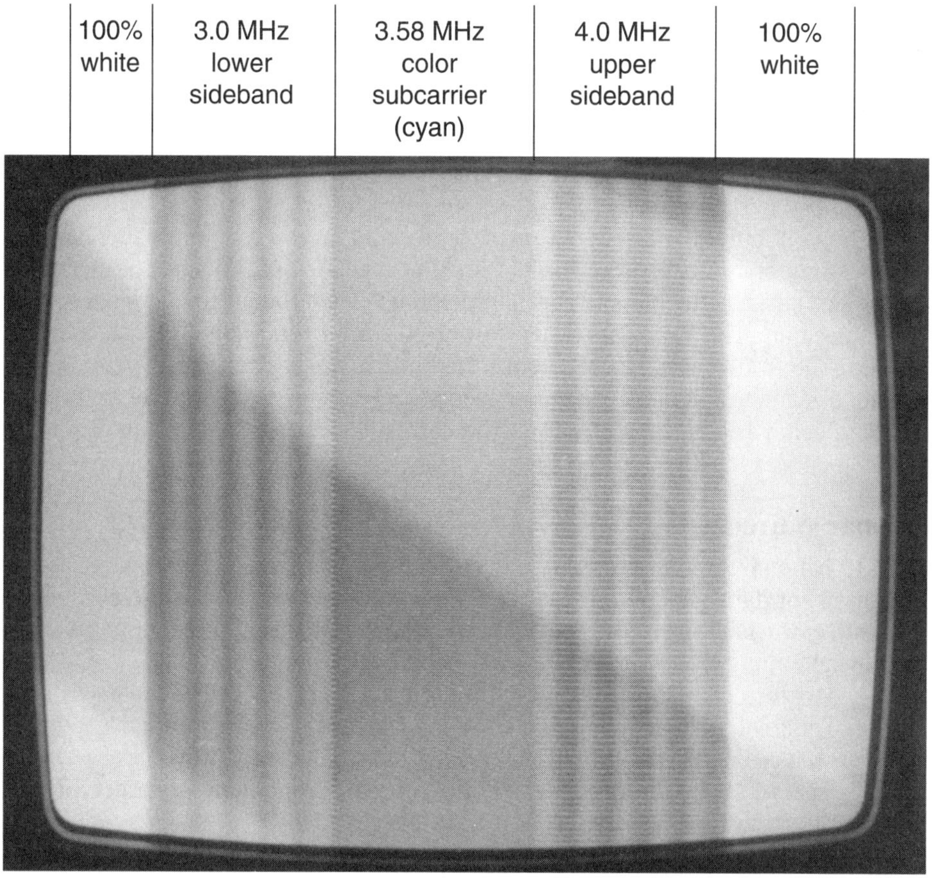

1-5 The chroma bar sweep pattern.

Multiburst bar sweep

The multiburst bar sweep pattern is used for troubleshooting or aligning the luminance circuits in a TV receiver or VCR. The pattern's main use is to isolate video problems related to frequency response. This pattern also provides a reliable method of aligning the video IF stages of a receiver for the best video frequency response with minimum picture distortion.

The multiburst bar sweep pattern consists of ten reference frequency bars beginning with effectively zero (96 kHz) and extending to 4.5 MHz. Each bar is 0.5 MHz higher in frequency than the one before it. This half-megahertz sequence follows the design of video systems.

Each bar is generated at the same amplitude, extending from pure black to peak white. This provides a dynamic test of each video stage through its entire range of operating frequencies. Any stage that restricts frequencies will reduce the amplitude or distort the shape of one or more of the bars.

Each bar is phase-locked to horizontal sync so the bars follow a predictable path through high-resolution comb filter systems. Nine of the ten bars are phased as luminance signals and should follow the luminance signal path. The remaining bar (the 3.5 MHz color carrier bar) is phased as a chroma signal to follow the chroma path in delay-line comb filters. (Charged-coupled-device, or CCD, comb filters require a color burst signal to properly separate chroma. The 3.5-MHz bar might not be properly combed in these systems.)

The multiburst bar sweep interrupt buttons allow you to identify any bar or to remove any of the bars from the pattern for special applications. The horizontal space occupied by the bar switches to pure black when the button is released to its "out" position. The VA62A produces a pure black raster when all buttons are released. The frequencies marked above the buttons are the *baseband frequencies*, or the frequencies of each bar in a demodulated composite video. The frequencies marked below the buttons represent the corresponding sideband frequency in the video IF stages for each bar.

10-bar staircase

The 10-bar staircase pattern is used to troubleshoot luminance problems related to amplitude or dynamic operating range. The pattern can also be used to align stages that affect the linearity of the composite video signal, such as the tuning adjustment of a synchronous video detector.

The pattern consists of 10 levels of brightness equally spaced between black and white. This results in 10 identifiable levels of brightness on a properly operating video receiver or monitor screen. An oscilloscope will show a waveform having 10 equally spaced steps. Circuit defects in the receiver or monitor can affect any portion of the stair-step linearity.

Interlace adder switch

The interlace adder switch allows any video pattern to be generated either fully interlaced or noninterlaced. The interlaced signal is necessary for troubleshooting vertical countdown systems that automatically select one of two operating modes for standard NTSC signals from a TV station or video tape; the interlaced signal is also necessary for nonstandard sync signals from video games or low-cost cameras. Some VIR-controlled receivers need a fully interlaced signal to properly identify the VIR signal. All VA62A patterns will be interlaced when the pushbutton is pressed in or noninterlaced when the button is released to the out position.

VIR adder switch

Some receivers use the VIR signal transmitted from the TV station to automatically control the color saturation and the tint phase of the picture. The VA62A provides a dynamic test for VIR systems. This signal is added to line 19 of the vertical blanking interval. The interlace adder must also be switched on for proper detection in both fields of the interlaced signal for most VIR-controlled receivers.

RF-IF generator

The RF-IF generator provides all the signals needed to inject before the audio or video detectors. The wide assortment of signals allows maximum troubleshooting effectiveness in the RF and IF stages. All signals are fed through a single attenuator and output jack to keep bench clutter to a minimum. Digitally preset functions combine with digitally programmable functions to greatly extend the uses of the generator for troubleshooting or alignment. Figure 1-6 shows a technician setting up an RF frequency.

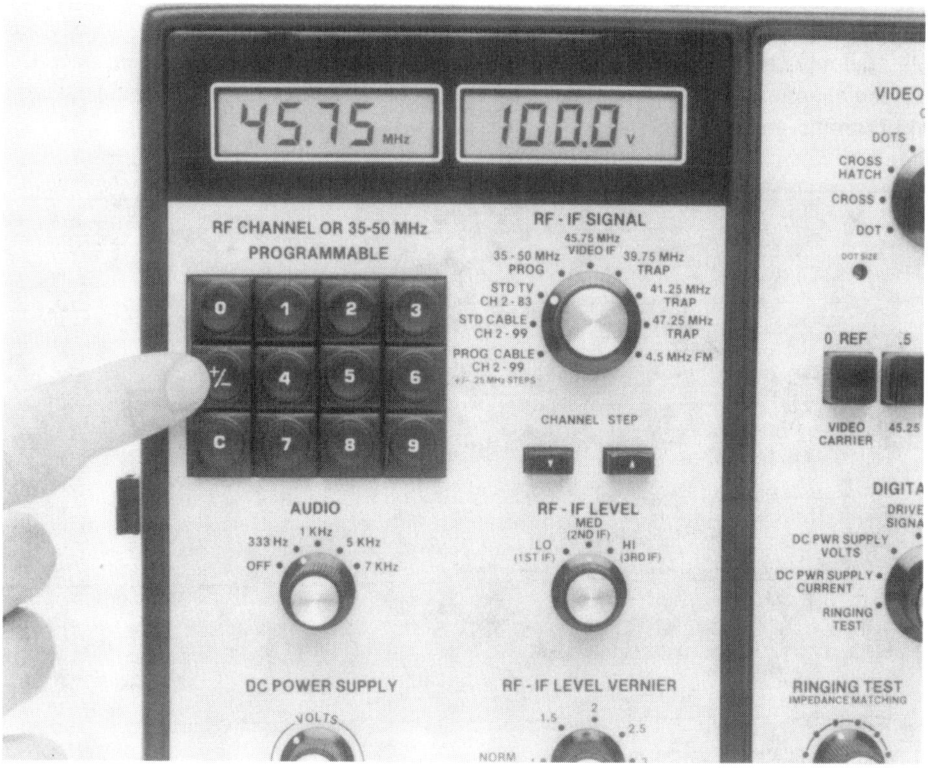

1-6 RF frequency being set up on the VA62A.

RF generator

The RF generator has three functions that closely match the functions of a controlled tuner. The first function provides standard VHF and UHF channel frequencies to test either conventional or digital tuners. The second function replaces UHF channels with cable channels. The VHF channels in this mode are the same as in the first mode. The third mode (or function) is the same as the second, except that any of the selected channels can be moved above or below the standard FCC frequencies to match frequency offsets used by many cable systems.

RF attenuator

The output attenuator affects all three RF generator functions the same. The output level is fully adjustable from 5 microvolts to 5000 microvolts (5 millivolts) in three steps. The attenuator is microprocessor controlled.

Note: The signal levels for the IF generator outputs are 100 times larger than the RF generator for each of the settings of the RF-IF level switch positions.

Selecting channels

Any of the RF channels can be selected using either of two methods. Channels can be entered randomly by entering the two numbers corresponding to the desired channel into the keyboard, or channels can be stepped up or down in sequence using the channel step buttons. Figure 1-7 illustrates the cable channels being entered in programmed steps.

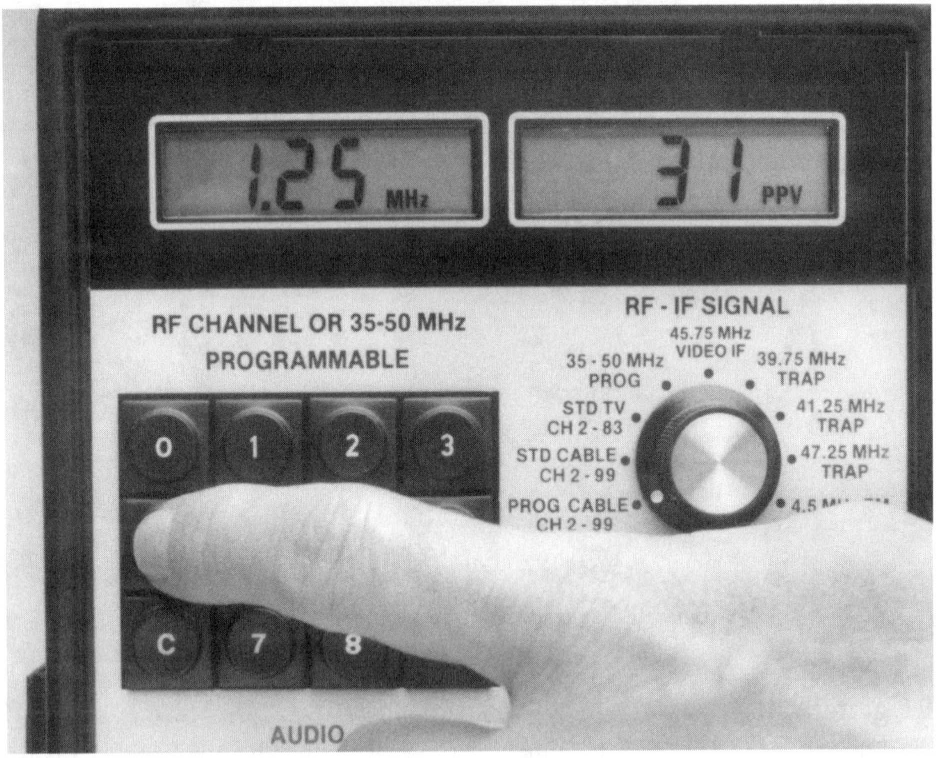

1-7 Programming the cable channels on the VA62A.

Both digits of the channel number must be entered when using the keyboard method. This means that the number zero must be entered before the channel number when selecting channels 2 through 9. The digital display guides you as you change numbers. For example, if you want to change the RF generator from channel

12 to 9, you'd begin by pressing the zero key. The left-hand digital display will then show "0-," indicating that number zero has been accepted and that one more number is needed to complete the program. The RF generator tunes off channel as soon as the first digit is entered. Pressing the 9 key causes the channel display to read "09." The generator immediately moves to channel 9.

Pressing the clear (C) button clears both digits from the RF generator. This is especially helpful if you forgot to enter the leading zero. The RF generator will automatically clear itself if you enter an invalid number. The display shows the same "-" as if you had cleared the generator manually, indicating that you must enter both digits to return to a valid channel.

The automatic return memory is erased each time the VA62A is turned off. Both automatic return memories will always start at channel two when power is first applied.

Any channel produced by the VA62A will be modulated with the video pattern selected with the video pattern switch. In addition, you have the option of adding a sound carrier (4.5 MHz greater than the video carrier) that's frequency modulated with any of the four audio frequencies. The frequencies are selected with the audio switch. When the audio switch is in the off position, there's no 4.5 MHz sound carrier or sound modulation. This provides clear patterns, as there's no beat between the sound carrier and the video carrier or its sidebands. When the audio switch is moved to one of the four frequency positions, you will hear a tone at the speaker of the TV receiver if the sound stages are all operating correctly.

Temporary direction shift

Certain applications might require that you test a given channel at frequencies greater than or less than the assigned FCC frequency. For example, you might need to know if the AFC alignment provides an equal amount of pull in both directions from the center frequency. The VA62A lets you change the direction of the shift on any channel without reprogramming the shift memory. The only limitation you have is that the amount of the carrier shift must be the same in both directions. For example, a channel programmed for a +.75 MHz shift might be alternated between the programmed +.75 MHz and a –.75 MHz shift.

The temporary shifting uses the special function "±" key. After the first time it's pressed to enter the program mode, the "±" key causes the direction of the shift to alternate each time it's pressed. The direction of the shift will be indicated by the polarity sign to the left of the shift frequency displayed in the digital display. The shift memory will not be affected unless a new three-digit shift number is entered after the RF generator is in the programming mode. If you want to permanently change the direction of the programmed shift, simply reenter the three digits representing the shift after the desired polarity is shown.

IF generator

The VA62A IF generator can be broken into four main functions:

1. The programmable generator.
2. The crystal-controlled, troubleshooting generator.

3. The special trap-setting signals.
4. The sound IF generator.

Each of the four sections have different methods of applying audio or video modulation that provide extra testing capabilities for a variety of applications. All signals are fed through the same attenuator and output jack as those used for the RF generator.

Programmable IF generator

The 35- to 50-MHz PROG position of the RF-IF signal switch provides a single carrier at any frequency between 35 and 50 MHz. The carrier can be programmed to within .01 MHz (10 kHz) by entering the required frequency to the keyboard. The function has an automatic return memory so you can move to any other function of the RF-IF signal switch and return to the same frequency last entered. The memory will always start at 35.00 when power is first applied to the video analyzer.

The programmable IF generator doesn't have video modulation. The carrier can be amplitude modulated with any of the audio tones selected by the audio switch. When the audio switch is in the off position, the programmable IF generator produces an unmodulated carrier.

Selecting a frequency requires that you enter four digits on the keyboard, as shown in Fig. 1-8. The microprocessor recognizes the digits from left to right as you enter the digits. The decimal point is automatically placed after the second digit. The digital display shows positions that need filling with a "-." Pressing the C clear button or entering an invalid number (such as 25.00 MHz or 51.25 MHz) causes the generator to clear. The digital display will show "--.--," indicating that four digits must be entered before the generator will put out a valid signal. This is only true when a "clear" function is pressed, not when moving from one frequency to another without going through a "clear" function.

The normal sequence (without clearing the display) is to enter the four digits for the next desired frequency. The digital display will show the first number entered in the left-hand position of the digital display as soon as the first button is pressed. The last three places will show a dash, indicating that you must enter three more numbers before the microprocessor accepts the new number. The IF generator continues to put out the old frequency until the fourth digit is entered.

IF troubleshooting generator

The position of the RF-IF signal switch (marked 45.75 MHz VIDEO IF) produces a crystal-controlled carrier that's fully modulated with the video information selected by the video pattern switch. In addition, this signal can have a 4.5 MHz sound carrier added, which can be modulated with any of the four frequencies selected by the audio switch.

The full modulation allows the signal to be injected into any IF stage for troubleshooting or alignment. A picture and (if desired) sound will appear at the receiver output if all video and audio stages work correctly. The RF-IF level switch selects

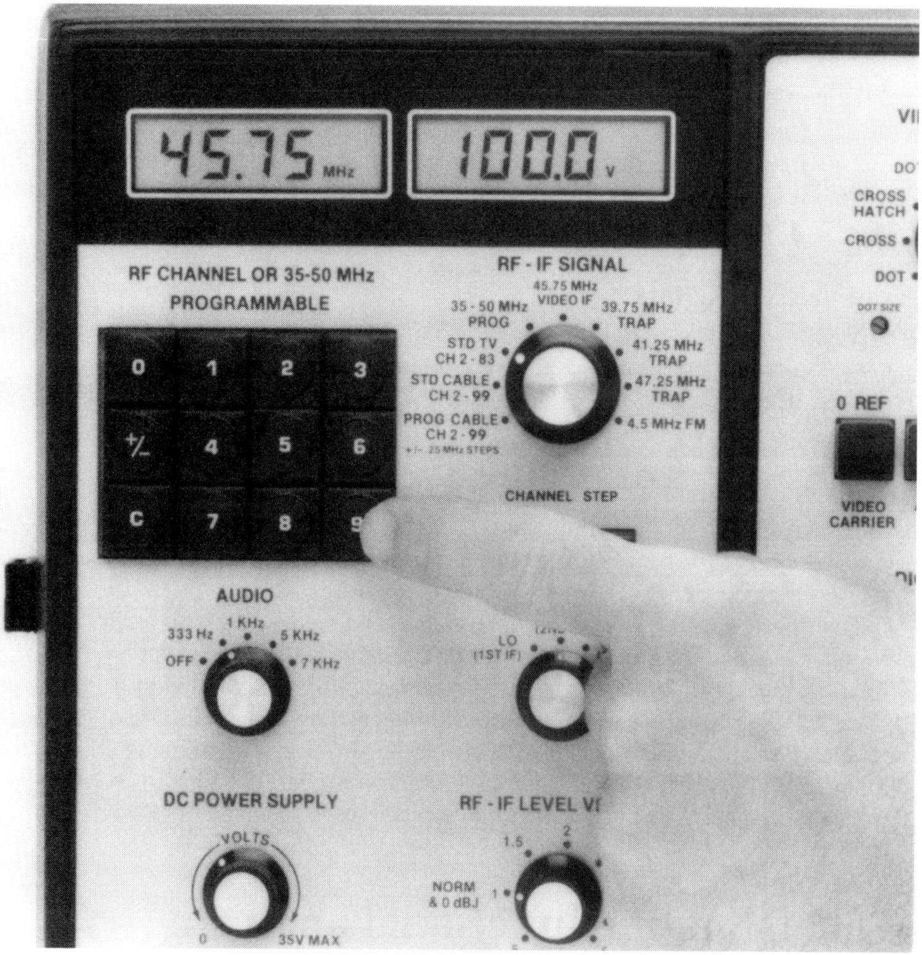

1-8 Programming the standard TV channels.

preset levels to match the levels normally found in the first, second, or third IF stages.The VA62A is supplied with adapters plus a special matching balun to use with the IF generator.

The audio modulation works exactly the same as the audio modulation for the RF generator. When the audio switch is off, there's no 4.5 MHz carrier. When the audio switch is set to any of the four positions marked with a frequency, the 4.5 MHz sound carrier is added to the video carrier, and frequency is modulated with the selected frequency.

Using the IF generator for troubleshooting

1. Move the RF-IF signal switch to the 45.75 MHz video IF position.
2. Set the video pattern switch to the desired video pattern.

3. Connect the RF-IF out cable to the desired test point using an appropriate connector or adapter.
4. Set the RF-IF level switch and the RF-IF Level vernier to the desired level.
5. Set the audio switch for the desired sound modulation.
   - ~ If no sound modulation is desired, set the audio switch to the off position.
   - ~ If sound modulation is desired, set the audio switch to the desired modulating frequency.

Trap-setting signals

The VA62A produces three totally unique signals that greatly simplify the adjustment of the three most common traps found in the video IF:

1. The 39.75 MHz adjacent video trap.
2. The 41.25 MHz sound trap.
3. The 47.25 MHz adjacent sound trap.

The VA62A dynamically duplicates the interference conditions the traps are designed to eliminate. The VA62A produces a fully modulated video carrier along with the interfering carrier that the trap should eliminate.

The interfering carrier can be unmodulated or modulated with any of the four frequencies selected by the AUDIO switch. The video reference carrier is supplied at a fixed amplitude of 1000 microvolts, which holds the keyed AGC at a fixed level that's unaffected by the level of the interfering carrier. The amplitude of the interfering carrier is fully adjustable with the RF-IF level switch and the RF-IF level vernier so that the ratio of the reference signal to the interfering carrier can be set for best contrast. This results in a usable signal for any receiver.

Understanding functional analyzing

There are generally two ways to test a TV receiver. You can signal trace or signal inject. *Signal tracing* is the age-old technique of connecting a scope to a test point and comparing the waveform on the schematic (if you have a schematic and it shows the waveforms). The idea is to move from one end of the circuit to the other until you spot a waveform or reading that doesn't match those shown on the schematic.

The success of this approach depends on your ability to analyze just what that waveform is telling you. This presents several problems. First, you must use a schematic. On many schematics the waveforms are small, dark photos that are difficult to read, and often don't show much detail. Second, your success can be hampered by the small changes in a circuit's operation! Third, you could be led astray, since the dc, peak-to-peak, and general waveshape can all appear to be correct.

A far more successful troubleshooting method is "functional analyzing," where you simply test the function of the suspected stages. If they work, they're good, and that's it; there's no need to measure voltages or check waveforms.

Functional analyzing eliminates the need for detailed measurements and concentrates, instead, on checking the functional operation of the circuits. You break the total TV into functional blocks. Your goal is to test the function of each block and isolate the problem to the defective block. Once you've isolated the problem to one small circuit block, you can find the defective components in a short time. The key to troubleshooting success is finding the defective block.

To find this block you need a known-good substitute signal. The VA62A Video Analyzer supplies all the basic signals needed to troubleshoot TVs and VCRs. You inject the same type of signals from the VA62A as the block normally receives for an input and monitor for proper operation. Let's now put the Analyzer to actual use.

You have a television with a video problem. You first connect the VA62A to the TV's antenna terminals and select the multiburst barsweep pattern. Normally you'd expect to see crisp, clean, vertical black-and-white bars. Instead, you see a washed-out picture with a loss of detail, as shown in Fig. 1-9.

The problem could be anywhere from the antenna to the CRT. Before you begin looking for components, you mentally break the overall TV up into blocks (as large as possible). For example, you can look at the TV as being composed of four functional blocks: the tuner, IF, video stages, and CRT. (See Fig. 1-10.)

There are three places you could inject a test signal in these blocks:

1. Between the tuner and IF stages.
2. Between the IF and video stages.
3. At the CRT.

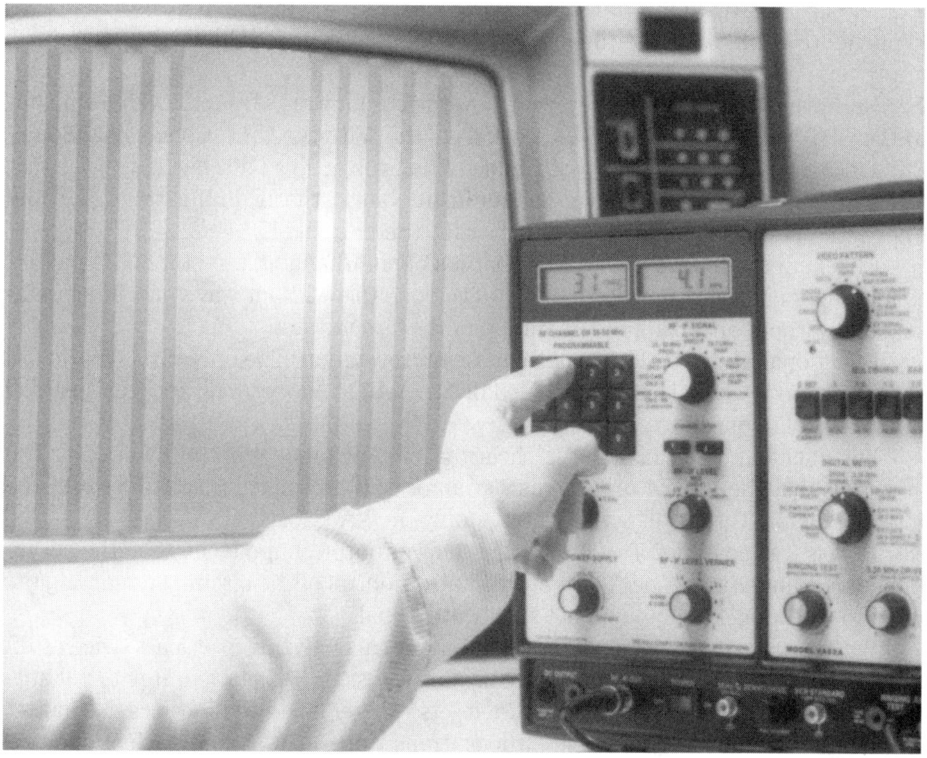

1-9 Washed-out picture with loss of detail.

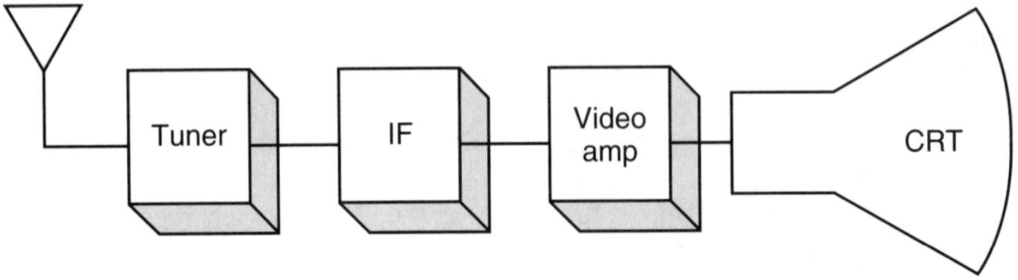

1-10 To isolate the problem, you can break the TV receiver into four blocks, any of which could be causing the problem.

Let's start by injecting between the IF stage and video stage. A good place to start, as you'll see later, is at the output of the video detector. The signal at this point is a video signal. Leave the VA62A Video Analyzer connected to the antenna terminals and select a video pattern. This signal is now identical to the signal that should be seen at the output of the video detector. Injecting this known-good signal at the output of the video detector, you'll see a clear, sharp picture with good contrast. Do you suspect a soft CRT? No! Do you need to do any more testing in the video circuits? No! Have you had to do any guesswork or make detailed measurements? No!

Let's move back one more step and inject a signal into the input of the IFs. Select the 45.75 MHz video IF signal on the VA62A's RF-IF signal switch and connect the RF cable from the tuner input to the input of the IF block . . . There's no improvement.

Now, what does this tell you? You know the circuit works from the video amp out to the CRT, yet the circuit doesn't work from the IF to the CRT. Conclusion: Something in the IF is defective. It's that simple. When you analyze this block, you see that there are only three major components in the block: the IF buffer transistor, the SAW filter, and the IF amp/video detector IC. (See Fig. 1-11.)

Checking the IF buffer, you find that it's open. Changing it restores the picture to its proper contrast and sharpness. You've done this without waveform interpretation and without analyzing the operation of the circuit.

Functional analyzing leaves nothing to interpretation. The only variable is in the signal selected, which is clearly labeled on the VA62A Video Analyzer. You just dial up the signal in question and inject it.

You might ask what happens to the signal already present in the TV circuit? The answer is that the VA62A drive signal "swamps" out any signal that's present at the injection point. The result is that you see only the VA62A signal.

How is this possible? The VA62A has a lower output impedance than the circuit impedance under test. The Analyzer thus takes control of the circuit, eliminating the need to disconnect components to substitute signals.

With rare exceptions, you'll find that functional analyzing is at least twice as effective as signal tracing. In addition to requiring less interpretation, functional analyzing makes all TV receivers look alike. It lets you get back to the basics. No matter what type of components are used, tubes, transistors, or ICs, you use just one universal troubleshooting approach.

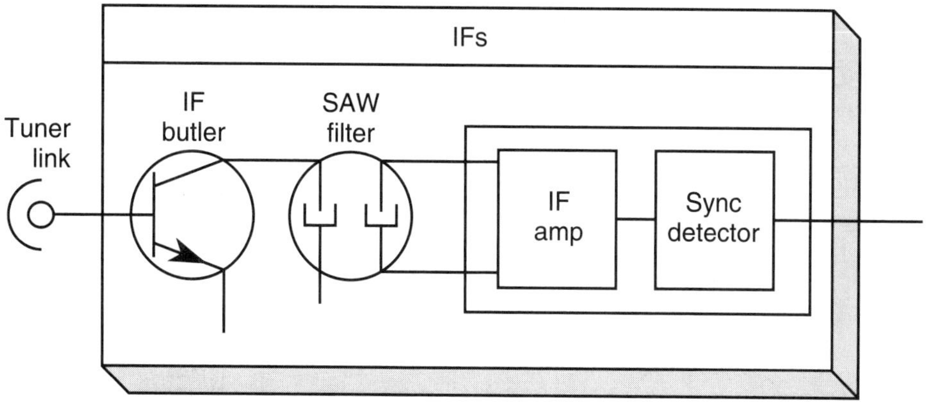

1-11 Once the problem is located in a major functional block, it can be divided into smaller functional circuits for more detailed testing.

Many servicers feel that the greatest benefit to functional analyzing, however, is that when you inject a substitute signal into a circuit, and the set operates, you know conclusively that every circuit is functioning from that point on.

Signal substitution

Signal substitution is a super timesaver in any circuit. However, signal substitution doesn't stop there. It solves one of the toughest problems of today's service technician who's checking ICs. Signal substitution is the only sure way to confirm the operation of ICs and to isolate problems associated with them. The only other alternative is expensive and time-consuming chip swapping. Let's see why this is so.

How do you tell if a chip is good? Many servicers measure the voltages on the various pins. That entails using a meter or scope to look at 14, 24, or even 48 IC pins. However, what if one of the voltages is slightly different than the voltage shown on the schematic? Chances are very good that you'll end up swapping ICs. You'll have to carefully unsolder and resolder every pin on the IC. Also, you've no guarantee that a new IC will solve the problem.

That's why signal substitution is so important. ICs are complete circuits. Each IC performs one or more functions. In fact, many of the chips used in TV sets today perform many functions. Signal substitution, using your VA62A Analyzer, confirms that the IC produces the correct output when a known-good signal is injected.

Divide-and-conquer technique

Let's take signal substitution one step further. If you're not careful when choosing injection points, you can end up jumping all over in your attempts to find the problem. By attacking the problem with a logical, systematic approach called divide and conquer, you become much more efficient.

Each functional block (previously shown in Fig. 1-10), is composed of several smaller functional blocks (shown in Fig. 1-12). You could have started at the driver transistors at the CRT and worked your way back, one functional block at a time. This would have required seven signal injections instead of three. By cutting the problem into halves, you cut your troubleshooting time by much more.

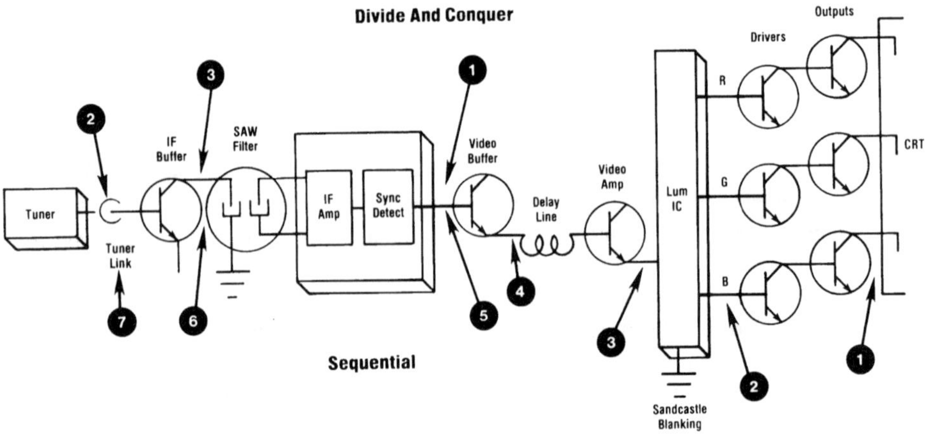

1-12 Divide-and-conquer functional analyzing minimizes the number of signal injections needed to find the defective stage.

With the divide-and-conquer technique, you divide the problem in half, then in fourths, then in eights, and so on. You quickly eliminate large sections of the TV set from suspicion, zero in on the defective block, and finally, find the defective circuit.

Detectors are at halfway point

The final way to make troubleshooting the most efficient is to make the first injection in the right place. Take a look at Fig. 1-13. The detectors cause the greatest change in the signals. The circuit before the detectors processes the signals much differently than the circuits after the detectors. Because of this, detectors are the best place to start the divide-and-conquer process. Injecting at the detectors cuts the number of signals that can cause a problem by at least half.

There are three types of detectors found in TV receivers: The video detector, the audio detector, and the chroma detector or demodulators. The *video detector* changes the amplitude-modulated IF signal into video; the *sound detector* changes the FM sound IF signal into audio, and the *chroma demodulators* change the phase-modulated color signals into the individual red, green, and blue drive signals. Sencore's proven "TV Functional Analyzing/Troubleshooting Guide," along with the TV Functional Block Diagram gives you some helpful points to start your troubleshooting.

Figure 1-14 shows the audio troubleshooting section of the Functional Analyzing/Troubleshooting Guide. It shows you what to look for, based on the symptoms, and where to start your divide-and-conquer signal injections based on the problem.

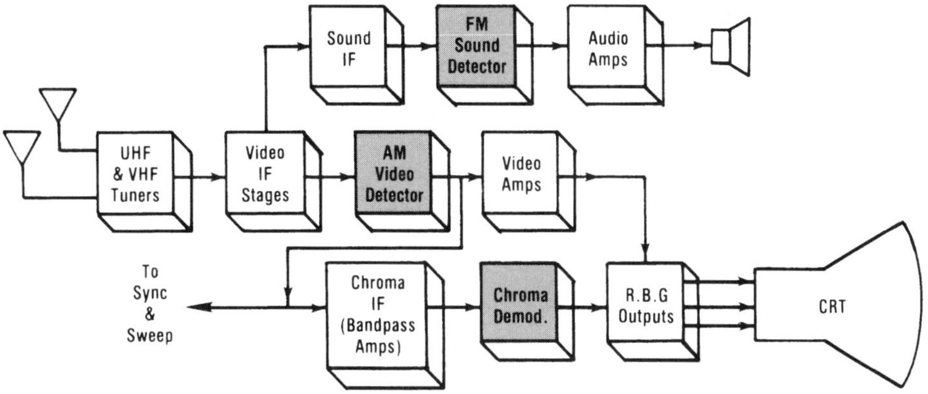

1-13 The detectors are the best place to start divide-and-conquer signal injection because they cause the greatest change in the signals.

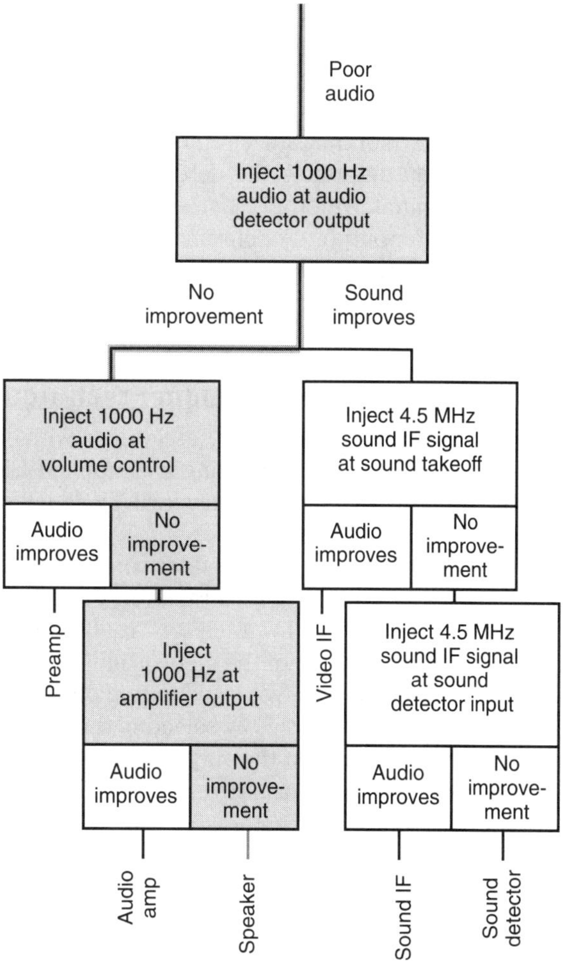

1-14 The TV functional analyzing/troubleshooting guide gives a systematic method of quickly locating the defective stage.

The numbers on the troubleshooting guide correspond with the numbers on the Universal TV Block diagram shown in Fig. 1-15.

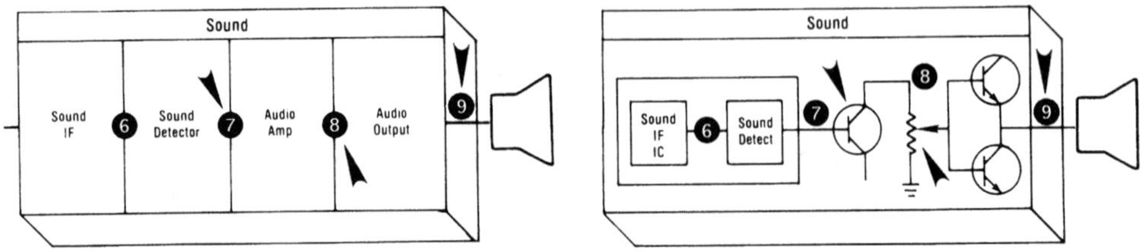

1-15 The TV block diagram has injection points keyed to the troubleshooting guide to help identify key injection points.

Using the guide

The TV Functional Analyzing/Troubleshooting Guide begins by telling you to inject an RF signal from the VA62A into the antenna terminals of the TV receiver. In this step, you look at the entire TV as one complete functional block and confirm that the receiver is truly defective. If it's defective, you'll observe one of the symptoms. Once you identify a specific symptom, you've already narrowed the problem down to about one-sixth of the total circuits.

Next, you determine the detector or other major circuit point that's involved with the general symptoms. For video and sound problems, inject at the video or sound detector output. For chroma problems, inject at the chroma demodulator's input. If the receiver operation returns to normal, you know that the problem is somewhere before the detector. If the operation remains bad, you know that the problem is somewhere between the detector and the output.

Using the divide-and-conquer technique

Let's look at one more example, a television with distorted sound, to see how divide-and-conquer functional analyzing is simplified using the TV Functional Analyzing /Troubleshooting Guide and Universal TV Block diagrams shown in Figs. 1-14 and 1-15.

The first thing you need to do is confirm the symptoms. When you inject an RF signal from the VA62A into the receiver, you confirm that the sound is, indeed, very distorted. Look at the TV Analyzer/Troubleshooting Guide and find the symptom "Poor Audio." A portion of the guide is illustrated in Fig. 1-14.

As the guide shows, you need to find out whether the problem is before or after the sound detector. Do this by selecting the audio drive signal on the VA62A and injecting the drive signal at the output of the sound detector (point 7 in Fig. 1-15). The sound is still garbled. This proves that the problem is between the detector and the speaker output.

When you follow the "no improvement" path of the Guide, you're instructed to inject the same VA62A audio signal into the volume control (point 8 in Fig. 1-15). When you do this, you find that the sound is still garbled. Following the "no improvement" path further, you inject the VA62A audio drive signal at the output of the audio amplifier (speaker leads). The sound is still distorted . . . You've narrowed the problem down to the speaker.

Confirming the symptoms first

The first thing you need to do is determine if there really is a problem in the TV set. Then, if there is, find out what the symptoms are. You can start by filling out a TV performance report card for the set in question. Filling out this report card lets you see exactly what's wrong with the set, plus it shows the customer what was done to repair the troubles.

You start by connecting the RF signal from the VA62A to the set's antenna and selecting the multiburst sweep pattern. On a good TV, you'd see a clean, stable picture with vertical black-and-white bars.

The set I've been describing, however, shows some very interesting symptoms. You can see the video in the picture, but the bottom half seems to be out of sync with the top half. There also seems to be some extra bars and bends present throughout the screen.

Adjusting the VA62A's variable RF levels or switching to different patterns doesn't affect the symptoms very much. A quick check of all the VHF channels and some random UHF channels confirms that the symptoms are similar on all channels. The VA62A's performance test has shown you that the TV's symptom remains the same under all varying conditions.

What has the VA62A told you? Plenty. You know you have a good signal going into the television's antenna, so you can eliminate cable or interference problems. Without a signal standard, you could chase a nonexistent problem all day when the problem is actually in the customer's antenna or cable system.

The VA62A has also told you exactly what the symptoms of the defective set are. By having the ability to choose different troubleshooting patterns and various channels, you've found out that the symptom is consistent for all channels with all patterns. You also know that the symptom is not RF-level related because the symptom didn't change when you adjusted the VA62A's RF level.

Finding the real-world problem

Great. You now know the problem could be anywhere from the antenna terminals to the CRT. That's a lot of space to cover. Let's get started.

Your VA62A can guide you to the defective stage so you don't end up troubleshooting good circuits or wasting time measuring voltages. Using the divide-and-conquer method, you simply inject the VA62A's signal into the TV's midpoint (the video detector) to divide the circuits in half.

The universal TV block diagram (Fig. 1-16) shows you that about half of the circuits are before the video detector, and about half come after. So, with the VA62A still connected to the antenna, you simply drive the detector output with the VA62A's video pattern drive signal and look for an improvement on the TV's CRT. Remember, the VA62A's antenna signal must be left connected to keep the good operating circuits in sync.

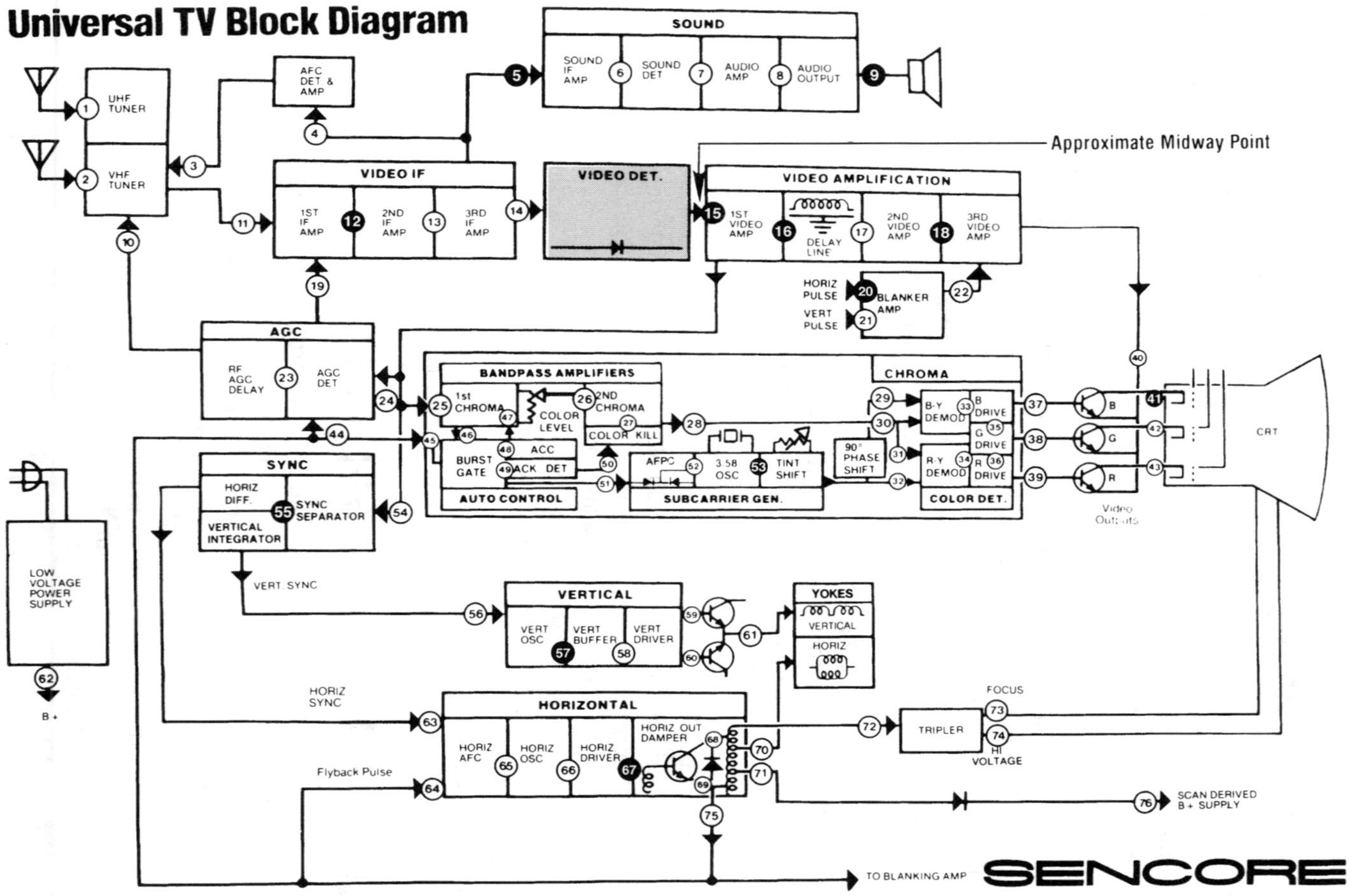

1-16 The universal TV block diagram shows you that the video detector is the approximate midpoint of a typical TV receiver. This makes the video detector an ideal point to begin divide-and-conquer analyzing.

If the symptom improves on the CRT, you know you're injecting into good circuits. If the symptom doesn't change, you're injecting into defective circuits. In either case, you've eliminated half of the TV's circuits with just one injection. You just keep dividing the circuits in half until you have the problem narrowed down to one stage. Then you use conventional troubleshooting methods to find the defective part.

You can see in Fig. 1-17 why you need to keep the antenna connected when you're driving circuits with the VA62A. At point 14, the signal divides into several different paths. In order to effectively troubleshoot any of these signal paths, all of them must be kept in sync with one another. For example, if you're driving point 35 with your VA62A, you must have the VA62A's RF antenna signal hooked up to point 1 in order to keep all of the branch circuits in sync with one another. If you drove point 35 without connecting the RF to the antenna, the video that you drive into point 35 might show up on the CRT, but it will be out of sync and without any color information because there's no reference signal to hold the branch circuits in sync with the injected signal.

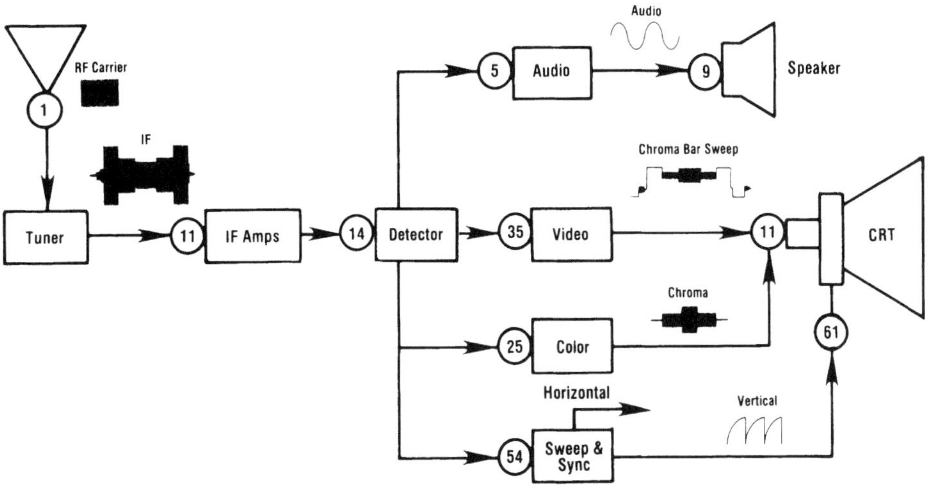

1-17 A simplified TV block diagram shows the parallel paths that must be kept in sync when driving a signal after point C.

A glance at the service information shows that the normal peak-to-peak voltage level at the video detector output on this TV set should be 1.6 volts. Use the VA62A's video pattern drive signal, since this is the signal present at the detector output. You now connect this point in the circuit with the VA62A's drive leads and turn the drive level to 1.6 volts peak to peak, which is monitored on the VA62A's LCD display. There's no need to unsolder any components because the in-circuit signal is swamped out and replaced by the good VA62A drive signal.

This makes it quite easy for you. One injection and you've eliminated over half of the TV set as the source of the problem. You don't have to interpret a waveform or even unsolder or substitute any parts.

You now know the problem is somewhere between the antenna and the output of the video detector. It's now time to eliminate more circuits using your VA62A to divide and conquer them. According to the block diagram, the next logical injection point is at the input to the first IF stage.

Figure 1-17 shows that any point before point 14 doesn't have any parallel paths to keep in sync, meaning you don't need to have the antenna signal feed connected at point 1 when you inject into these circuits. Since you're injecting into point 11, the signal has only one path to take, so all of the circuits will be kept in sync with one another.

Because this tuner has a cable feeding the IF section from the tuner, you can use the VA62A's IF adapter cables to inject the known-good signal into the circuit. This substitute IF signal also cleaned up the picture on the CRT. This indicates that the problem is between the antenna terminals and the first IF stage, which leaves little else but the tuner.

Upon further inspection of the schematic, you notice one power supply voltage feeding the tuner. It should be –15 volts. A quick check with the scope shows the dc voltage is at about –4.6 volts, with a ripple voltage of 40 volts peak to peak. This must be the trouble!

The defective –15-volt line traces back to a single leg off the front end of the power supply. The circuit is straightforward, with rectified ac being filtered by a 4.7 MFD, 250-volt electrolytic capacitor. Based on the amount of ripple on the –15-volt line, you suspect the 4.7 MFD filter right away. A quick check of the capacitor value with the Z meter proves that the capacitor is defective.

In three troubleshooting steps, you've located the defective stage without waveform interpretation or parts removal. Instead, you injected the proper drive signal and watched for a good pattern on the picture tube. In other chapters, we'll go into much more troubleshooting detail.(See Fig. 1-18)

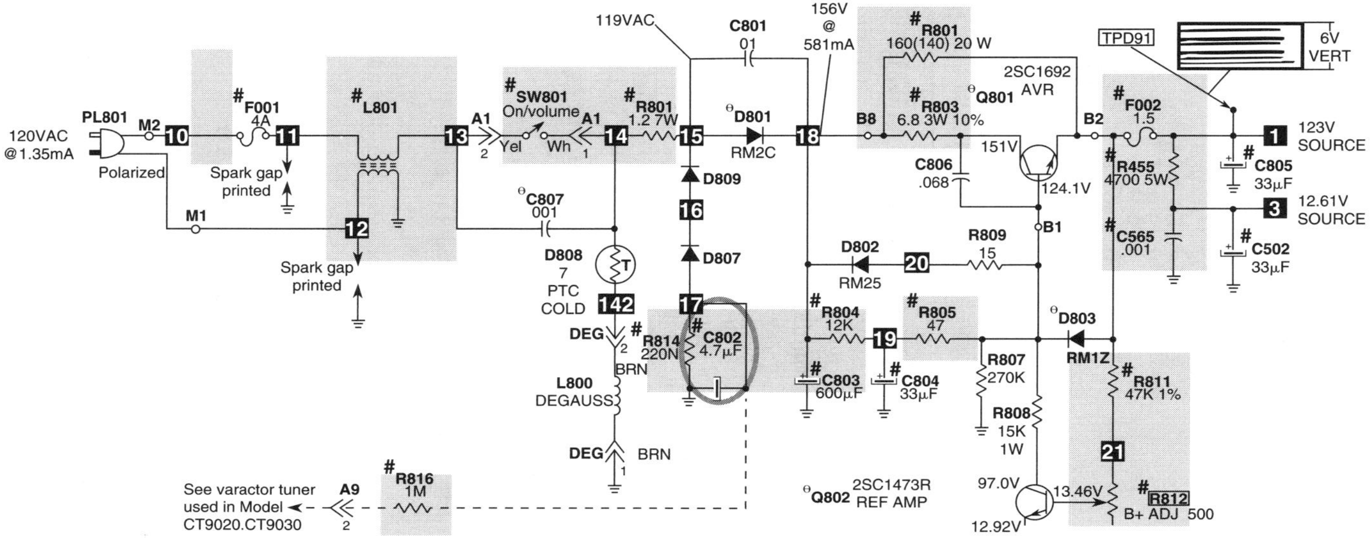

1-18 Capacitor C802 measured open, causing a ripple on the –15 volt line and a distorted signal to be coming from the tuner.

2

TV and VCR tuner servicing *VIRS signals*

THE MULTIBURST SWEEP BAR PATTERN FROM THE VA62A PROVIDES A REFERENCE for accurate testing of VHF and UHF tuner operation and alignment. The Analyzer can be used to align and troubleshoot conventional tuners that contain tubes or transistors as well as the newer electronic tuners that use PLL feedback control systems. Also, where there are minor tuner defects or link mismatch, the multiburst bar sweep signal allows the IF amplifiers to be adjusted to compensate for these faults.

All signals supplied by the VA62A are standard to the TV signal itself as specified by the FCC or EIA. Any special signals, such as signals needed for VCRs, cameras, or video disc systems, will be supplied through accessories that plug into the special connector in back of the VA62A. This prevents the VA62A itself from becoming obsolete if a new format or new customer signal finds wide usage. The interface jack supplies all signals and power supplies needed to accommodate any video or audio accessory dictated by later technology.

RF-IF generator section

The RF-IF generator of the VA62A is tuned by digital push-button operation and readout. The digital system results in much better frequency accuracy and many more functions for RF-IF system troubleshooting. The VA62A also provides a much wider range of signal levels because it has a three-step attenuator that controls the course levels of all RF and IF signals.

Standard TV channels

The RF generator produces any VHF or UHF channel, which you can select with the standard keyboard or by using the "up/down" buttons to step one channel up or down each time you press the button. All carriers are modulated with the video pattern selected by the video pattern switch. Audio modulation can be added by changing the audio switch to any of the four positions marked with a frequency. The selected audio frequency then frequency modulates the 4.5 MHz audio carrier that's added to the video carrier. (Note: To prevent interference with the video signals, the

4.5 MHz sound carrier is automatically turned off when the audio switch is in the off position.

A special feature of the VA62A RF functions is that the internal microprocessor remembers the last RF channel selected as long as power is applied to the Analyzer, no matter what other RF-IF functions are used in between the times that the switch is moved to the STD TV position.

The desired TV channel can be selected in any of three different ways. The procedures are similar to the entry made with most direct address TV receivers:

1. The two-digit channel number can be entered directly. The left-hand digital display shows a "-" in the place of the second digit of the channel number when a second digit must be entered.

   Note: The microprocessor requires two digits to be entered for any channel. Remember to enter a leading zero for channels 2 through 9. If you enter an invalid channel (such as 01 or 96) when using the keyboard, the display will show "--."

2. The "C" button can be pressed first to clear the old channel out of the display. The two-digit number for the desired channel is then entered. Remember to enter the leading zero for channels 2 through 9.

3. The RF generator increases or decreases by one standard TV channel each time the channel up or down button is pressed.

   Note: When using the up/down buttons, the microprocessor automatically "wraps around" the end of the band, jumping from 83 to 2 when stepping up or from 2 to 83 when stepping down.

Standard cable channels

A total of 73 cable channel frequencies are available when the switch is moved to the STD cable position. Channels 2 through 13 are standard FCC-specified VHF channels. Channels 14 through 73 are standard cable channels. The channel numbers for cable channels 14 through 73 agree with the numbered sequence used by most TV receivers that have cable-ready tuners.

Note that the frequencies produced by this position of the RF-IF signal switch are all FCC-specified frequencies. You can switch back and forth between this position and the PROG cable position when it's necessary to alternate between the FCC-assigned frequencies and a programmed cable frequency shift. The microprocessor keeps the RF generator on the same cable channel for the two cable functions, alternating between the unshifted and the shifted carrier.

Cable channel entry procedures are identical to the STD TV functions, including invalid channel display and wraparound between channels 2 and 73. The microprocessor remembers the last cable channel selected in a separate memory location, as with the STD TV channel, so that switching between the STD TV and STD cable functions often shows two different channels. The return memory is erased when the power is turned off.

PROG cable channels

The PROG cable function produces the same cable channels as the STD cable function. The difference is that any of the cable channels can be shifted above or below the FCC-assigned frequencies. This is necessary in order to duplicate the carrier shifts used by your local cable TV systems. There are many different frequency shifting schemes used. The VA62A can be programmed to produce any combination of shifts so that the cable shifting takes place by simply selecting the PROG cable function and selecting the desired cable channel.

The VA62A stores the programmed cable shifts in a special nonvolatile programmable memory. You can change the frequency shifts as often as desired. The memory will not forget the shifts when the power is removed from the Analyzer. The circuit uses a special type of memory IC that doesn't require internal batteries to maintain memory. This eliminates the problem of a dead memory battery that occurs with some microprocessor-controlled units.

You can check the frequency shift of each channel without changing its programming, or you can reverse the direction of the shift (from above the FCC-specified frequency to below, or vice-versa) without changing the memory. As an example, if channel 2 is programmed for a 1.00-MHz shift, you can alternate between –1 and +1 MHz without changing the program memory. This is especially helpful for checking the lock-in range of the AFT circuits from the antenna terminals. The following instructions explain how to use the three options discussed:

1. Checking the shift.
2. Temporarily shifting directions.
3. Reprogramming the memory.

Selecting any option affects only one channel at a time, leaving all other channel shifts unchanged. All options use the special-function "±" key on the VA62 keyboard.

Programmed shift checks

1. Set the RF-IF signal switch to the PROG cable position. Note: You can't check the shift in any other function of the RF-IF signal switch.
2. Select the desired channel with the keyboard or with the up/down buttons.
3. Press the "±" key.

The left-hand digital display will show the amount of frequency shift. Shifts to a frequency higher than the FCC-specified frequency will not show a polarity sign, and shifts to a lower frequency will indicate a "–" sign.

The RF generator continues to produce the same RF carrier frequency whether the digital display is showing the cable channel number or the shifted frequency number. You can return the display to the channel number by momentarily switching the RF-IF signal switch to the STD cable position and then back to the PROG cable position.

Direction shift change

Use the following procedure to change direction of shift without reprogramming the memory:

1. Set the RF-IF signal switch to the PROG cable position and select the desired channel.
2. Press the "±" key. The left-hand digital display will show the programmed shift.
3. Press the "±" key again. The polarity sign in the digital display will change from blank to "–" or vice-versa, and the RF generator will change its direction of shift. Note: The program shift memory is not changed with this special function. You can return to the preprogrammed shift by switching the RF-IF signal switch to the STD cable position and back to the PROG cable position.

Programmed shift-any channel

To change the programmed shift of any channel, use the following procedure.

1. Set the RF-IF signal switch to the PROG cable position and select the desired channel.
2. Press the "±" key. The left-hand digital display will show the shift presently programmed into the memory.
3. If the display shows no sign (a positive shift), and you want a negative shift, or if the display shows a negative sign, and you want a positive shift, press the "±" key again. You can shift the direction as often as you want before entering the programmed change.
4. Enter any shift you want up to 9.75 MHz. You must enter all three digits (for example, enter "100" for a 1-MHz shift). The microprocessor places a "-" in any position that needs another digit entered. The decimal is automatically placed after the first digit.

After the third digit is entered on the keyboard, the display will blank out for approximately one second and then return to the channel number. The memory has been reprogrammed when the channel number reappears. The RF generator produces the desired shift as soon as the channel number reappears.

Other operational tips

1. The RF generator can only be shifted in multiples of .25 MHz (250 kHz). If you enter a number that doesn't end in "00," "25," "50," or "75," the microprocessor will automatically round off the entered number to the nearest quarter of a megahertz. The correct number appears in the display the next time you check the shifted value.
2. If you make an error while entering the shift, enter all three digits and wait for the channel number to reappear. Then repeat the programming sequence. If

you press the "C" (clear) button, the display will show "--". You must then reenter the desired channel number before reprogramming the shift.

RF-IF attenuator notations

The levels for the RF signals are different than the IF signals because of the differences in applications of the two signals. Table 2-1 shows the different signal levels available between the RF and IF functions. The RF-IF level vernier is used to select any level in a range of the drive signal switch. For example, the RF-IF level vernier produces 100 microvolts on a range of the RF-IF level switch covering 50 to 500 microvolts.

Table 2-1. Signal levels

Switch position	RF signals	IF signals
LO	5–50 microvolts	500–5000 microvolts
MEDIMUM	50–500 microvolts	500–5000 microvolts
HIGH	500–5000 microvolts	50–500 megavolts

The reason you need shifted CTV channels is because in order to prevent co-channel interference or poor maintenance, some cable systems don't always transmit their signals at the FCC-specified frequencies. Without a full set of test cable channels, you often can't tell if there's a problem in the TV receiver, the cable signals, or the cable converter box.

Overall tuner operation check

Use the following procedure to test for overall tuner operation:

1. Make sure the AFT and IF circuits are aligned properly.
2. Feed the multiburst bar sweep pattern into each channel of the VHF tuner while observing the video detector output with an oscilloscope.
3. Should each of the VHF channels show a rolloff in the same area of the multiburst bar pattern, the IF stages can be readjusted to compensate for tuner frequency loss:
   - Feed the multiburst bar sweep pattern into any of the VHF channels.
   - Reset the IF alignment coils for even bars at the output of the detector.
   - Check the other VHF channels and adjust for the best balance of frequency response from channel to channel.
4. If only a few of the channels show a frequency rolloff (as indicated by uneven bars) the tuner tank circuits for these channels are out of alignment. If the rolloff is severe, the tuner might have to be replaced.

You can also use the VA62A as a tuner subber to determine if the set's tuner or other systems (such as the IF or AGC stages) are at fault.

AFT alignment checks

Note that all AFT alignment begins with the same preliminary steps. At the conclusion of these procedures, locate the final alignment information for the particular type of circuitry used. The specific instructions are based on the number of AFT adjustments that the TV receiver will require.

Figure 2-1 shows the VA62A control setups and TV set connections for AFT alignment. Now use the following procedure checks:

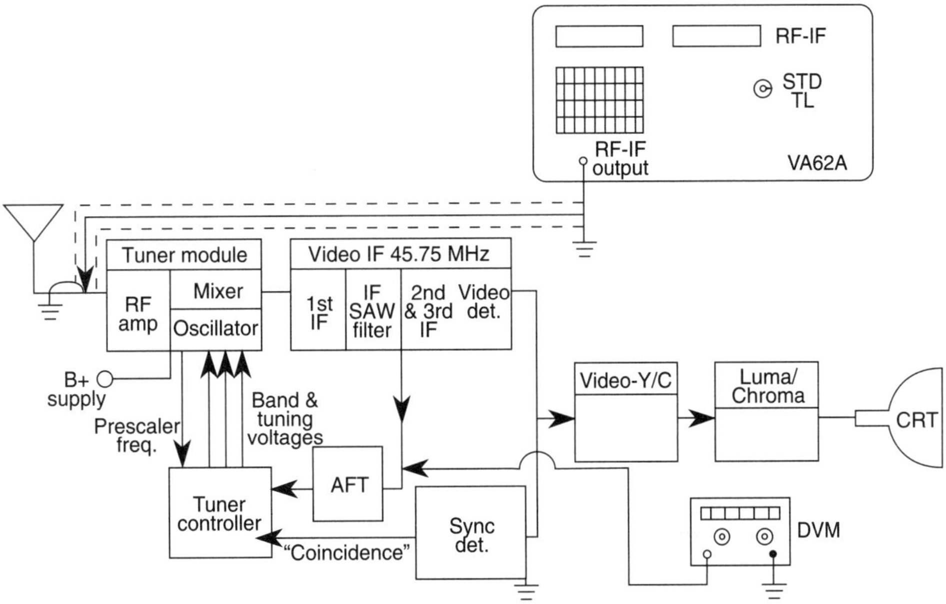

2-1 Control setup for making AFT adjustments. AFT circuits are adjusted to produce the same AFT control voltages with AFT off or on.

1. Connect the tuner subsignal to the input of the IF stages, as illustrated in Fig. 2-1.
2. Connect a dc voltmeter or dc-coupled scope to the AFT control line that goes to the VHF tuner.
3. Turn the AFT off and note the dc voltage reading.
   - If a dc meter without a zero-center scale is used, note the dc reading.
   - If a dc meter with a zero-center scale is used, set the meter for zero center with the AFT voltage applied.
   - If a dc-coupled scope is used, set the vertical position control to place the trace at the CRT center line, with the AFT voltage applied. The scope can now be used as a zero-center dc voltmeter.

4. Turn the AFT on. If the voltage reading from step 3 changes, the AFT circuits require realignment. If there's no change, the AFT circuits are operating properly.
5. If the need for realignment is indicated, note the number of adjustments in the AFT circuit, and refer to the following instructions.

Aligning AFT circuits with three controls

Refer to the complete RF-IF alignment information data given in the set manufacturer's service literature to locate the control used to center the 45.75 MHz point of the position marker on the "S" curve, and the control that adjusts the position of the negative marker on the "S" curve.

1. Adjust the coil for the 45.75 MHz marker (detector secondary) for the highest positive dc reading possible.
2. Adjust the control that affects the positive portion of the "S" curve for the highest positive dc reading.
3. Readjust the detector secondary coil for the greatest negative reading possible.
4. Adjust the control that affects the negative portion of the "S" curve for the greatest negative dc reading.
5. Readjust the detector secondary coil for the same reference voltage found in step 3 of the preliminary alignment instructions.
   - If a dc meter without a zero-center scale is used, set this adjustment for the same voltage.
   - If a dc meter with a zero-center scale is used, reset this adjustment until the meter reads center scale.
   - If a dc-coupled scope is used, reset this control until the trace is centered.

AFT alignment comments

The microprocessor-controlled 45.75 MHz carrier used to align the video IF amplifiers provides a reference for testing or realigning the automatic fine tuning (AFT) or automatic frequency control (AFC) circuits. These adjustments should be performed after the IF stages are properly aligned.

Understanding AFT alignment

The microprocessor-controlled IF generator in the Analyzer provides an accurate reference for alignment of the AFT circuits. Generally, the Analyzer procedure will result in a more accurate alignment of the AFT circuits than by making these adjustments with a sweep and marker generator. The main difference is that the alignment is compared with the AFT control turned off and then with the AFT correction circuit turned on to be sure that the tuner's fine-tuning point doesn't change when the AFT control is activated.

When a sweep and marker generator is used for AFT alignment, a variable-frequency signal is applied at the input to the AFT detector. The circuitry is then adjusted to center the 45.75 MHz center-frequency point between the two extremes of the AFT lock-in range.

AFT operations check

Use the following check points to test for proper AFT operation:

1. Connect the RF-IF output of the Analyzer to the TV set's VHF antenna input terminals.
2. Turn the set's VHF tuner to an unused channel and punch up the same channel number on the VA62A.
3. Turn the RF-IF signals switch to the standard channel position. Set the RF-IF level control for a snow-free picture.
4. Select the multiburst bar video pattern, the 3.5 MHz color carrier, and the 41.15 MHz for the audio carrier.
5. Turn off the AFT and manually tune the VHF tuner for color and the least amount of 920 kHz beat produced by the 41.15 MHz sound carrier beating with the 3.58 MHz color carrier in the IF stages.
6. Turn the AFT on. There shouldn't be a noticeable change in the color level or any increase in the amount of 920 kHz beat. If this is not the case, the AFT will have to be realigned.

Note: On most TV sets that have microprocessor (PLL) tuner controls, the AFC special mode should be able to search for the TV carrier above and below the FCC standard by 1 MHz. To check this out, set the VA62A in the PROG cable mode, step the frequency up or down (±1 MHz), and see if the tuner will lock onto the signal.

Tuner substitution

The video IF position on the VA62 lets you feed the microprocessor-controlled 45.75 MHz signal directly into the first IF stage for troubleshooting and alignment. This signal can be modulated with any of the video patterns. The tuner IF link cable can be disconnected from the input to the IF amplifiers and the IF signal injected directly as a substitution for the VHF tuner. This test signal should be injected into the UHF input of the VHF tuner for general troubleshooting and alignment. This procedure injects the signal directly into the VHF mixer circuits. The cable coming from the UHF tuner is disconnected and the connector adapter is put in its place. When using this type of signal injection, the VHF tuner must be placed in the UHF position.

If the tuner doesn't have a separate UHF input jack, the tuner substitution signal can be fed directly to the mixer test point input by using the 75/300-ohm balun transformer for matching. To inject a signal in place of the VHF tuner:

1. Connect the output of the RF-IF output jack to the UHF input of the VHF tuner, the mixer test point, or the input to the first IF.
2. Select the desired video pattern with the video pattern selector switch. If a sound carrier is required, turn on the audio switch.

3. Select the 45.75 MHz video IF position of the RF-IF signals selector switch.
4. Adjust the RF-IF level control for a snow-free picture.
5. Perform any of the desired tests that might be required.

Use as a tuner subber

You can substitute for any tuner (TV, VCR, etc.) with the VA62A, be it an electronic tuner, tube, or solid-state, and you can localize the problem to the tuner or the IF stages. All you'll need to do is inject a calibrated signal directly into the IF stage, bypassing the tuner. You can check the operation of the set right on the TV set's screen, eliminating all guesswork. You might also want to take the Analyzer on home service calls to check the set and know what repairs are required before shop work is performed. This will allow you to give much more accurate estimates before any shop work is performed. You can also check out each and every channel on the tuner to determine if all channels or just certain channels are defective. This also helps you to determine if you have cable or TV-set troubles. Figure 2-2 shows you how to set up the controls on the VA62 Analyzer for the tuner sub-mode of operation.

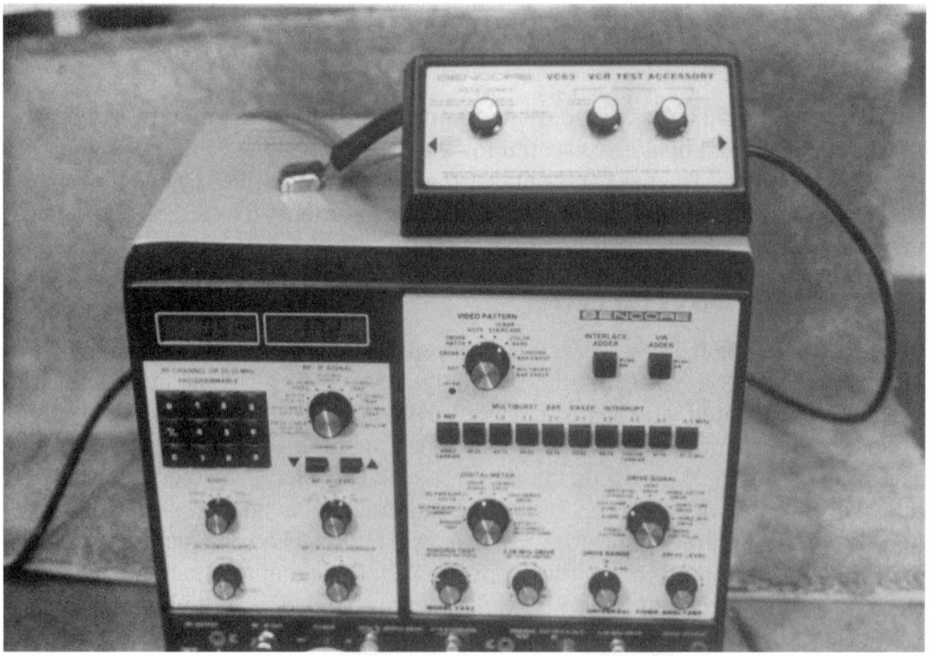

2-2 Using the VA62A as a tuner subber.

MATV and CATV systems check

The VA62A Analyzer provides you with both 75-ohm and 300-ohm outputs for matching into MATV and CATV systems without having any standing waves. The multiburst bar sweep patterns let you check the frequency response of splitters and amplifiers from the head end right down to the last tap-off. The calibrated RF output gives you a dynamic check of the system's gain and signal required at the amplifier input for a good, snow-free picture. When servicing MATV or CATV systems, feed the analyzer test signal into an unused channel. This allows you to check out the system of individual TV sets without disrupting the other channels or the system's TV receivers.

The variable calibrated RF output control on the VA62 allows you to test the system at various signal levels. The fastest method of testing an MATV or CATV system is to feed the signal at the head end and view the resulting pattern on the different sets that are normally connected to the system. The multiburst bar sweep pattern let's you check each section of the system for overall frequency response.

Overall system operation can be checked with the addition of a field-strength meter (such as the Sencore FS134) or a TV monitor that has been modified to display the output of the video detector through a test point. This test point, when fed to an oscilloscope, will provide a method of verifying the operation of the system at each section of the distribution system.

The VA62A also provides reference signals for aligning translators (channel converters) in the system. The multiburst bar sweep pattern can be used as the reference pattern to align the receiver portion of a converter for maximum flat response. The transmitter portion can then be aligned to provide equivalent response by using a properly operating TV receiver to monitor the transmitted output signal. The signal from the VA62 should be fed into the TV monitor on the same channel that the translator is designed to provide. (The IF stages should be adjusted to provide a good alignment pattern at the video detector by viewing a scope display.) The output of the translator should then be fed into the translator's receiver section. The transmitter portion should then be aligned for best overall response at the video detector of the TV monitor. The use of a frequency counter or SC61 scope and a field-strength meter will also aid in setting RF output levels and frequencies.

The VIR test signal

VIRS is an abbreviation for the vertical interval reference signal and is a reference signal inserted during the vertical blanking interval of a color TV program transmission. A VIRS is used to smooth out variations in color throughout the TV network system by helping TV engineers and operators adjust various signal parameters. This is used in order to help keep the color information similar when station programs change or channels are switched.

The VIR signal is intended to be used with reference to a particular TV program. The VIR signal was not intended to be used for test data on transmission distortions. Figure 2-3 is a drawing of the VIR signal as it appears on line 19 of the picture scan. This complete VIR signal is put on one scan line, which has a duration of 63.5 microseconds. A breakdown of the different test components are as follows: Starting at the left side, the interval from the leading edge of the horizontal sync pulse to the trailing edge of the blanking pedestal is 12 microseconds, which also includes color burst. Now the signal takes a sharp rise to the 50 IRE units level, where you'll find

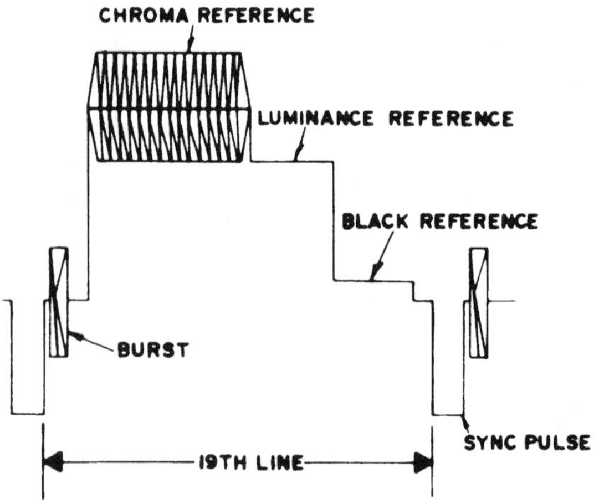

2-3 VIR waveform with a breakdown of components.

the chroma reference bar located between the 50 and 90 IRE units. The chroma reference bar is 24 microseconds long, followed by the luminance (LUMA) reference with a duration of 12 microseconds. Now you see a fast drop to 7.5 IRE units (black reference level) that also has a duration of 12 microseconds. At this point the signal goes into the blanking pedestal and starts over again with the sync pulse and the 3.58 MHz color burst pulse. For correct color level, the chroma reference bar should fill the IRE units between lines 50 and 90. These IRE units have been set by the Institute of Radio Engineers.

A VIR test signal stays with a TV program from the point of insertion to its final destination. A VIT signal is generally used only over some portions of a television system and will usually be taken out and reinserted point by point. The vertical interval test (VIT) signal, however, is used as a diagnostic test signal primarily intended to monitor and measure the characteristics of a transmission system or facility and not for any one individual program. Thus, the VIT signal would be used to check the condition of the color transmitter, video tape unit, or the (STL) studio transmitter microwave link.

The VA62A doesn't have a VITS signal adder. However, it has some signals that can be used as VITS test signals such as the multiburst bar sweep and staircase (ramp), which can be used for video circuit checks.

The VITS signal is strictly a TV transmitter test signal. There are many different video patterns that can be inserted on the VITS lines by the network or station, so these signals aren't predictable standards that can be used by a TV receiver. The VIR signal, on the other hand, is strictly specified by the FCC, allowing the TV receiver manufacturers to use the signal as a standard to reference automatic circuits. Some sets now have VIR circuits that are used to automatically adjust the color tint and level. The VIR is a standard FCC signal that all networks and most local TV stations now use on every program. The TV sets you service now cannot have VIR capabilities, but it's a good bet that in the near future you'll have to service sets with these circuits, and the VA62A will be ready for you to troubleshoot them with.

GE's VIR color control system

The General Electric Color Controlled System provides the TV receiver with automatic color intensity and tint adjustments. The system does this by detecting and processing a VIR signal transmitted on the 19th line of each field of the composite video information.

The results of the processed VIR signal are dc voltages that automatically replace the dc voltages developed across the manual tint and color level controls. Thus, the system automatically maintains the parameters of chroma gain and phase as controlled by the VIR signal transmitted from the TV station.

VIR system analysis

Five basic blocks make up this VIR system used in the GE PC chassis. These are the line recognizer, tint controller, color controller, chassis interface, and VIR sensor. Most all of the components are located in the AN5330 chip. The VIR system provides four functions, as shown in the IC block diagram of Fig. 2-4.

The line recognizer receives horizontal and composite sync pulses. The composite sync is used by the line recognizer to identify the 19th line. It then produces a turn-on pulse for the VIR sensor circuit.

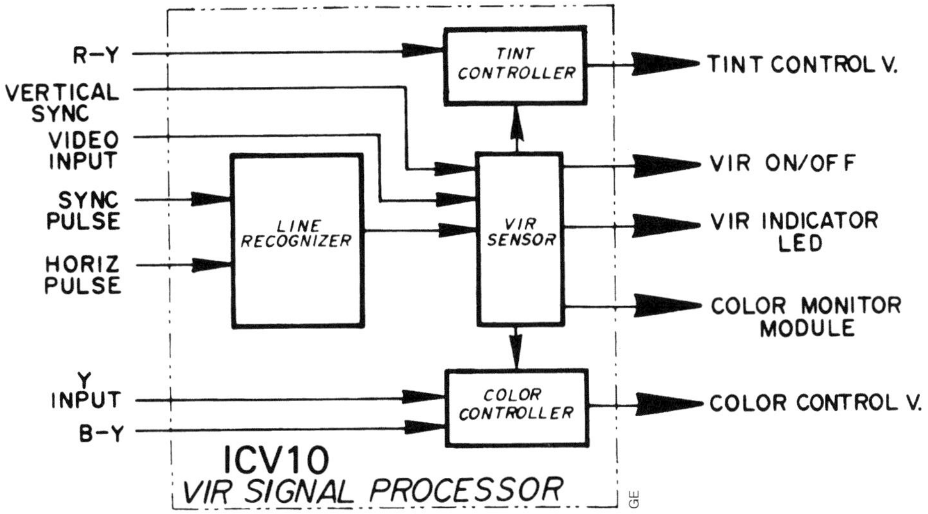

2-4 Block diagram for a GE VIR system.

Inputs from video, vertical sync, and the line recognizer are applied to the VIR sensor. When both video information and a line recognizer pulse are applied, the VIR sensor knows that line 19 has VIR information. It then turns on the sensor and produces outputs to turn on the tint and color controller circuits, the VIR indicator, VIR on/off, and color monitor on/off. The vertical sync pulse is used to trigger the "reset" circuit for the VIR sensor.

The color controller samples the matrixed sum of B-Y and Y signals during the VIR chroma reference signal (note the waveform in Fig. 2-5) and compares the B-Y and Y matrixed sum to the black level. If any differences are found in this comparison, a control voltage is produced and fed to the voltage-controlled color-level circuitry on the signal module to cause B-Y to equal Y.

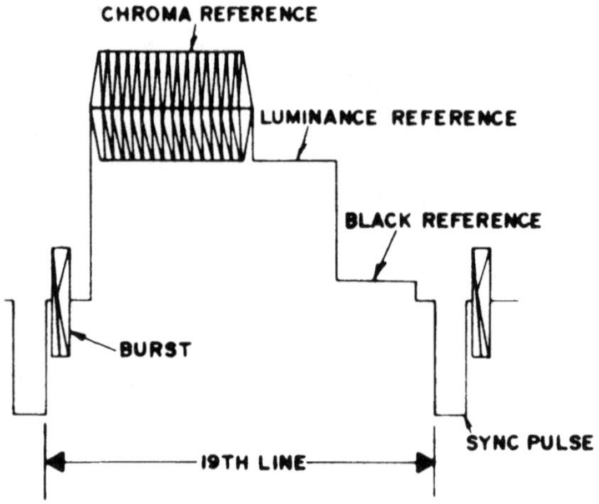

2-5 VIR signal showing black level, chrominance, and luminance information.

The tint controller samples R-Y output signals during the VIR chroma reference interval and adjusts it to black level. At this time the R-Y signal should be equal to black level (zero). If not, a control voltage output is generated to control the tint circuitry on the signal module to bring R-Y output to zero or black level. As I discuss how these color-control circuits operate, refer to Fig. 2-6 for the complete VIR module diagram.

VIR sensor

The VIR sensor produces the control voltages that will correspond to the presence or absence of the VIR signal and switches the TV set to automatic or manual control. The fixed amplitude luminance signal is fed into pin 2, clamped by the clamper to black level, and then fed to the VIR keyer. The VIR keyer picks up the luminance portion of the line 19 chrominance reference signal. When the signal level exceeds the threshold level (0.38 volts lower than pedestal level), the VIR sensor circuit recognizes the presence of a VIR signal and sets the R-S flip-flop.

When the R-S flip-flop is set, the voltage of pin 3 becomes approximately 3.3 volts (0 volts when not set), and the LED indicator circuit connected to pin 5 is turned on. Also, the switch that controls the tint and color controller circuits is switched to the automatic mode.

The vertical blanking pulse fed into pin 7 is used in conjunction with the line 17 through 24 pulse to reset the R-S flip-flop before the 16th line and after the 24th line.

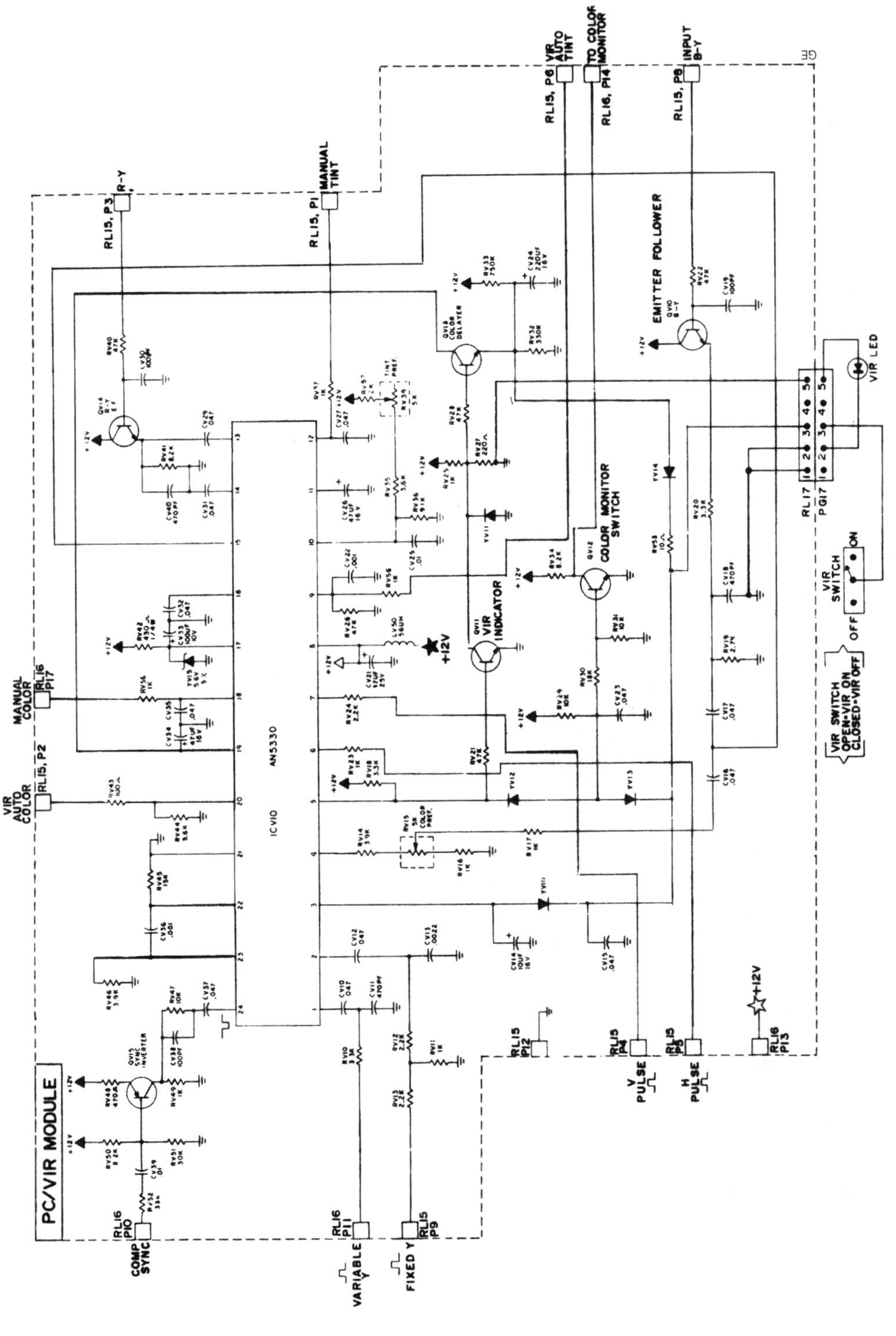

2-6 A GE VIR module circuit diagram.

This pulse refreshes the VIR sensor in every field. The vertical blanking pulse must be in the low state before the 24th line.

Color controller

The Y and B-Y signals are fed into pin 15 and are clamped so that the pedestal level of the B-Y and Y signals coincides with the voltage of the clamp bias source. The line 19 keying pulse samples the signal during the chrominance reference interval, and the extracted signal is held by the holding capacitor connected to pin 16. The output of the sample-and-hold circuit is compared with the voltage of the clamp bias source at the next stage, and the difference is amplified to produce the color control voltage. The output from the comparator and amplifier is refined by the low-pass filter after the automatic switch and fed to pin 20 as the color control voltage. A smoothing capacitor connected to pin 19 composes the low-pass filter.

The color control voltage from pin 20 is applied to the color gain control terminal of the TV receiver and adjusts the amplitude of the B-Y signal. As shown in Fig. 2-7, the negative feedback loop controls the combined B-Y and Y signals during the chrominance reference interval to coincide with the black level. If the color saturation is low, as seen in the Fig. 2-7 waveform, the comparator and amplifier generate a positive voltage. This increasing positive voltage applied to the color control causes an increase in the amplitude of the B-Y signal and makes the combined B-Y and Y signals zero during the chrominance reference interval.

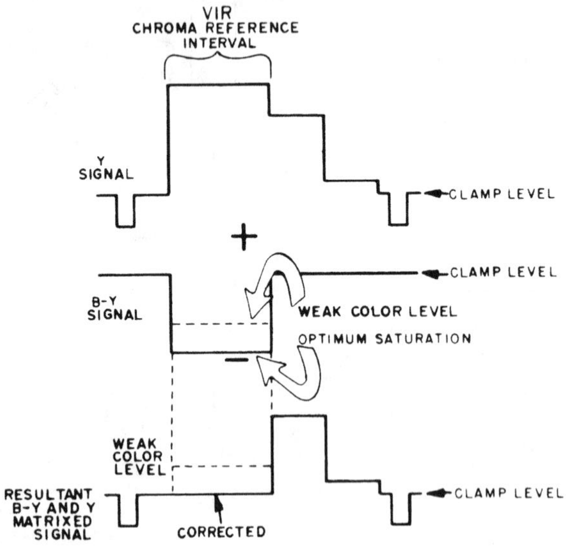

2-7 Color control effect on a signal waveform.

The auto-manual discriminating voltage generated by the pulse amplifier is applied to two switches. When the VIR signal is present, D is low, and the automatic color level correction circuit is activated. When the VIR signal is absent or when the VIR switch connected to pin 3 is closed, D becomes low, and the manual color control voltage fed into pin 18 appears at pin 20. When in manual, the filtering capacitor of the low-pass filter is also applied to the manual control voltage. This setup minimizes the transient when switching from auto to manual, and vice versa.

Tint controller

The tint controller circuit is very similar to the color control circuit, except for the addition of the tint preference circuit. The R-Y color demodulator signal during the chrominance reference interval is fed to pin 13, and the pedestal level of this signal is clamped at the clamp bias source by the clamper. The chrominance reference signal is extracted by the sample-and-hold circuit and is held by the tint-holding capacitor connected to pin 14. The voltage held in this capacitor is compared with the clamp source voltage at the comparator and amplifier circuit. The difference is amplified and goes to pin 9 through the auto-manual switcher, low-pass filter, and the tint preference circuit. The low-pass filter is composed of a smoothing capacitor connected to pin 11. This serves as a ripple filter for the tint control error voltage.

Use the VA62A Analyzer in the VIR adder mode to see if the set will go into the automatic VIR mode of operation. If you have VIR problems, then use the VIRs from the VA62A to inject or trace signals with the scope in the VIR system.

VHF electronic tuner

The VHF varactor tuner is a four-transistor device using a dual-insulated gate field-effect transistor in the RF stage. (See Fig. 2-8.) The tuner has five switching diodes that change the circuit characteristics so that the tuner will cover both the low-VHF and high-VHF bands with approximately the same varactor tuning voltage. Let's now look at some of the RF amplifier stages.

The 75-ohm antenna input is coupled to the antenna filter assembly, T1. The assembly contains two separate filter sections:

1. Blocking capacitor C1 and inductors L1, L2, and L3 provide trapping for FM frequencies.
2. Blocking capacitor C2 and inductors L4, L5 and L6 provide IF trapping. This prevents any possible 40 MHz signal, present at the IF module, from beating with a channel carrier.

The output of the filter assembly, via pin 2 of capacitor C2, is then coupled to the antenna preselector circuit.

From the antenna preselector, the signal is coupled via C6 to gate 1 of the RF amplifier. In this circuit design, the dual gate FET is operated as a common source (similar to a common emitter) amplifier. Gate 2, used for gain control, is biased to +12.8 volts (maximum gain) by a resistor divider network mounted outside the tuner. Under medium-strength signal conditions, the receiver produces a negative-going AGC voltage that opposes the positive gate voltage. With a very strong signal, the gate voltage is reduced to +3 volts for approximately 50-dB gain reduction. Any additional negative voltage will have no gain reduction effect. The ferrite bead, L11, on the drain (output) lead suppresses parasitic oscillations that can possibly fall in the UHF spectrum.

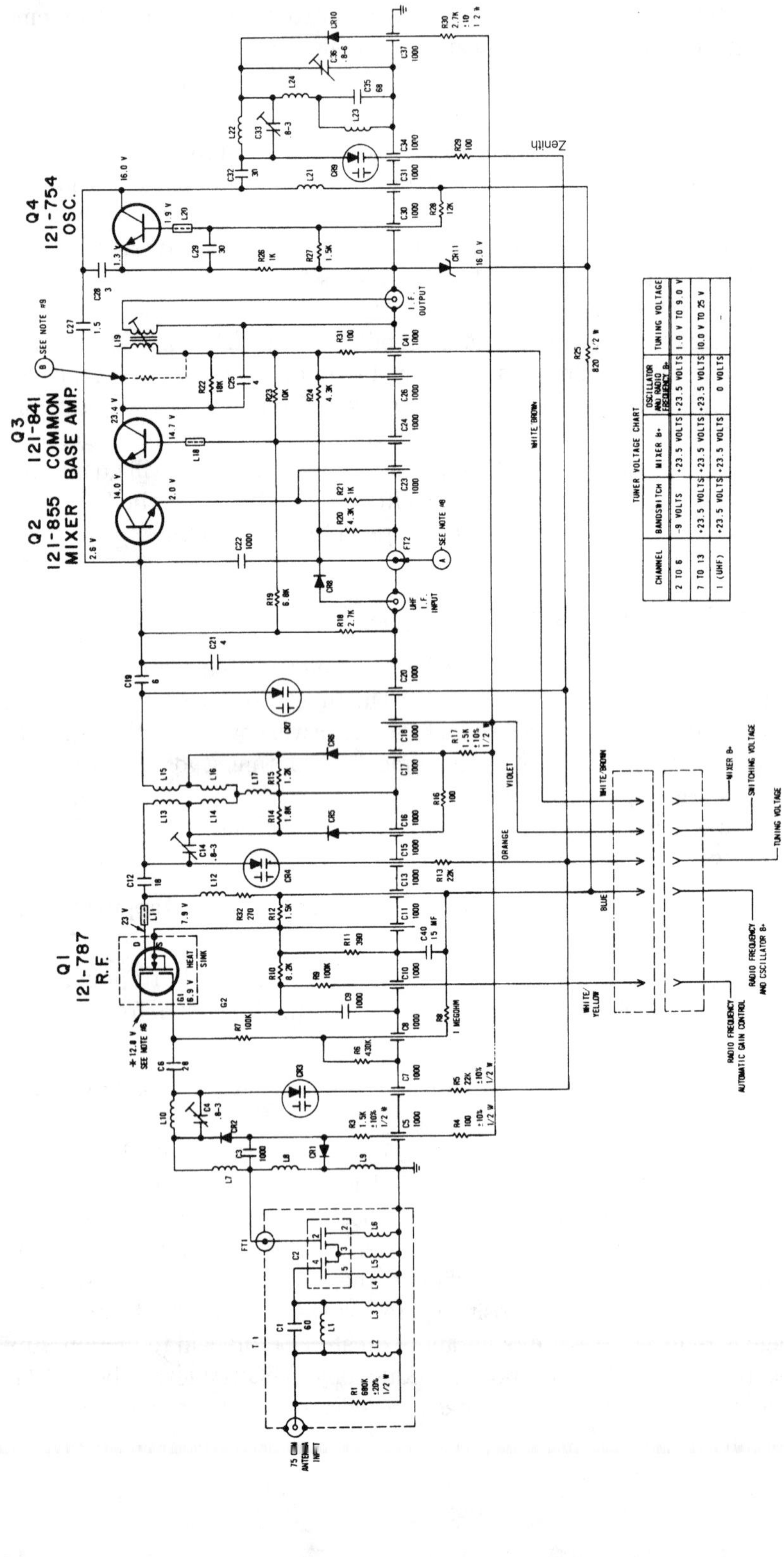

TUNER VOLTAGE CHART

CHANNEL	BANDSWITCH	MIXER B+	OSCILLATOR AND RADIO FREQUENCY B+	TUNING VOLTAGE
2 TO 6	-9 VOLTS	+23.5 VOLTS	+23.5 VOLTS	1.0 V TO 9.0 V
7 TO 13	+23.5 VOLTS	+23.5 VOLTS	+23.5 VOLTS	10.0 V TO 25 V
1 (UHF)	+23.5 VOLTS	+23.5 VOLTS	0 VOLTS	-

2-8 Schematic for a Zenith electronic tuner.

The amplified signal is coupled from the drain to the RF-to-mixer interstage network through capacitor C12. The drain-tuned circuit is composed of varactor CR4 and inductors L13 and L14. The mixer-tuned circuit consists of varactor CR7 and inductors L15 and L16.

The signal is coupled from the interstage network to the mixer transistor base by C19. Oscillator injection, coupled through gimmick capacitor C27, is also applied to the mixer base. The resultant 40-MHz signal appearing at the collector (high impedance) is directly coupled to the emitter (low impedance) of the common base amplifier. The output at the collector of the common base amplifier is high impedance and matches the IF output coil, L19. The purposeful mismatch between the two stages provides the mixer with stability and isolation. This isolation allows the IF output coil to be tuned during the IF alignment procedure without affecting the mixer input circuit.

Transistor Q4 is operated as a Colpitts oscillator. The feedback circuit is used to sustain oscillation and consists of capacitor C28 from the collector, in series with C30 to "ground." The emitter is connected to the midpoint of this capacitive divider, thus providing the proper feedback level.

You can check out the RF stages of tuners by setting the VA62A Analyzer to the STD TV channel mode and injecting the RF signal into various test points. Make sure you set the TV set's tuner and the VA62A Analyzer to the same channels for this check. This can be used if you have a weak, snowy picture or suspect a dead RF tuner stage. Referring to the varactor electronic tuner diagram in Fig. 2-9, just inject the RF signal from the VA62A to the points marked with arrows in the RF stages. Start at the balun coil and work your way up to the mixer stage. When the signal clears up on the set's screen, you've just passed the faulty stage.

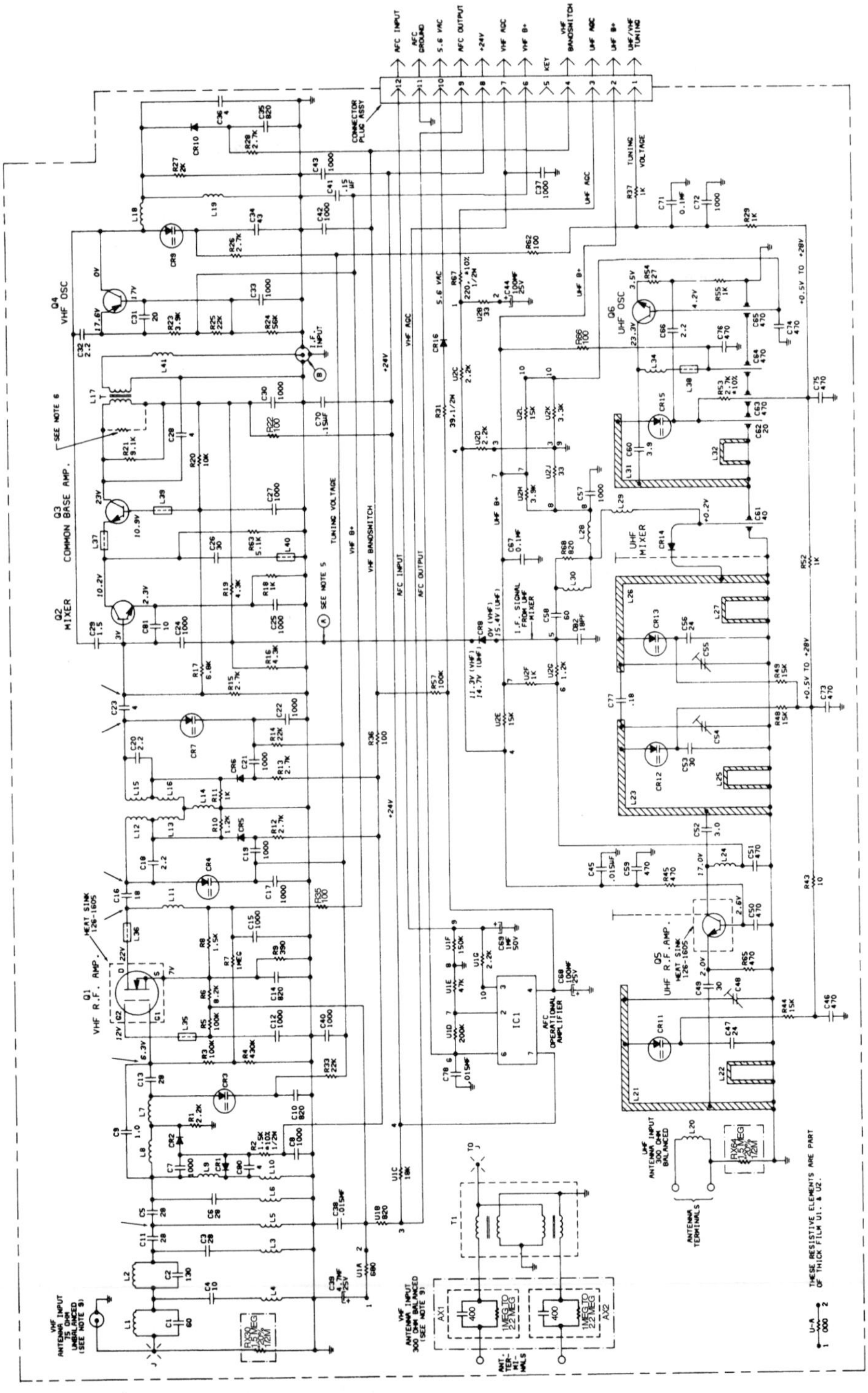

2-9 Zenith varactor tuner schematic.

3 Troubleshooting video amplifiers

IN THIS CHAPTER I LOOK AT TECHNIQUES FOR TROUBLESHOOTING VARIOUS VIDEO amplifier systems using the multiburst bar sweep and staircase pattern injected from the VA62A Video Analyzer.

The video pattern position of the drive signal output provides reference signals for analyzing the various video stages of all TV receivers and VCRs. The patterns provided by this signal are selected with the video pattern switch. These patterns are the same ones that are used to modulate the RF-IF test signals. Since this same video information is available from both the RF-IF generator and the drive signals output, the RF or IF signal provides a picture-locking reference while the video amplifier is being analyzed.

The multiburst bar sweep and staircase patterns provide a method to test the frequency response of the video amplifier. These signals will test the overall response to locate any stages that might be restricting the video resolution or to locate a completely dead video stage. The video drive test signal from the VA62 provides sufficient amplitude to substitute for any signal, including the output fed to the picture tube.

When the video drive signal is used for video amplifier troubleshooting, the RF cable should be attached to the antenna terminals of the set. This provides sync information to the deflection circuits to lock in the test patterns.

Video amplifier comments

A video signal is composed of a combination of transient voltages, and a good TV picture depends on the transient response of the video amplifier. The multiburst bar sweep and staircase pattern can be put to good use for making frequency-response and overall performance checks of color TV video amplifiers. With these test patterns, clues to circuit defects that cause poor picture reproduction can be uncovered.

Sometimes marginal defects are caused by cumulative drift of component values. Capacitance, inductance, and resistance can be crucial in obtaining peak-picture quality from video-amplifier circuits. Peaking coils develop shorts, opens, or leakage between layers, thus changing their inductance. Resistors tend to increase in value as the circuits age and capacitors (common troublemakers) develop various degrees of leakage and loss of capacitance. Of course ICs, transistors, and tubes also become defective. Tolerances are cumulative, and as picture quality deteriorates, it's very difficult to determine which components are defective. Video circuit faults will usually show up as preshoot, overshoot, or unsymmetrical distortion of the staircase pattern.

Bar and staircase patterns

The VA62A multiburst bar sweep has 10 bars. (See the TV screen pattern in Fig. 3-1.) These bars fall every .5 MHz all the way to 4.5 MHz, so you have a complete check of the horizontal resolution of any high-tech video system. There are no jumps in the frequencies that might cause you to miss important information between bars.

The EIA staircase signal (see TV screen pattern shown in Fig. 3-2) is a standard test pattern with equally spaced steps (note scope waveform pattern in Fig. 3-3) so you can accurately test or align synchronous video detectors or check the dynamic range of any video amplifier, video tape system, or other luminance circuit. This is the first time this handy test pattern has been available outside the broadcast studio.

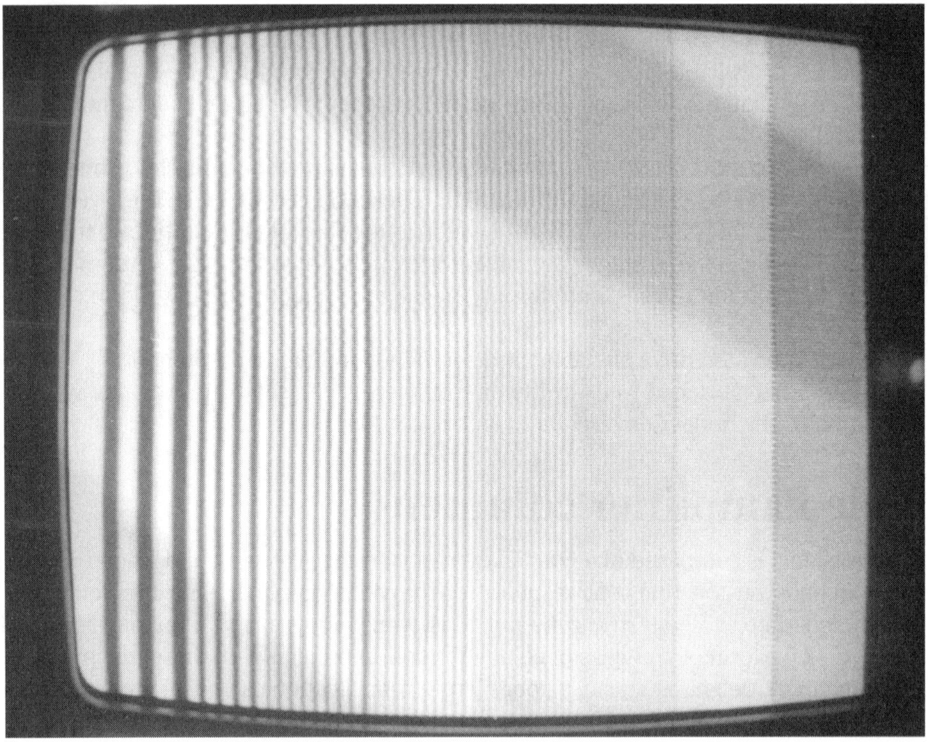

3-1 Correct multiburst bar sweep pattern.

Locating faulty video stages

The video drive signal can be used to locate faulty video stages. The video test signal is injected at the input and output of the suspected faulty video stage. If the opera-

3-2 Staircase pattern as seen on the TV screen.

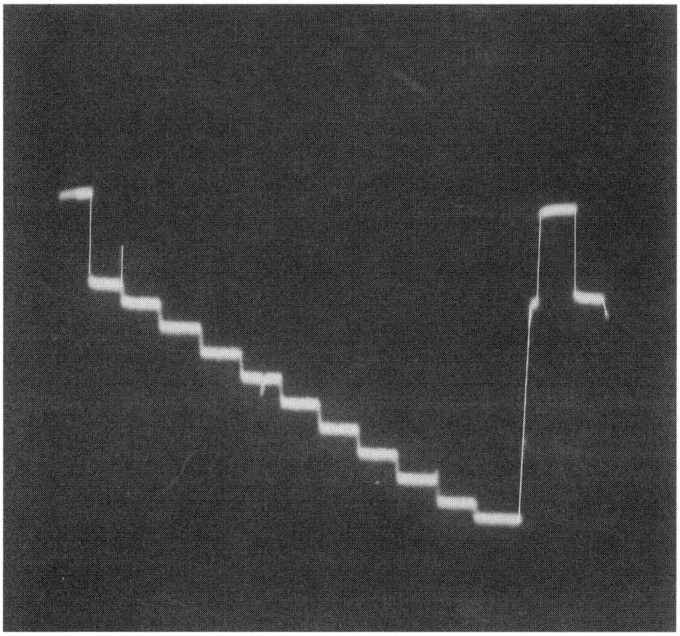

3-3 EIA staircase scope waveform.

tion returns when the signal is applied at the output of a stage but not at the input, then you've located the defective stage. If the video amplifiers are direct-coupled, an external bias signal might be necessary to get the video stages to operate normally. Refer to the VA62 control setup and various stage test points illustrated in Fig. 3-4.

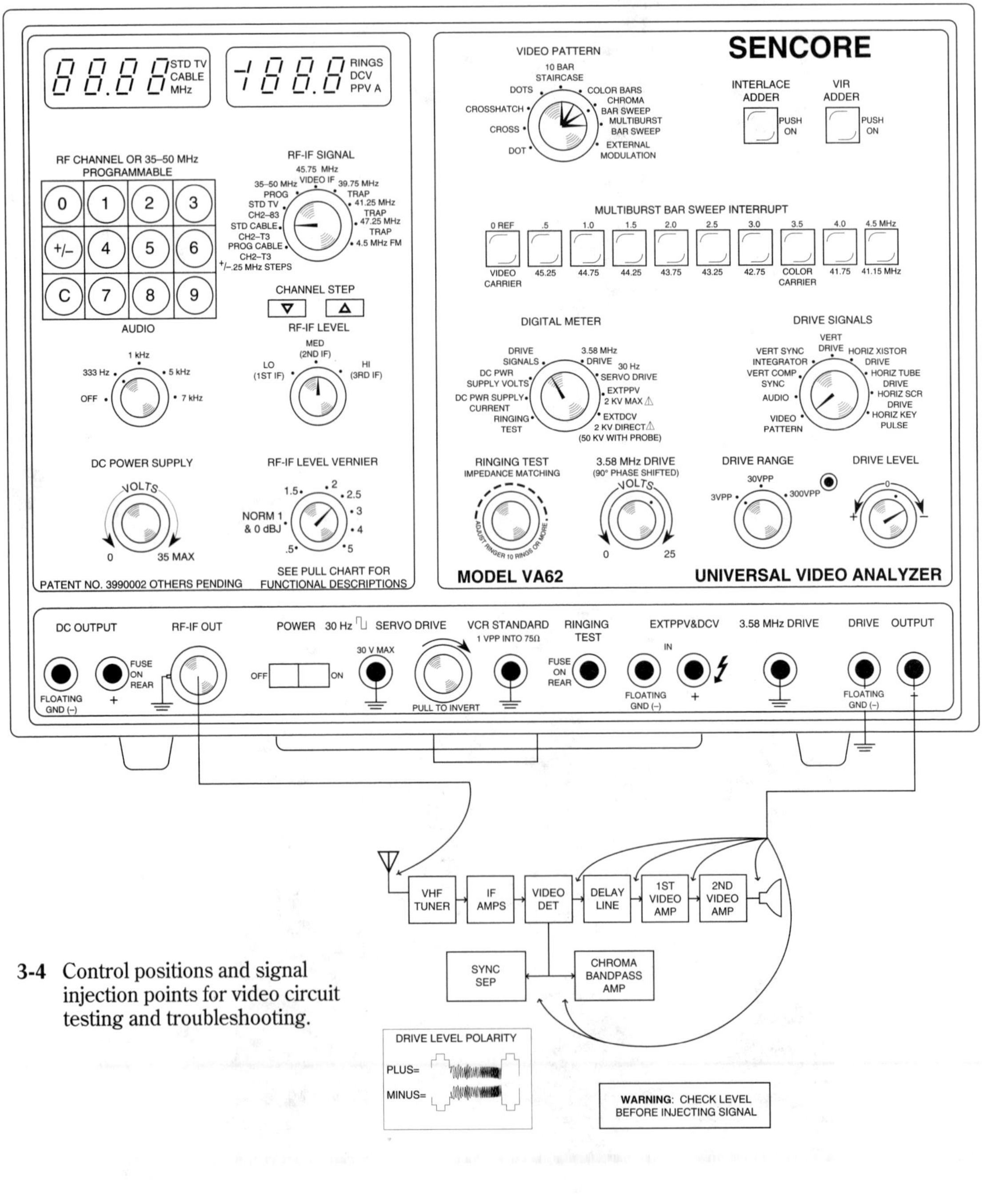

3-4 Control positions and signal injection points for video circuit testing and troubleshooting.

Use the following procedure to locate a faulty video stage by signal substitution:

1. Locate the output of the suspected faulty video stage.
2. From waveforms given in the service data, notice both the amplitude and polarity of the signal that should be present at the amplifier output.
3. Connect the RF-IF output to the VHF antenna input of the TV receiver. Set both the Analyzer and TV receiver to the same channel.
4. Set the digital meter switch to the drive signal's monitor position.
5. Adjust the drive level control for the proper signal level and polarity. To prevent possible damage to solid-state circuits, always observe proper polarity and level of the drive signal before injecting it.
6. Substitute the drive signal for the output of the suspected defective stage. If the picture appears on the screen, the defective stage is before the injection point. If there's no change, the defective stage is after the injection point.

Video amplifier frequency response check

The 10 multiburst bar sweep allows a complete frequency response check of the video amplifiers. With this test signal you can check video amplifier operation throughout the entire normal frequency range for these stages. The amplitude of these multiburst bars at the drive signal's output is the same, which means that an amplifier with flat frequency response will provide an output with all bars at the same level.

The video amplifier normally has a trap for the 3.58-MHz color subcarrier frequency. This means that the 3.5-MHz bar will normally be of lower amplitude than the other bars.

Here is the video amplifier frequency response test procedure:

1. Locate the input to the video amplifiers.
2. By checking the scope waveforms in the service information, notice both the amplitude and polarity of the signal that's normally present at the video amplifier input.
3. Adjust the drive level control for the proper signal level and polarity.
4. Substitute the drive signal for the input of the video amplifier.
5. Monitor the output of the video amplifier with an oscilloscope or observe the picture on the CRT.
   - Connect an oscilloscope to the output of the video amplifier. Set the scope for proper triggering. All bars (except the 3.5-MHz bar) should be about the same level.
   - Or, turn down the brightness control until the bars just disappear; the 3.5-MHz bar should be the first to disappear. The 3.0-MHz bar might disappear before the others because of the effects of the 3.58-MHz trap.

- Black-and-white receivers might normally roll off the high-frequency video response, especially on small-screen-size sets. The 3.0- to 4.5-MHz bars should be turned off for these checks. The .5-, 1.0-, and 1.5-MHz bars should be of equal amplitude and disappear simultaneously as the brightness is reduced.

Setting the 3.58-MHz trap

The color subcarrier trap in the video amplifier can be set by observing the effects of the trap on the video amplifier response. The chroma bar sweep provides the best signal for setting this trap.

The 3.58-MHz trap is set as follows:

1. Feed the chroma bar sweep signal into the video amplifiers. If the IF stages are working properly, the easiest place to inject this signal is through the VHF antenna input or the IF input. If desired, this signal can also be injected at the input to the video amplifier.
2. Select the chroma bar sweep position of the video pattern switch.
3. Connect a scope to the output of the video amplifier.
4. Adjust the 3.58-MHz trap for minimum amplification of the (3.5-MHz) carrier of the chroma bar sweep pattern.

Troubleshooting direct-coupled (dc) video circuits

Many video circuits contain one or more stages of direct coupled (dc) amplification. Modern circuit design can include dc circuits in the video amplifiers, vertical output stages, or the audio amplifiers. Direct-coupled stages are easily identified on a schematic because there are no blocking capacitors located in series with the output of one stage and the input of the next stage.

Some technicians might find these direct-coupled amplifier stages difficult to troubleshoot. This difficulty is compounded when there are more than two stages of dc amplification.

Since each stage of dc amplification depends on both its input and output connections to properly bias the transistor, a defect in either the input (source) or output (load) will cause the bias voltages of the suspected stage to vary. A single defective component in any of the dc stages can cause improper voltages in all other stages. In other circuits, a single defective component can actually cause damage to transistors several stages before or after the original defect.

An in-circuit transistor tester can be quite helpful for troubleshooting dc amplifiers, since the transistors can be tested in circuit. If all of the transistors test well, the next step is to measure the voltages and inject dc substitute bias and ac signals to locate the defective stage.

When substituting for the required bias signals, remember that the important parameter is the amount of emitter/base current. Simply applying the voltage indicated on the service data with a direct connection from the dc power supply output might not restore normal circuit operation. A direct connection from the VA62 Analyzer dc voltage supply might cause additional damage if too much current is allowed to pass through a small or medium transistor.

To apply external bias voltage to a dc stage:

1. Determine the proper bias polarity from the drawings shown in Fig. 3-5. As you'll note, the bias depends on the type of circuit under test.

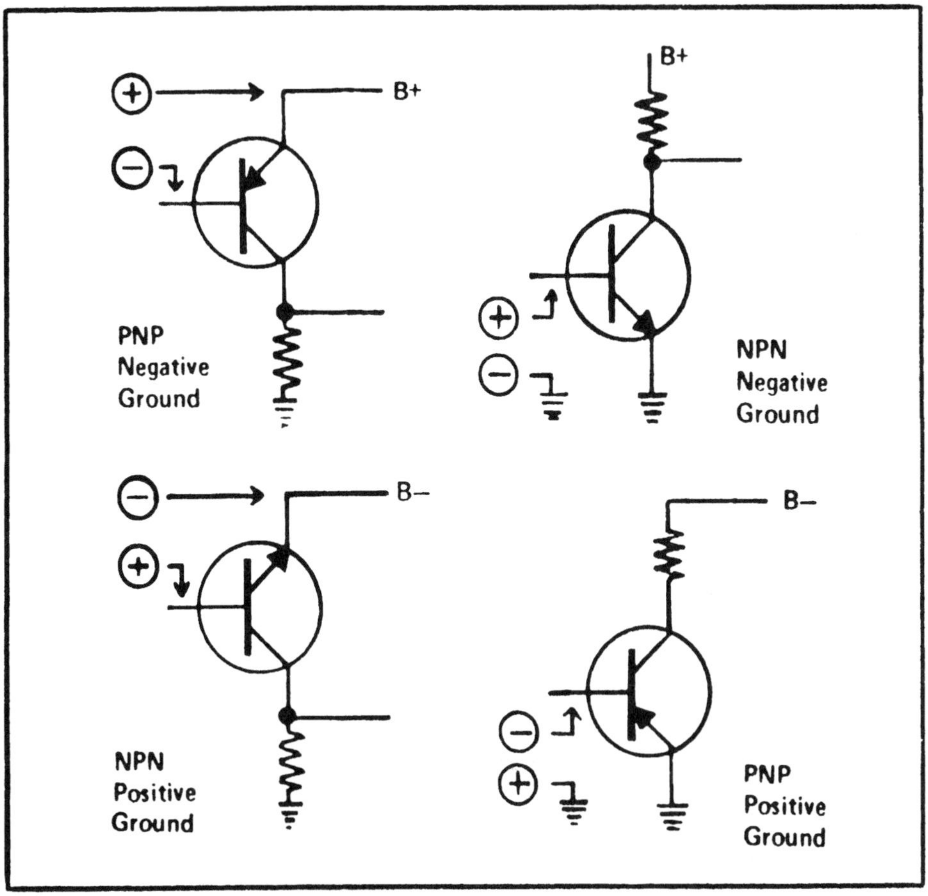

3-5 Proper connections of the bias supply leads for substituting bias voltages in various direct-coupled circuits.

2. Insert a current-limiting resistor in series with one of the VA62 supply leads before making the connection to the circuit being tested. Generally, a small-power (signal) transistor will require a 1K resistor. A medium-power (driver) transistor will require a 100-ohm resistor. A high-power (output) transistor will not require any limiting resistor.
3. Monitor the transistor base voltage with a dc voltmeter.

4. Apply the substitute bias voltage from the dc power supply output from the Analyzer.
5. Increase the output voltage of the dc supply until the proper bias voltage is indicated on the monitoring voltmeter. Note: The voltage drop across the transistor is dependent on the E-B current, not the applied voltage. For this reason, the output voltage of the B+ dc supply might be much higher than the base voltage indicated on the schematic.
6. If the proper base voltage is not obtained with the full output voltage of the supply, make the following checks:
   - Make sure the output of the transistor is not loaded and causing the low base voltage.
   - Decrease the value of the limiting resistor to the next smaller value listed in step 2. Use care when reducing the limiting resistor value, as the chances of causing circuit damage is increased.

Video circuitry

Let's now look at some actual video circuit operations. The video circuit in Fig. 3-6 for the GE PC color TV chassis shows the 1.5-volt, negative-going video signal from the emitter of Q110 entering the video processing stages via R216. The signal then goes through a 3.58-MHz trap onto the video delay line, L220. Two bias resistors terminate the delay line's output.

Transistor Q206 is an emitter follower that isolates the delay line termination from the variable input impedance of Q204, the video peaking stage. The high frequency gain of Q204 is adjusted by the sharpness control, which is connected to its emitter. High-frequency video is coupled from the collector of Q204 to a matrix network. The fixed-amplitude, low-frequency video is taken from the emitter of Q204 via L200 and applied to the same matrix network. The values of R209 and R210 determine the video drive level that's coupled through C216 into pin 27 of the video processor, IC300.

The IC300 chip processes the video signal through six internal stages. The amount of signal peaking is first determined by the value of C210 connected to pin 28. Next, a dc voltage adjustable picture level control stage receives the signal. The dc level from the picture control is present at pin 26 and controls this stage. An internal connection in IC301 also ties the picture control to the chroma gain section, thus allowing the chroma amplitude to track with the contrast level. A black level-clamping circuit with an external connection to C206 is the next component encountered by the video signal. A video amplifier stage, with its output at pin 24, provides a positive-going video signal for the VIR circuitry. The final two stages of IC300 provide brightness-level control and blanking.

The brightness control, R4220, is connected to pin 23 and is adjustable from +8.5 volts at minimum brightness to +8.9 volts at maximum brightness. Minor adjustment to this voltage range is accomplished with the brightness centering control, R262.

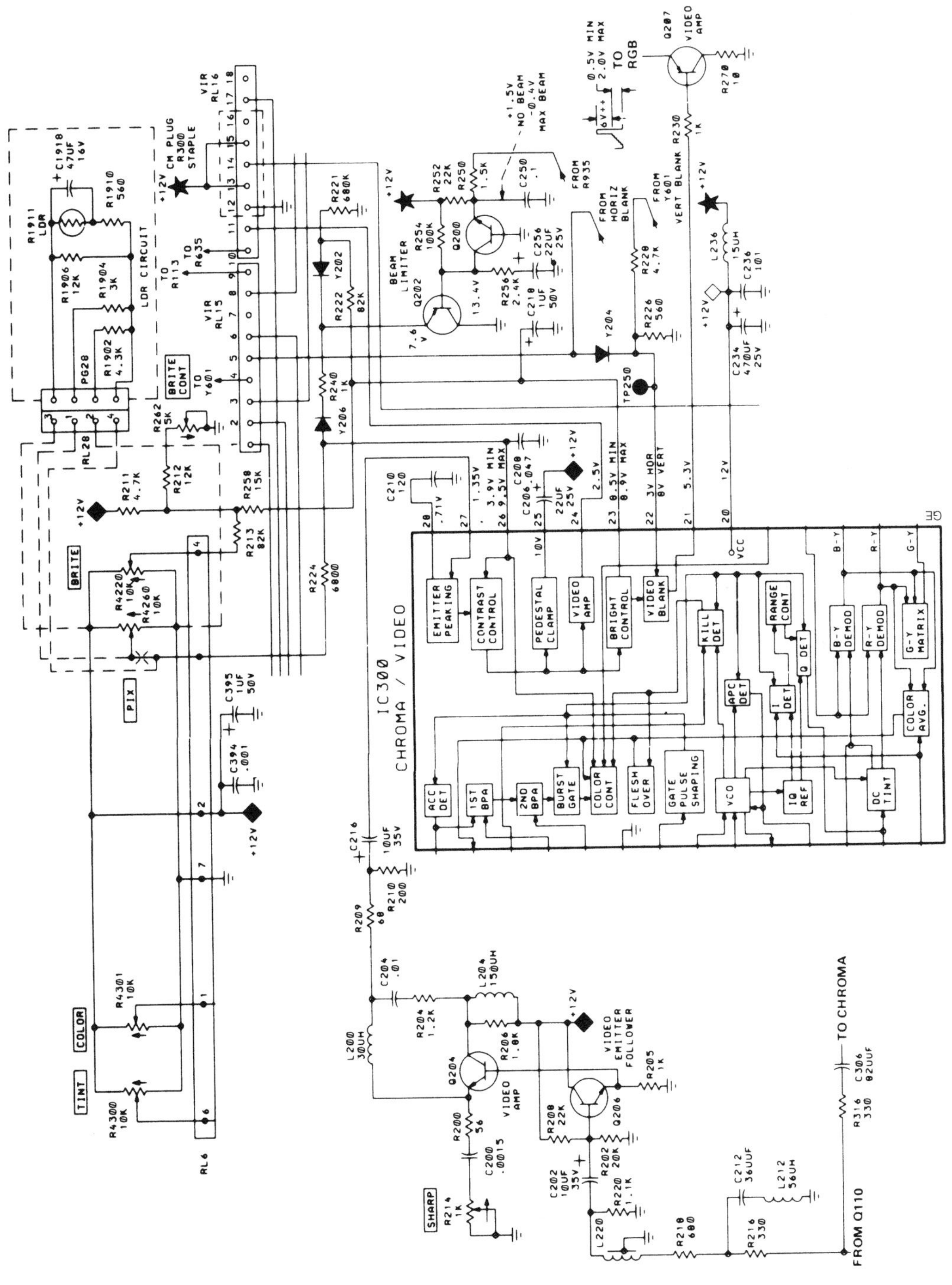

3-6 Video signal processing circuitry used in a GE PC color TV chassis.

A beam-current limiter circuit is connected to both the brightness control and picture control through diodes Y202 and Y206. These diodes permit the controls to operate normally when picture-tube beam current is below the maximum limit. Should maximum beam current be exceeded, diodes Y202 and Y206 become forward biased, clamping the picture control and brightness control voltages to a level that reduces beam current to acceptable limits.

The circuitry controlling Y202 and Y206 is made up of Q200, Q202, and their associated components. Increased picture tube current causes a negative voltage to be developed across R935, the beam current sampling resistor. Resistor R935 is connected to the emitter of Q200 by R250. Increased beam current will lower the voltage on the emitter of Q200 until it's driven into conduction. Conduction by Q200 in turn drives Q202 into conduction. With Q202 turned on, Y202 and Y206 have a path to ground that lowers the brightness and picture control voltages through a dividing action.

Note that capacitor C250 is used to filter the beam current sampling voltage from R935, C256, and R256, which form an antihunt damping circuit to prevent the pulsating effect of low-frequency oscillation that might occur in this circuit.

Horizontal and vertical blanking pulses are added to the video signal in the final stage of IC300. The horizontal pulse is derived from a capacitive divider on the collector of the horizontal output transistor. The vertical blanking pulse is taken from pin 22 of IC300 through isolating diodes Y204 and Y601. If you retrace lines in the picture, use the scope to make sure that you have these H&V blanking pulses at pin 22 of IC300.

Certain failures in the vertical oscillator can cause pin 17 of IC501 to produce a constant +5-volt dc output rather than the normal blanking pulse. This condition can blank the raster out completely. This is a good troubleshooting clue if you have a set with a blank screen. Use the scope to see if you have a vertical pulse at pin 17 of IC501. Many other brands of sets use this same CRT shutdown technique. So, if you find this symptom, check for a missing pulse or scope the vertical oscillator and sweep output stages for proper operation.

The video output from IC300 (pin 21) connects through R230 to Q207, the luminance driver transistor. From this point the composite video (Y signal) enters plug PG4 of the RGB section via pin 8 and is then applied to the emitters of the output amplifiers. Pin 8 is a good test point to inject a video signal from the Analyzer to check for proper operation of the RGB video output stages. (See Fig. 3-7.)

Video processing (black-and-white TV)

The circuit in Fig. 3-8 is for a black-and-white TV signal processing system that uses discrete transistors. These transistor stages operate on 18 volts B+.

Beginning at the video detector, the video signal is elevated above ground by resistors in order to provide proper bias for the NPN video driver. The video signal is fed to the base of Q801, the video driver, which also functions as a buffer. At the collector of Q801, the amplified video waveform is coupled into the base of Q802 (sync separator) by means of a series of resistors and capacitors selected to filter out any noise or improper signals that might affect sync. The output at the collector of Q802

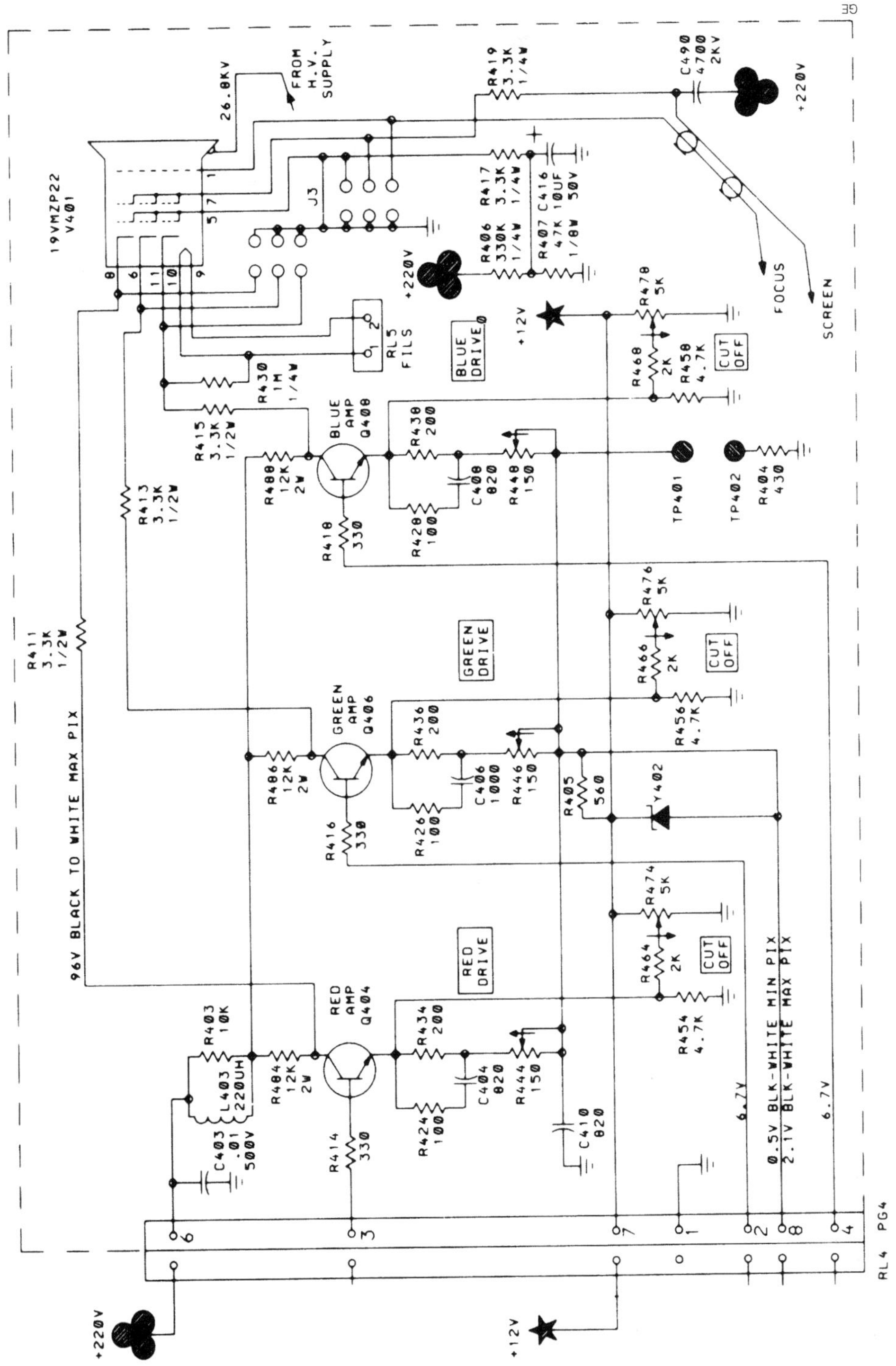

3-7 RGB video output circuit for GE PC color TV set.

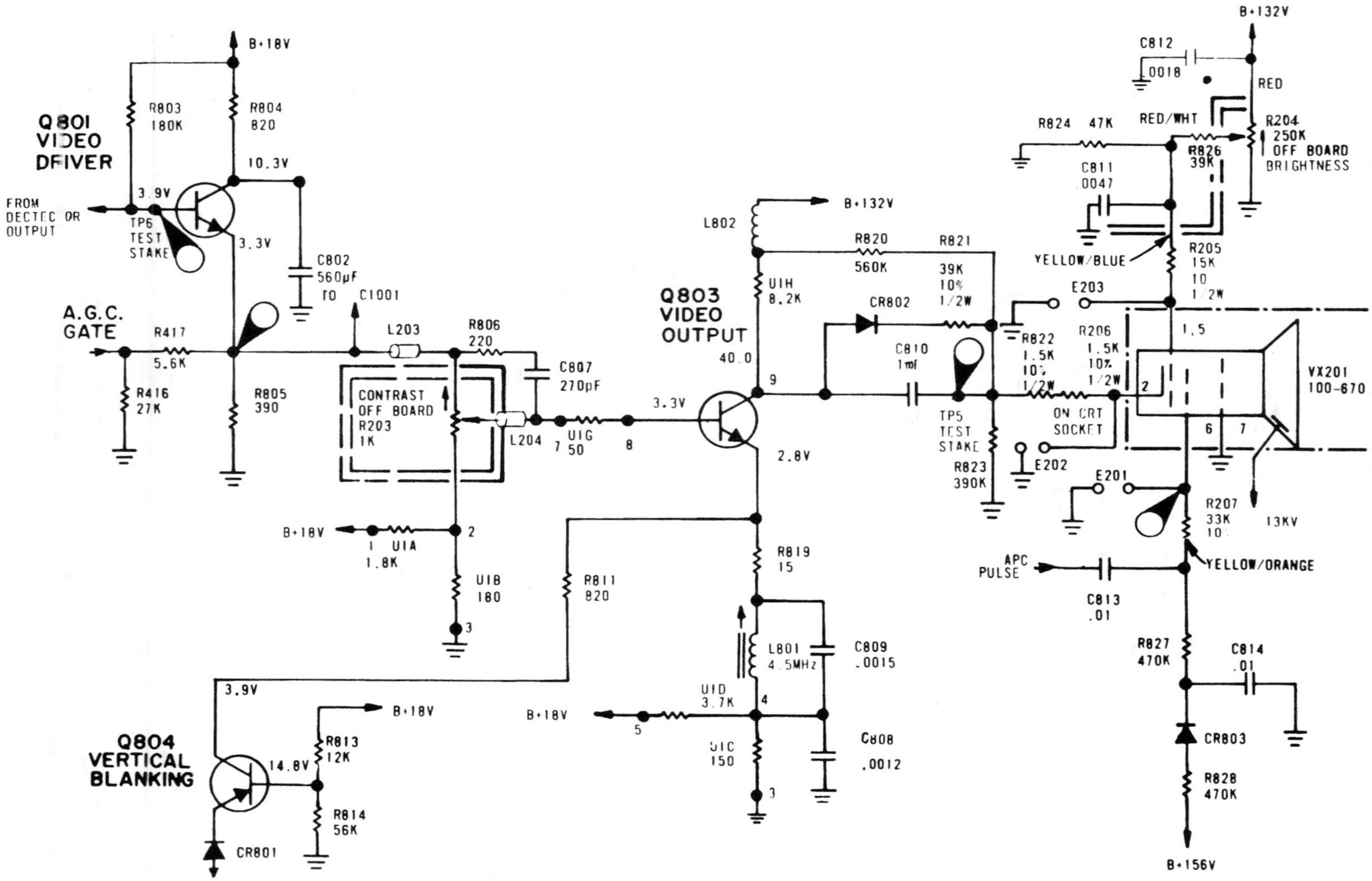

3-8 Black-and-white TV video circuit. GE

noise or improper signals that might affect sync. The output at the collector of Q802 is approximately 18 volts peak-to-peak of composite sync.

The emitter of Q801, the video driver transistor, provides a low-impedance signal to the sound circuit and the video output transistor Q803. The base of Q803, the video output transistor, is a dc-coupled circuit from the video driver through the contrast control. Divider resistors, using the +18-volt supply, provide bias to prevent cutoff of the video output stage at minimum contrast setting. Resistor R806 and capacitor C807 minimize the change in video peaking with various contrast control settings. Resistors and capacitors in the emitter and collector of the video output stage provide frequency and phase compensation. The video signal is coupled by CR802, C810, R823, and R821 to the picture-tube cathode. This network provides approximately 60 percent dc coupling to the CRT, and limits beam current to a maximum of 300 micro amp. Brightness is controlled by R204, which provides G1 bias for the picture tube.

The vertical retrace blanking is injected into the emitter of transistor Q803, the video output stage, from the collector of transistor Q804, which is the vertical blanker stage. A portion of the composite video signal is also coupled from the emitter of Q801, the video driver, by resistor R417 to the base of Q403, the AGC gate transistor. (Some circuit information in this chapter courtesy of Zenith Radio Corp. and General Electric Co.)

Poor video problems

Many problems affect the picture of a TV receiver. Troubles that cause a poor picture, such as smear, weak contrast, harsh edges, or poor detail, are tougher to find than problems caused by a dead stage. These problems often don't cause a noticeable difference in either dc bias or peak-to-peak voltage. They only affect bandwidth or waveshape.

Video frequency problems can affect a single frequency or a whole band of frequencies. To make matters worse, the problem could be in the video detector, in the IF stages, or in one of the video amplifier's circuits. The VA62A Video Analyzer has a special video pattern called the multiburst barsweep to help isolate these tricky problems.

The multiburst barsweep video pattern (refer back to Fig. 3-1) has samples of ten different video frequencies, ranging from 0 (dc) to 4.5 MHz. You judge the performance of all the video circuits, from antenna to the picture tube, by simply looking at the CRT screen. To understand why this is possible calls for an understanding of how different video signals relate to frequency.

Video signal frequency relationship

A video circuit must have enough frequency response (bandwidth) because a video signal contains many different frequencies. The video circuits must treat all the frequencies the same to produce a good picture. However, how do video frequencies relate to picture quality?

A TV picture is made up of 525 horizontal lines, repeating at a rate of 15,734 Hz. Each line changes in amplitude to form the picture as the beam moves across the front of the screen. The video frequency produced depends on how often the amplitude changes in a scan line. You can calculate the exact frequency if you know how often the beam is interrupted as it moves across the screen. (Note the waveform shown in Fig. 3-9.)

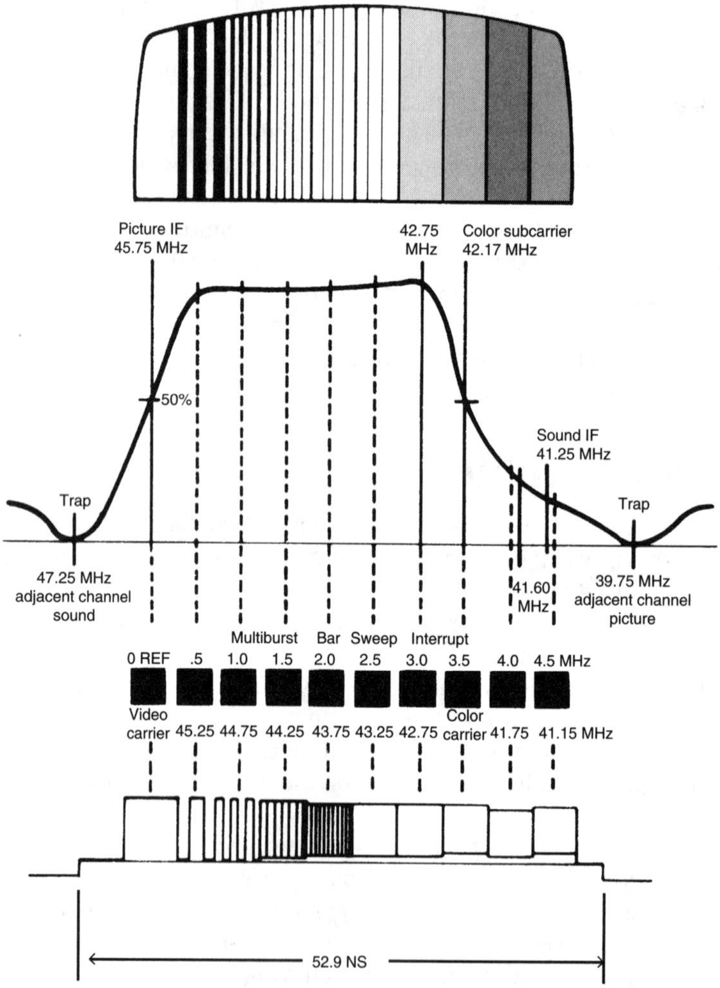

3-9 The composite video signal is "active" for 52.9 microseconds, so this time is used for frequency calculations.

To make these calculations, you need to know the "active time" for one horizontal line. This is the time between blanking and sync pulses, but doesn't include the blanking interval itself. To determine active time, you start with the time for the entire line, including blanking. Now invert 15,734 (divide it into one) to get 63.5 microseconds. Now subtract the 10.6 microseconds used for blanking and sync to get 52.9 microseconds for trace time. (Refer to Fig. 3-10 for waveform trace time.)

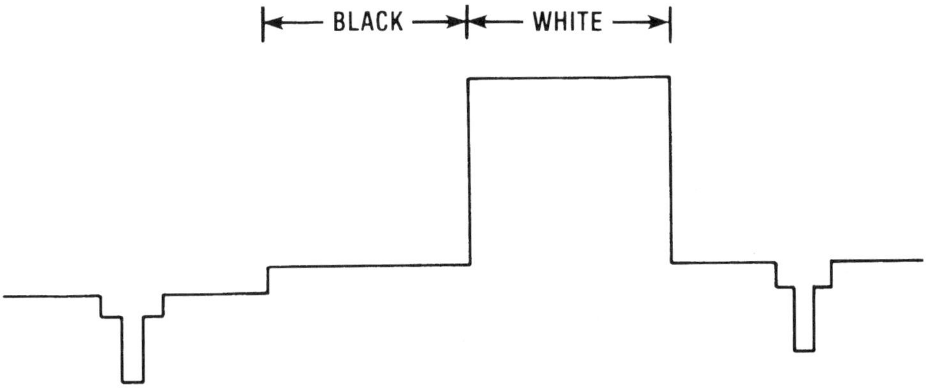

3-10 If the left half of the screen is black and the right half is white, then the video forms a square wave with a period equal to the "active" time of the horizontal line—18.9 kHz.

Now, let's see what happens as we apply video to interrupt the beam. We'll start when the screen is completely white. The video signal is now interrupted only by horizontal blanking. Since the signal is at a fixed level for 83 percent of the time, it's more like a dc signal than an ac signal.

Next, let's see what happens when we change the picture by making the left black and the right half white (see Fig. 3-11). The video signal now becomes a square wave, which forms one complete cycle during the active time of each horizontal line. Earlier, I noted that this period is 52.9 microseconds. Inverting 52.9 microseconds tells us that the video frequency is now 18.9 kHz.

Now let's double the interruption rate to form two white stripes of equal size. The time for each cycle is half as long as our first example. This calculates to twice the first frequency, or 37.8 kHz. As we interrupt the screen more often, we can translate the interruptions into frequency by multiplying the interruptions times 18.9 kHz.

Most TV pictures have a mixture of large and small objects, instead of a row of identically sized objects. This causes the video signal to be a mixture of many different frequencies. Each different-sized picture element produces a signal of the same frequency as it would if it were repeated all the way across the TV screen. Large objects create high frequencies, and small objects create high frequencies. The drawing in Fig. 3-11 puts this into a better perspective.

The ten soldiers across the screen are each one-twentieth the width of the screen. If you think of the soldier as a high-level and the in-between space as a low-level, the electron beam is interrupted ten times across the screen. The ten soldiers then calculate to 10 times 18.9 kHz, or 189 kHz.

If our video circuits only had 189 kHz of bandwidth, We would not see the soldiers. We would only see their outline; and not a very clear one at that. We would need more bandwidth to see smaller details.

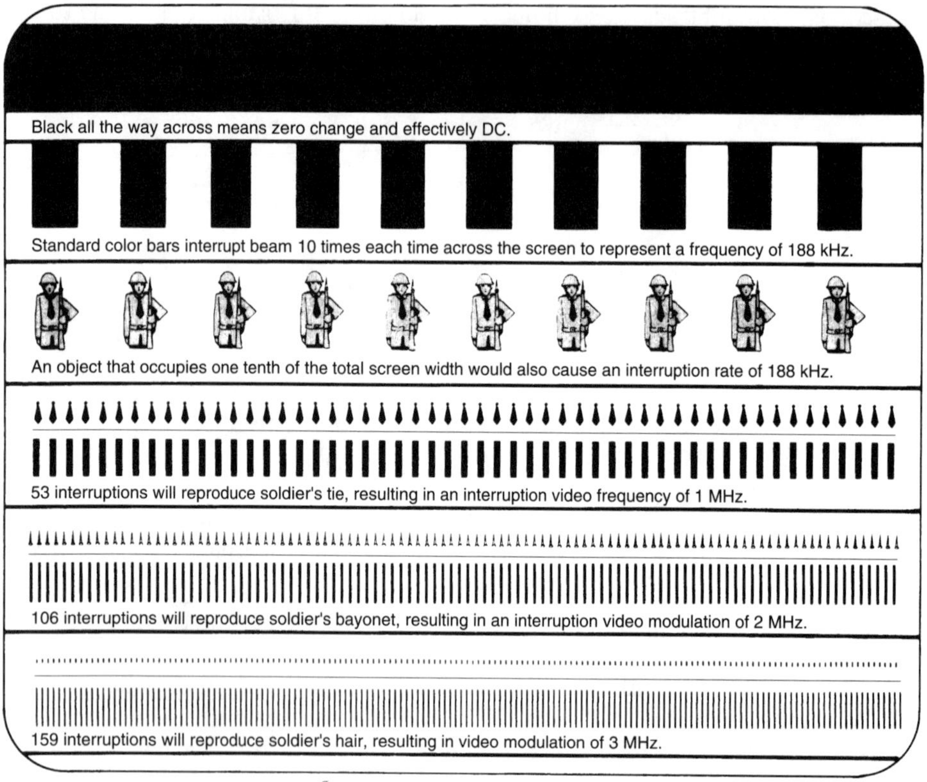

3-11 A TV picture is made up of many video frequencies, all mixed together, as these soldiers represent.

If the soldiers' ties are one-fifth the width of the soldier, we would need enough bandwidth to place 50 necktie-sized objects (plus equal blank spaces) on the screen. This needs 50 times 18.9 kHz, or 945 kHz of video bandwidth. We'll round this to 1 MHz, which would allow 53 neckties across the screen.

A 2-MHz signal could represent 106 objects across the screen, or it could represent one object $\frac{1}{106}$ the screen's width. This might be the bayonet on the end of the soldiers' rifles.

Increasing the frequency to 3 MHz allows you to form objects $\frac{1}{159}$ the width of the screen, such as the hair on the soldier's head. Since most TV receivers have 3-MHz video response, this is the smallest object they can show on the picture.

A TV monitor with a direct video input (which bypasses the tuner and IF circuits) might provide a 4.5-MHz response. If so, the monitor will be able to place 238 objects across the screen, with spaces of the same size. This would be objects 24 times smaller than the outline of the soldier.

Now you know how different-sized objects create different video frequencies. Let's next examine how the multiburst bar sweep pattern relates to this, starting with video processing ahead of the video detector.

Using the multiburst bar sweep for IF circuit tests

The multiburst bar sweep pattern provides ten bars (bursts) of video, each with a different frequency. Any rolloff or distortion in the video circuits will affect the corresponding bars. You simply view the screen to see if a video problem has restricted one or more bars.

Sencore chose half-megahertz spacing of the test frequencies because everything in video happens at multiples of 500 kHz. To understand this, we need to look at the way video signals are processed in the IF stages. They must filter unwanted interference while feeding as much video bandwidth as possible to the video amplifiers.

Part C of Fig. 3-12 shows the ideal frequency response at the video detector output. It doesn't matter what type of circuits are used in the IF stages. Variations from this ideal response curve will affect the picture, whether the TV has adjustable IF amplifiers or a fixed-tuned SAW filter.

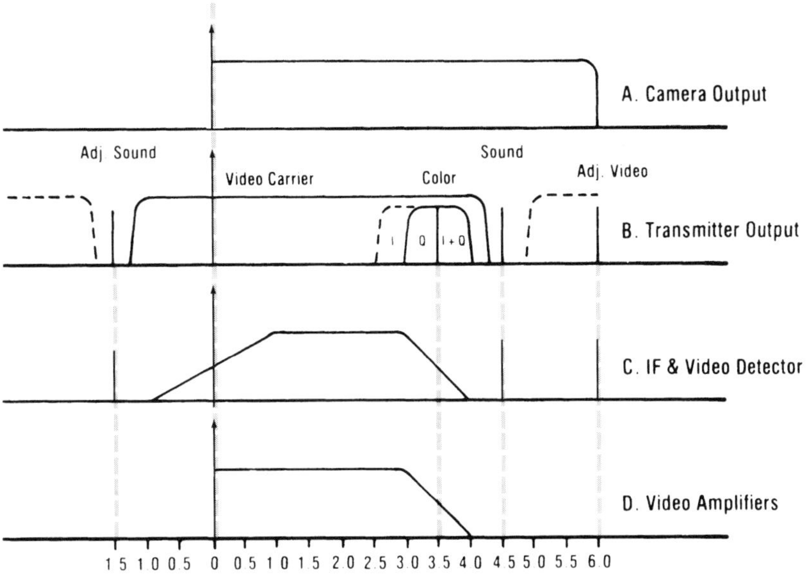

3-12 The relationship of video frequencies: A. The output of the TV camera. B. The TV transmitter output. C. The combination of the IF stages and video detector. D. The video amplifier response. A frequency lost in any part cannot be recovered in a later group of circuits.

First, notice that there are three carriers that call for trapping to prevent interference with the video carrier:

1. The adjacent audio.
2. The on-channel audio.
3. The adjacent video.

The carrier for the adjacent audio (the next lower channel) is 1.5 MHz below the video carrier. The sound carrier for the channel we're viewing is 4.5 MHz higher than the video carrier. Then, the adjacent video carrier (the next higher channel) is 6.0 MHz above the video carrier. Are you beginning to see a trend? Each of these frequencies is displaced by a 500 kHz multiple from the video carrier. Next, let's see what happens at video frequencies below 1.5 MHz.

As discussed earlier, large objects in the picture cause low video frequencies. When these low frequencies are used to amplitude-modulate the video carrier, they produce sidebands close to the carrier. As Fig. 3-12 shows, the video transmitter uses full double-sideband modulation (standard AM) when transmitting these frequencies.

Smaller objects produce higher video frequencies that, in turn, form the carrier. Sidebands more than 1.25 MHz from the carrier have only one (the upper) sideband. This is the special vestigial-sideband transmission that allows you to squeeze reasonable picture detail into a 6-MHz TV channel.

The IF circuits must compensate for the gradual loss of the upper sidebands as the modulation frequency increases. This compensation gives signals with only one sideband twice as much gain as those with two sidebands. This, in turn, places the video carrier at the 50 percent point on the bandpass curve. If the carrier is too high or too low on the curve, the picture develops harsh edges or ringing. Checking for proper gain at 0, 0.5, 1.0, and 1.5 MHz ensures that the rolloff is smooth and balanced.

The sideband frequencies represented by 2.0 and 2.5 MHz correspond to the flat area of the response curve. At the highest video frequencies, the gain must be reduced to make room for the 4.5-MHz audio carrier. To prevent ringing, the response must have a smooth rolloff from 3.0 to 4.0 MHz, with the 3.58-MHz color signal at the 50 percent level.

Now you know the significance of half-megahertz stepping. However, how can we be certain that there's not a problem between each half-megahertz test frequency? We might miss problems if we tested with sine waves. However, square-wave testing eliminates this, since a square wave needs correct phase and gain response of all its odd harmonics for proper shape. The harmonics effectively fill all the gaps between the fundamental test frequencies. This is why Sencore uses square waves instead of the sine waves used on generic multiburst generators.

The first use of the bar sweep is to troubleshoot any kind of video response problem. The bar sweep also, by the way, lets you realign the IF stages if you want. However, its first use is for troubleshooting.

How the video amplifiers affect the picture

If a video amplifier restricts gain of some frequencies, the picture will have the same poor quality as when you have a problem in the IF stages or the video detector. The same VA62A video pattern that dynamically tests the IF stages tests video amplifiers too. This isolates troubles like bad peaking coils or compensation capacitors. Note the waveform shown in Fig. 3-13.

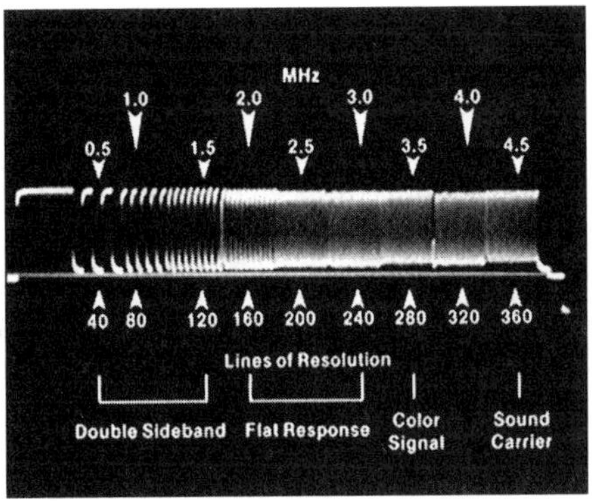

3-13 The half-megahertz steps of the VA62A's multiburst bar sweep relate directly to the video circuits.

As you can see, the multiburst bar sweep pattern lets you check the video response every half megahertz to confirm that all the video stages have the correct response. You simply look at the CRT screen to see if all the bars are getting through to the output. Next, you'll see how to interpret the results.

How to test response by looking at the CRT

You can look at the TV screen and tell if each multiburst bar sweep frequency has made it to the CRT. You know a frequency has reached the CRT if the bar shows detail in the form of black-and-white vertical lines. In addition, the brightness (white level) of each bar shows how the gain at one frequency compares to the others. Simply lower the brightness or contrast controls to see whether all bars have the normal brightness level. Restricted bars will turn dark before the others.

To identify frequencies, use the "interrupt" buttons. Just reach up to one of the buttons and watch the CRT as you turn a bar off or on. If the bar was not the one that you wanted to identify, move over a button or two until you're interrupting the one you want. Then, just look above the button and read the video frequency of the test bar. Now, let's see what to consider normal response.

A color TV receiver should show video response to 3.0 MHz when feeding through the antenna. The 3.5-MHz bar must always be filtered (show no detail) to confirm that the color trap is working correctly. The 4.5 MHz should be trapped as well.

A high-resolution monitor with a direct video input (that bypasses the tuner) might show video response as high as 4.0 or even 4.5 MHz. Even in these monitors, however, the 3.5-MHz (color) signal should show no detail, since the color signal still must be trapped from the video circuits.

Most black-and-white sets only show video response to 2.0 or 2.5 MHz. The lower bandwidth gives the appearance of added sensitivity, since high-frequency noise is filtered in the video circuits.

In each case, the multiburst bar sweep tests every video circuit from the input, right to the CRT. Simply look at the CRT to see whether the circuits show normal response.

Isolating problems

Use the VA62A's signal injection features to isolate the cause of the poor picture. The RF/IF generator lets you inject modulated signals into the antenna or any IF stage. The drive signals section lets you inject composite video into any stage after detection, using the "video pattern" drive signal.

You need only to watch the screen as you inject substitute signals. If the pattern returns to normal, you know that you're injecting the signal after the defect. If the screen still shows a problem, you know that the VA62A's good signal is affected by the defect, proving that you're injecting ahead of the problem. The defective stage is the one that gives good results when injecting at its output, and bad results when injecting at its input.

You don't have to disconnect components when substituting signals from the VA62A because it swamps out the existing signal. When injecting into stages after the video detector, remember to leave the VA62A RF cable connected to the antenna terminal. This provides sync to the other stages, allowing the phase-locked substituted signal to lock properly onto the screen.

4

Chroma circuit alignment and troubleshooting techniques

FOR MANY YEARS NOW, ELECTRONICS TECHNICIANS HAVE FOUND THAT SIGNAL substitution or signal injection is the preferred method for locating circuit faults. Many have indicated that this technique is much faster than other methods they've used. Signal substitution is easier, more accurate, quicker, and it keeps you from replacing components that aren't defective.

The VA62A Analyzer was designed with these advantages in mind. The VA62A and VC63 accessories provide all of the required signals for substitution into all stages of a TV set or VCR, from stem to stern. The Analyzer provides the basic patterns most technicians have used, plus new and unique patterns that will make your servicing faster and easier.

The Analyzer has the following patterns available as modulated RF or IF or variable video from the drive signal output. These patterns are standard, have been used for many years, and include:

- Single cross. Used to center the picture with the yoke and for purity adjustments of the color CRT.
- Single dot. Used to set the static convergence magnets.
- Crosshatch. Used in setting the vertical linearity and dynamic convergence controls. This pattern has 21 vertical and 15 horizontal lines so that each section forms a square for easier height and linearity adjustments.
- White dots. The white dots are used in dynamic convergence.
- Standard color bars. The 10-NTSC-phase color bars allow you to match the waveforms found in the TV service manuals. Also used for TV and VCR chroma circuit troubleshooting and alignment.

Special patterns

Let's now look at the special test patterns found on the VA62A Analyzer.

Chroma bar sweep Used for chroma circuit troubleshooting and color circuit alignment.

10-bar staircase The ETA staircase signal is a standard test pattern with equally spaced steps so you can accurately test or align synchronous video detectors or the dynamic range of any video amplifier, videotape system, or other luminance circuit.

The staircase signal is ideal for synchronous video detectors, which require a linearity pattern for testing and alignment. Misalignment causes poor contrast or brightness, which is especially noticeable on large-screen units. Other video circuits can cause compression at any point of the operating curve (white, grey, or black) which can only be found with a test of the full operating range of the system.

The VA62A also has an Interlace signal adder and a VIR signal adder.

Multiburst bar sweep

The multiburst bar sweep pattern is unique to TV and video servicing. You'll find this pattern is similar to the multiburst pattern used at the TV stations to check out the transmitter and video equipment monitors. The bar sweep pattern contains 10 different frequency bars that provide you with a complete check of the response of the IF and video amplifiers in conventional TV sets, projection TVs, VCRs, and high-tech video equipment. The drawing of the IF response curve and the multiburst bar sweep pattern in Fig. 4-1 shows where each bar is located with respect to the IF frequency on the curve. The multiburst bars fall every .5 MHz all the way to 4.5 MHz, so you have a complete check of the horizontal resolution of any high-tech resolution video system. There are no jumps in the frequencies that might cause you to miss important information between bars.

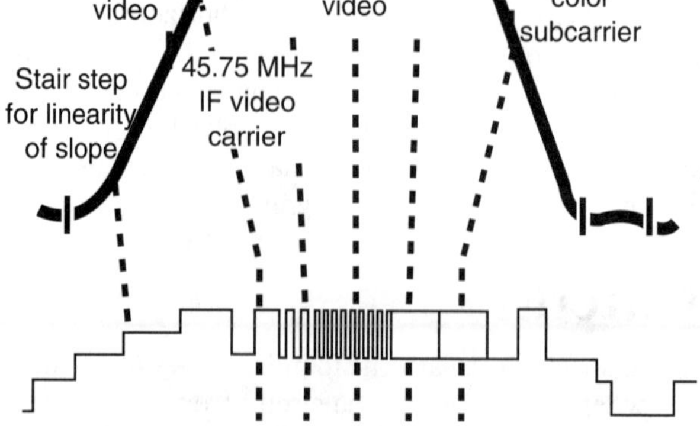

4-1 IF response curve showing the key frequency points. The multiburst bar waveform is shown at the bottom.

Each of the 10 bar frequencies is controlled by a pushbutton on the front panel of the VA62A. Depress the button of the desired bar to add or remove it from the test pattern. All of the bars are phase-locked to the sync pulse for a clear and stable pattern as viewed on the oscilloscope.

Figure 4-1 shows the details of the multiburst bar sweep pattern as viewed on a scope set to the horizontal sweep rate. There are 10 sections or bursts to this pattern. Each one has a specific purpose for analyzing problems in the IF and video amplifier stages, plus alignment quick-checks.

Staircase pattern

As shown in Fig. 4-2, the amplitude of each step in the staircase pattern increases in equal steps to provide a check of the linearity of the IF amplifiers, video detector, and video amplifier stages. The TV or VCR will have poor contrast or unstable sync if these steps aren't linear at the video detector output when the Analyzer signal is injected into the tuner or first IF stages. Poor alignment or a defect in the third IF stage or video detector are the most common causes of poor linearity.

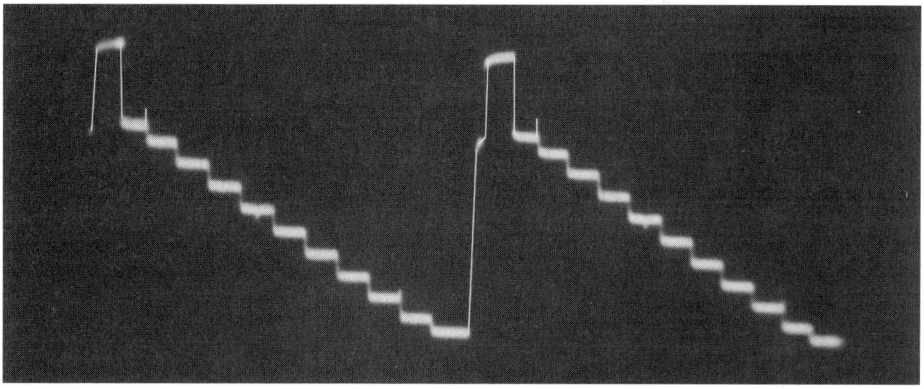

4-2 Linear steps of the stair-steps pattern are used for aligment of the synchronous detector reference coil.

The stair-step pattern can be used for the alignment of a synchronous detector. The 45.75-MHz reference coil must be adjusted for the best linearity of the detected signal to obtain proper set operation. This coil is adjusted until the stair steps are all equal in amplitude, with no overshoot or ringing on the leading or trailing edges of the steps. Poor linearity will cause sync compression and poor picture contrast.

The synchronous detector requires the 45.75-MHz carrier to be present continuously, as it is with the TV-station signal, for proper operation. This continuous 45.75-MHz carrier is not available on a sweep/marker generator. With a sweep/marker generator, the detector will operate only once every $\frac{1}{60}$ of a second when the sweep signal passes through 45.75-MHz. The resulting curve will not allow you to adjust the IF stages or synchronous detector reference coil. The multiburst bar sweep pattern has the continuous 45.75-MHz carrier to allow the synchronous detector to operate as if it were receiving a TV station signal.

The 1.0-MHz bar will also show ringing or multiple images in the picture. This will show up as ringing or overshoot on the leading or trailing edges of the square waves that make up this bar. The sine wave output of a sweep/marker will not show this type of problem because a square wave is required to produce the overshoot or ringing effect.

A key element of the bar sweep pattern is the 1.50-MHz bar. This bar produces an IF frequency of 44.25 MHz. The amplitude of this bar should be very close to the amplitude of the first two bars for good detail on smaller objects in an off-the-air picture. Reduced amplitude of this bar usually indicates a defect or poor alignment in the second IF stage.

This bar is made up of square waves like the first two. The square waves usually appear as sine waves when viewed at the output of the video detector because the response of the IF stages is restricted to 6 MHz. Therefore, you should concentrate on the amplitude of the 1.50-MHz bar, rather then its shape, when aligning or troubleshooting IF problems.

An important point to remember about ringing on square waves is that they're often caused by the scope or the scope probes and not the IF stages of the set under test. You can often reduce the amount of ringing by carefully adjusting the compensation capacitor in the scope probe. The easiest way to check your scope is to connect the scope probe directly to the VTR standard output jack on the Analyzer and select the bar sweep pattern. Be sure to use a 75-ohm resistor to terminate the VTR standard output to provide proper termination.

Some of the newer sets are using the comb filter. The comb filter separates the video from the color after detection. This reduces the 920-kHz beat and allows the video amplifiers to have a wider bandwidth for greater picture detail. With the multiburst bar pattern, you can check both types of sets, with and without the comb filter, in the same manner and in the same amount of time. The multiburst bar sweep pattern will become more important to you as the technology and change in video equipment continues to become greater.

The last bar sweep covered in this section is the 3.50-MHz bar. This bar represents the color carrier in the IF stages. The 3.50-MHz bar, just like the 3.0-MHz bar, should be adjusted for maximum amplitude in alignment and should be less than 50 percent of the amplitude of the lower frequency bar.

The 3.50-MHz bar in the multiburst sweep pattern is ideal for setting the color killer and APC. You should adjust the color killer until there's color running in the 3.50-MHz bar on the screen. Adjust the APC control until the bar locks into a single color. Then turn the color killer control until the color just disappears. You have now adjusted the color oscillator and color killer dynamically without shorting out a test point or making any voltage measurements.

The 3.50-MHz bars are ideal for those sets having a 3.58-MHz trap in the video. Connect a scope to the video amplifier just after the 3.58-MHz trap point and adjust the trap until the 3.50-MHz bars are at a minimum as viewed on the scope trace.

IF alignment tips

Now that I've put the bar sweep pattern together with some of the sweep frequency bars, you can see that that will be ideal for IF alignment. The different bars allow you to see just what's going on in the circuits. There's no need to connect a bias supply to the AGC line, since the AGC will function normally because the sync pulses of the multiburst sweep pattern are the proper amplitude and width as specified by the EIA. Alignment of the IF amplifiers with the bar sweep pattern is easy: simply adjust the IF coils for the best possible bar sweep pattern.

There are three important points that you should look for during IF alignment:

1. Be sure that all the bars are as close to equal in amplitude as possible. The .5-MHz to 3.0-MHz bars should be equal. In some sets, the 3.0-and 3.5-MHz bars will be lower. Also, in some older sets the 4.0- and 4.5-MHz bars will probably not be seen at all.
2. Be sure that there's no ringing or overshoot on the leading and trailing edges of the bars or on the stair steps when you switch over to the staircase pattern.
3. Be sure that none of the bars fall below the reference line created by the video carrier reference mode.

When aligning the IF stages, the polarity of the multiburst sweep pattern viewed at the video detector depends on the polarity of the diode. If the information is taken from the anode of the diode, the waveform will appear as shown in Fig. 4-3A, as a negative-going signal. If the video is taken from the cathode of the diode, the waveform will be like that shown in Fig. 4-3B, a positive-going signal. The waveforms in Fig. 4-3A and B were taken from properly aligned TV receivers.

A

B

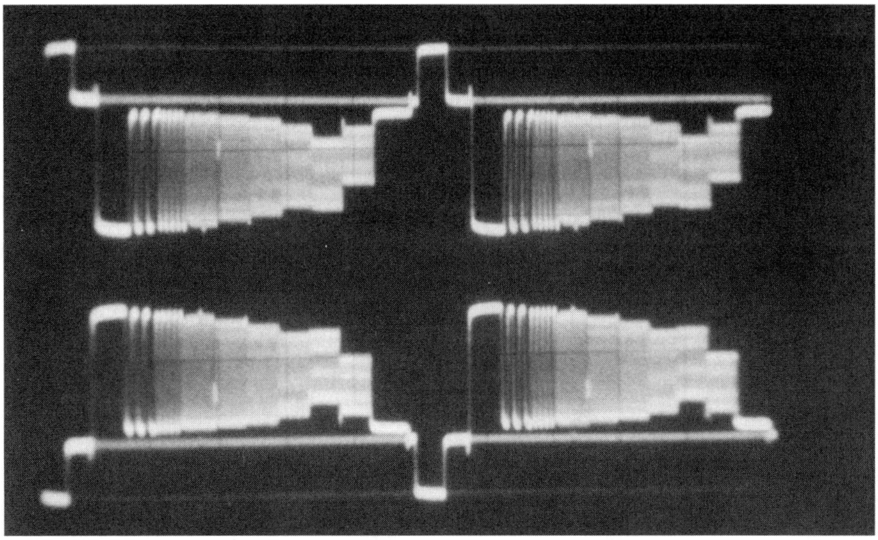

4-3 The polarity of the signal at the video detector is determined by the detector diode. Waveform A shows a negative phase, while waveform B is a positive phase.

The VA62A has crystal-accurate trap signals so you can set the receiver traps right on the nose. With the multiburst bar sweep signals, the crystal-controlled trap signals and the stationlike sync signals, you can quickly align any TV receiver or VCR.

Trap alignment notes

The trap alignment notes, a new feature on the VA62A, enable the TV or VCR technician to align all standard video IF traps with crystal accuracy, from the top of the TV chassis, by just viewing the set's picture. This is the only system that dynamically duplicates the interfering signals that the IF traps must eliminate. The VA62A supplies a reference-level, fully modulated video carrier to lock the AGC to a fixed level and then add adjustable-level-control, crystal-controlled interfering signals, one at a time. The CRT on the set will then show the same type of interference as when there's interference from an adjacent channel.

The VA62A dynamically duplicates the interference the traps must eliminate when your customer uses the TV set. Thus, the interfering signal is being duplicated and not just simulated.

Chroma bar sweep alignment and checks

The chroma bar sweep pattern, just like the multiburst bar sweep pattern, is a new TV servicing technique. The chroma bar sweep pattern has three different square-wave bar frequencies for checking out and aligning the chroma bandpass amplifiers and chroma circuits in TV sets and VCRs.

Each chroma bar is controlled by a front-panel pushbutton with the bar frequency labeled right above it. The chroma bar sweep also has a color burst on the back porch of the horizontal blanking pulse. This burst is the proper amplitude and number of cycles as specified by EIA for the TV broadcast industry. The first of the three bars, on the left side when viewing the TV screen, is the 3.00-MHz bar. This bar represents the lower color sideband and is 500 kHz below the color carrier. This bar will normally be the largest of the three at the output of the video detector, due to the nonlinear amplification of the chroma signal in the IF stages. This bar lets you determine how well the receiver will amplify the color sideband information.

The right-hand bar of the chroma bar sweep is the 4.00-MHz signal that represents the upper-color sideband. This bar is 500 kHz above the color carrier and will appear as the smallest bar at the video detector output. Both bars should, however, be the same amplitude after passing through the chroma bandpass amplifier.

The last part of the chroma bar sweep pattern is the 3.50-MHz, (or color carrier bar), which appears in the center of the chroma bar sweep pattern as shown in Fig. 4-4. This bar will appear halfway between the amplitude of the 3.00- and 4.00-MHz bars when viewed at the output of the video detector. When viewed with the scope at the output of the chroma bandpass amplifiers, the bar should be equal or slightly higher in amplitude.

The three chroma bar sweep bars are designed for both alignment and troubleshooting of the chroma bandpass amplifier stages. Alignment is done by injecting the signal at the input to the first video IF amplifier. This is the same point used for alignment of the IF amplifiers with the multiburst bar sweep pattern. The bandpass amplifiers are then adjusted until the amplitude of the three bars at the bandpass output is as close as possible to equal. You can also inject the chroma bar sweep signal directly into the bandpass amplifiers for isolating a color problem to a single stage in less time than signal tracing with an oscilloscope. Note that the standard 10-bar color pattern will not indicate problems in the color stages that are related to frequency.

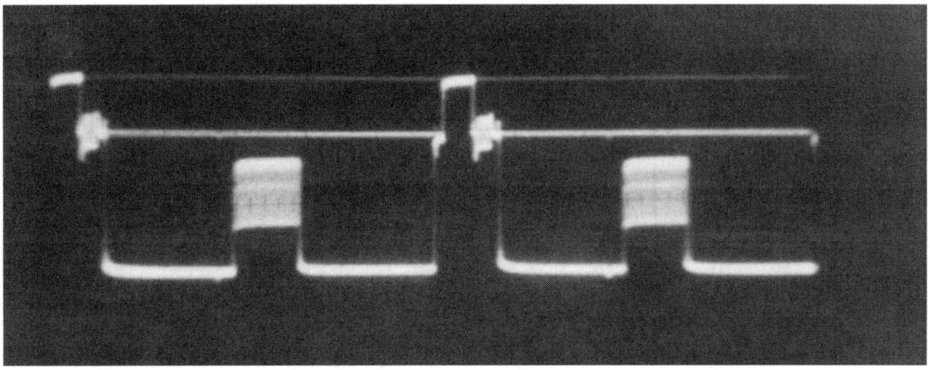

4-4 The 3.50-MHz color carrier bar appears in the center of the chroma bar sweep pattern.

Pinpointing chroma bandpass troubles

The primary function of the chroma bandpass amplifier is to separate the color information from the composite video signal. The bandpass amplifiers are actually filters tuned to accept the 1 MHz of color information that extends 500 kHz above and below the 3.58-MHz color subcarrier. The bandpass amplifiers have two outputs, one that goes to the burst gate to operate the 3.58-MHz local oscillator, and the second that goes directly to the chroma demodulators. There will be a complete loss of color if either of these outputs aren't present.

Figure 4-5 is a block diagram of a typical color circuit. Notice that the output of the second bandpass amplifier is fed to the burst gate for separation of the color burst signal. The output of the second bandpass amplifier stage is also fed to the color level control, which feeds the third bandpass amplifier. The third bandpass amplifier boosts the signal to a level that will operate the demodulators.

Let's now look at a technique for dividing and conquering bandpass amplifier troubleshooting. A defective bandpass amplifier will result in a total loss of color because the chroma signal doesn't reach the demodulators. If you have no color picture, you don't know whether the trouble is in the bandpass amplifiers or if the 3.58-MHz local color oscillator is dead.

You can quickly eliminate the 3.58-MHz local oscillator (and associated stages) as the defect by substituting the Analyzer's 3.58-MHz signal at the input of the demodulators and noting whether the color returns to the set. If the color does return, you know that the trouble has to be in the 3.58 MHz signal path because subbing in this signal returned the color information. Thus, you know that the chroma bandpass amplifiers are passing a signal to the demodulators because, if the chroma signal was not present, you'd not obtain color when subbing the 3.58-MHz signal. The demodulators require both signals to produce color.

The VA62A Analyzer has a separate 3.58-MHz color test signal, phase-locked 90 degrees from the color pattern, for you to inject the two color signals simultaneously

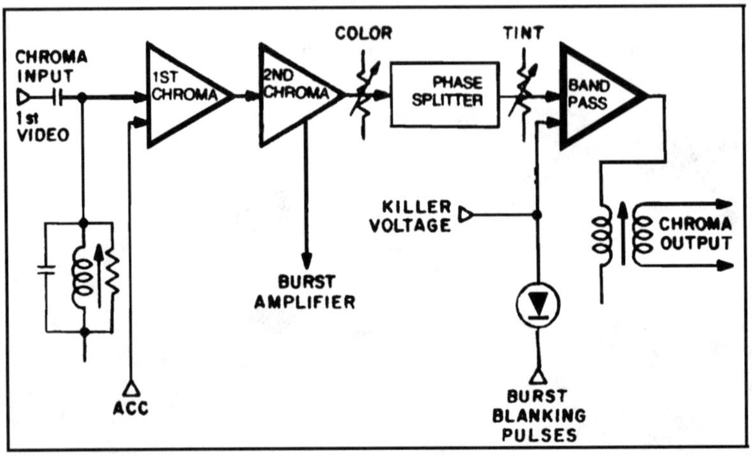

4-5 Block diagram of a typical three-stage solid-state chroma bandpass amplifier system.

into the color demodulator to simulate the color mixing action that takes place within the TV receiver. This enables you to divide and conquer the color circuits at the demodulator.

A defective bandpass amplifier will be indicated by a loss of colors, even when the 3.58-MHz signal is substituted at the demodulators. To locate a specific stage, switch to the video pattern mode and punch up the chroma bar sweep buttons. This now gives you a substitute chroma signal to inject for each chroma bandpass amplifier stage. The reason for using the chroma bar pattern (instead of the color bars) is that this pattern will indicate how well the bandpass amplifiers are passing the full chroma frequency range.

Start by injecting the composite chroma signal at the output of the chroma bandpass amplifiers. This is a good point to start because it tells you more about the operation of the bandpass amplifier stages than any other injection point.

Let's say, for example, that subbing in at the output of the last bandpass amplifier returns solid, locked-in colors on the set's CRT. This tells you that the 3.58-MHz local oscillator is locked onto the color signal being fed into the set's antenna input. This also indicates that the correct signal is appearing at the input to the burst gate amplifier. This tells you that the bandpass amplifier signal is present up to the point where the burst gate signal is separated from the composite chroma signal. This point is usually just before the color level control, so you know the defect is between the color level control and the demodulators. You can now inject across each component from the demodulators back to the color control until you find a point that doesn't return the locked-in color to the CRT. At this point you'll know where the fault is located.

What if injecting either the chroma bar signal or the 3.58-MHz signal at the demodulator inputs doesn't produce color? This tells you that neither signal is present. Troubles after the demodulator will cause no color only if both demodulators are defective, otherwise, you'd see some color. The trouble must then be in the first or second bandpass amplifier because they're the only stages that are common to both signals.

Now move the injection point to the input of the first bandpass amplifier to see if the problem is actually in the color stages. If you now have good color when injecting at the input to the first bandpass amplifier, you know that the color stages are good. The problem would now be a loss of high-frequency (color) response in the RF, IF, or video detector stages. You can now use the IF signal generator mode to check the IF alignment and make sure that the color portions of the composite video response are being passed by the IF stages.

Finally, if you don't obtain color when the signal is injected at the input to the first bandpass amplifier, you know that the problem is between the input to the bandpass amplifier and the demodulators. Now move the injection point one stage at a time toward the demodulators until the color appears on the set's CRT. For example, if you don't obtain locked-in colors when injecting at the input to the first bandpass amplifier, move to the output of the same stage. If color now appears, you know that the defect is in the first amplifier stage. If color doesn't appear, repeat the injection process at the input and output of the second bandpass amplifier and on upstream until you find an injection point that returns the color.

You might find a TV chassis where test signal injection at each test point and stage doesn't return the color. Generally, this indicates that there's more than one defect present and you'll need to tie down one defect and then go onto the other one. To be absolutely sure that the chroma bandpass amplifier is operating properly, connect a scope to the output of the last bandpass amplifier stage and watch the waveform amplitude as you inject at each stage. Keep the level of the injection signal at about the same amplitude as shown on the service data to make sure that you aren't forcing a signal through a defective stage. The combination of signal substitution and signal tracing with a scope lets you tie down the most complex bandpass amplifier faults as you divide and conquer the number of possible defective stages down to a single one and then go to conventional troubleshooting techniques on that stage.

Chroma troubleshooting

Let's look at our typical color circuits with a different problem. This time, you find that substituting the 3.58-MHz signal at the demodulator does correct the loss of color. Remember that feeding in the 3.58-MHz signal with a defective bandpass amplifier gave no improvement because there's no composite chroma signal from the bandpass amplifier. The loss of 3.58-MHz could be in any of the circuits, from the burst gate to the injection point of the demodulators.

The 3.58-MHz oscillator is the first point that gives a continuous wave signal and should be the first item to suspect. Injection of the 3.58-MHz phase-locked test signal at the oscillator output will tell you whether the trouble is somewhere between the oscillator and the demodulator, such as the tint circuits or 3.58 MHz amplifiers. Refer to this circuit and the various test points shown in Fig. 4-6. If the color returns when you inject at this point, you know the trouble has to be in one of the remaining 3.58-MHz circuits, which are the oscillator, the phase correction circuit, or the burst gate.

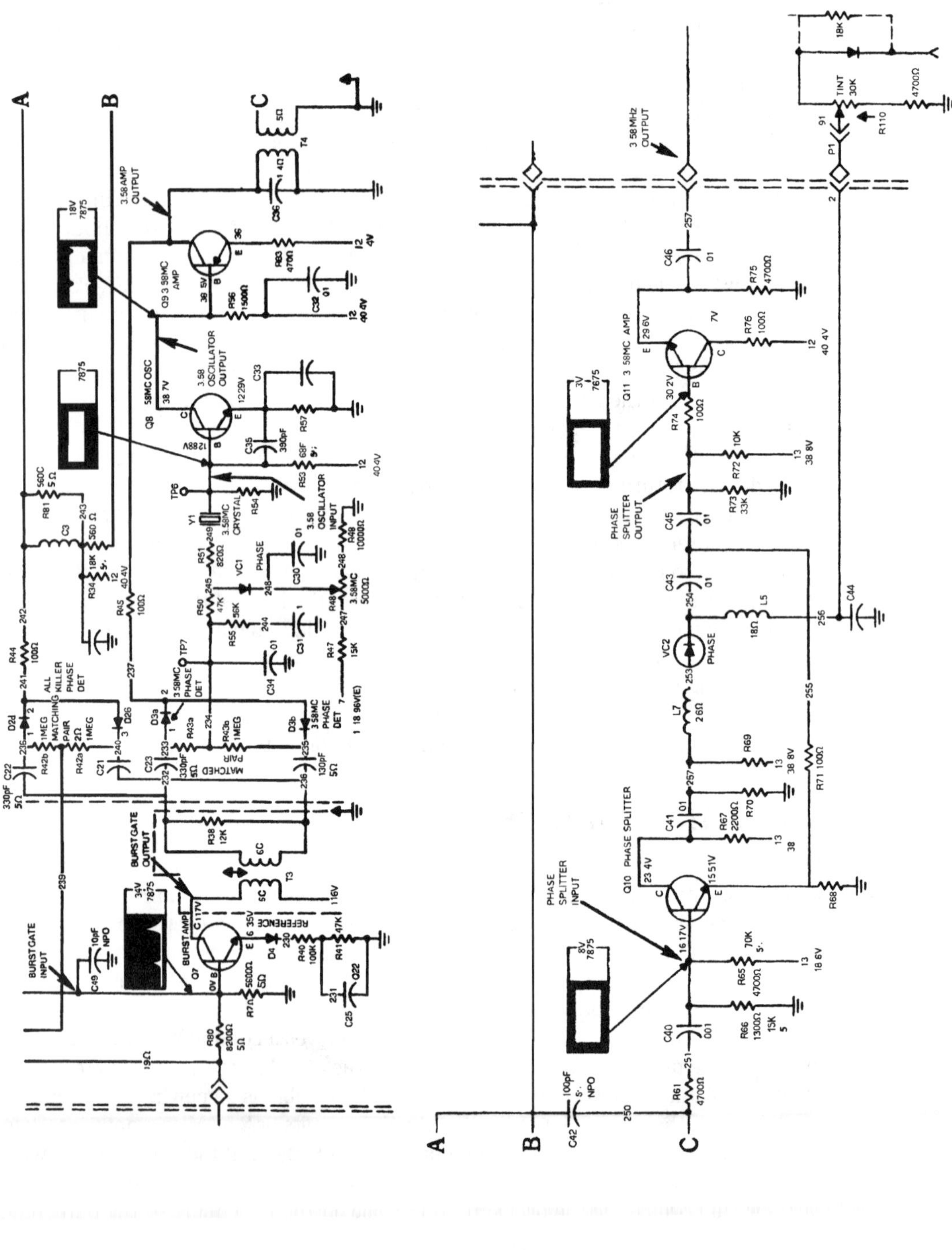

4-6 Test points where the 3.50-MHz signal can be injected in the color circuits.

A defective burst gate can cause either of two different problems. A burst gate that's open (or is missing the keying pulse) will result in no color at all because the color killer will switch to the black-and-white mode since no color burst is detected. A shorted burst gate, on the other hand, will produce color that's not in sync because all color signals, not just the burst, are passed to the 3.58-MHz phase detectors. The 3.58-MHz oscillator no longer has a reference, so the result is rainbowing color.

To check this stage, inject the 3.58-MHz signal at the input to the burst gate and see if any color is displayed on the CRT. It doesn't matter if the injected signal is CW because the burst gate will pass the signal only during the time when the keying pulse is present at the burst gate. If you find that injecting at the burst gate input gives color, you know that the signal from the second bandpass amplifier output is not reaching the burst gate, even though the chroma signal is getting to the demodulators.

If you don't obtain color, the 3.58-MHz signal should be substituted at the output of the burst amplifier to check the 3.58-MHz phase detector. Note that a continuous 3.58-MHz signal is being subbed for a 9-cycle burst pulse. This means that the phase detector will operate continuously and might improve stability; it will also serve as a good test on the phase detector operation. A positive check on the 3.58-MHz phase detector circuit is to use the filtered B+ supply from the VA62A to substitute for the dc output of the phase detector to see if the color can be synchronized. If it can, the phase detector is defective.

Color killer analysis

The automatic color killer (ACK) circuits depend on the presence of a color burst signal to indicate whether the picture should be shown in color or black and white. The bar sweep pattern provides a dynamic test of the color killer circuits because it doesn't have a color burst, but it does have a 3.56-MHz bar. (See the color killer circuit shown in Fig. 4-7.) You shouldn't have any color at all with this test. If you get any color, it will show on the 3.56-MHz bar because this frequency will pass through the chroma circuits and be demodulated. You should set the color killer control until the 3.56-MHz bar just loses its color. When you then switch to the color bar pattern or to the chroma bar sweep pattern, the color should return because both of these signals have a standard color burst signal whose relative amplitude is the same as the color burst sent out by the TV station.

Should the color killer not operate properly, it can be isolated to the actual problem quickly by substituting the dc B+ supply from the Analyzer. For this check, increase the dc potential, positive or negative (depending on the circuit used), to see if the color bars switch from black and white to color as you change the dc voltage.

Chroma demodulator checks

Up to this point, we've been working with defects that occur before the color demodulators. The VA62A Analyzer provides a full range of signals for injection at the output of the demodulators and the matrix amplifiers.

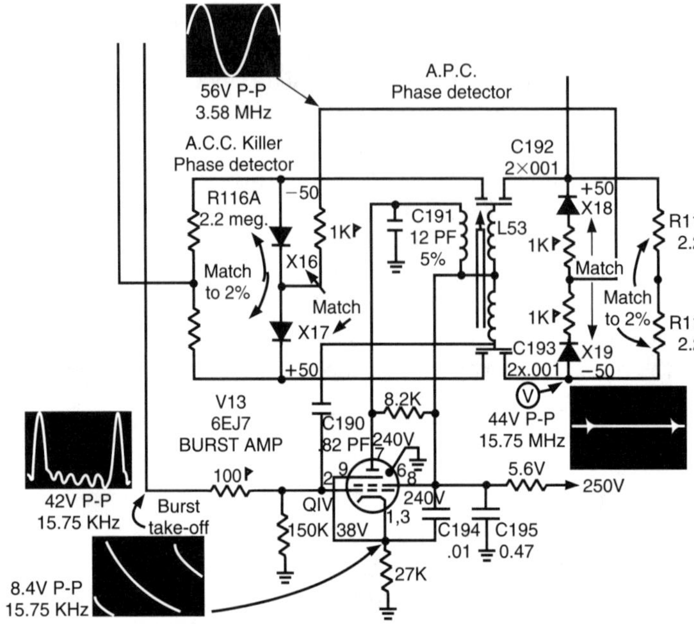

4-7 The phase-locked 3.50-MHz test signal can be used to inject either signal that operates the color killer phase detector.

A defect in the demodulators, matrix amplifiers, or the R-Y, G-Y, or B-Y output stages results in a different symptom than defects in earlier stages. Instead of a total loss of color, you now find that only one (or two) of the colors is missing. For example, if the R-Y demodulator is defective, you'll have blue-green signals but no red signal. A defect in the demodulator is not the only stage, however, that will result in a similar symptom. See the block diagram of a typical demodulator system in Fig. 4-8. Also, let's not forget a faulty red gun in the CRT.

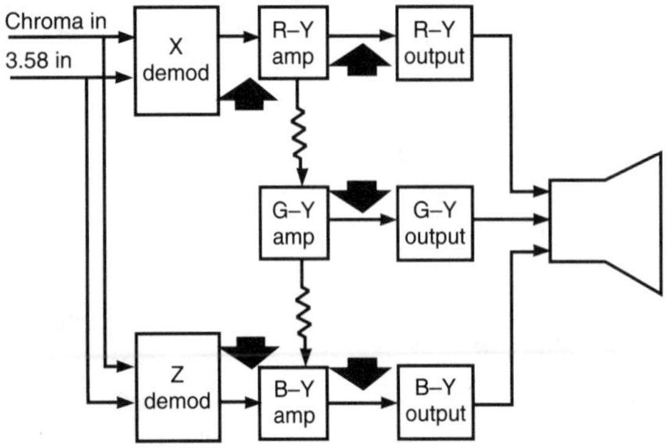

4-8 Block diagram of the color demodulator matrix and output stages.

To locate this type of defect, inject the phase-locked color bar signal at each point in the red color circuits. You can start at the input to the R-Y output stage and see whether or not red returns to the screen. You'll see all of the color bars turn red because the signal is injected after demodulation. If there's no change, you know the trouble is in the output stage itself, in the wiring, or in the CRT socket. The CRT can be driven with the substitute signal in some solid-state sets, or the CRT can be tested with a CRT checker.

Should you obtain a red screen when injecting at the input to the R-Y output stage, just move the injection point back one stage at a time until reaching the output of the R-Y demodulator. If the reds are present when you inject at the output of the demodulators, you know that the demodulator itself is defective. Thus, you're able to isolate the fault to a single stage to reduce the color circuit troubleshooting time.

Color circuit adjustments

The bar sweep pattern allows you to make adjustments like the 3.58-MHz oscillator (APC) and color tracking adjustments much faster because it's not necessary to disable the APC circuit or even flip any switch.

First, the color oscillator is adjusted by turning up the color killer adjustment until the 3.50-MHz bar of the bar sweep pattern displays color. The color will not be locked in because this pattern doesn't have a color burst. Now, adjust the 3.58-MHz APC adjustment until the bar shows a solid color from top to bottom. This action will zero beat the set's local oscillator with the master oscillator in the Analyzer. Remember to readjust the color killer, using this same pattern, when you've completed these color adjustments.

The CRT screen and drive controls that determine black-and-white tracking can be performed without flipping setup switches to adjust for a white line and then obtaining a black-and-white picture from an off-the-air signal. The bar sweep pattern has a special grey scale that's used as a black-and-white reference. The CRT drive and screen controls are then adjusted until there's no hint of color in any portion of the grey-scale pattern.

Finally, the VA62A Analyzer has all of the standard color bar generator patterns, including a single cross, a single dot, a crosshatch and a dot pattern so that you can make all of the set's convergence adjustments.

Chroma processing circuit (chip type)

The RCA CTC97 chassis chroma processing system is combined into one IC chip, eliminating many external parts. Some external parts are required, such as the 3.58-MHz crystal and coupling capacitors. Thus, the chroma IC can be the prime cause of many color defects. However, external components can still cause color problems, and the signal injection technique can be used to isolate these faults.

In many cases, the chroma chip is changed to determine if the IC or outboard components are at fault, and this can be quite time-consuming. Also, if the chip must be unsoldered to be removed, this can damage a good chip. You'll find that signal injection will take less time than changing the IC to determine if the chip or other components are at fault.

To find a no-color fault, for example, you must first establish a reference. To do this, inject the IF signal from the Analyzer into the input of the IF module. This provides the necessary reference signal to keep the other circuits operating properly. The chroma signals that will be injected will now be phase-locked to the reference signal and thus provide a meaningful test.

Inject the chroma bar sweep signal from the driver output into the chroma IC, using the first test point shown on the block diagram in Fig. 4-9. If color returns at this point, check out the chroma take-off. If color doesn't return, the IC or circuit components are at fault. Then inject the signal into the demodulator input to bypass the chroma amplifiers. If color returns, the problem could be the IC or a missing keying pulse. Inject the keying test pulse from the Analyzer into the keying pulse input circuit to determine if the IC is defective.

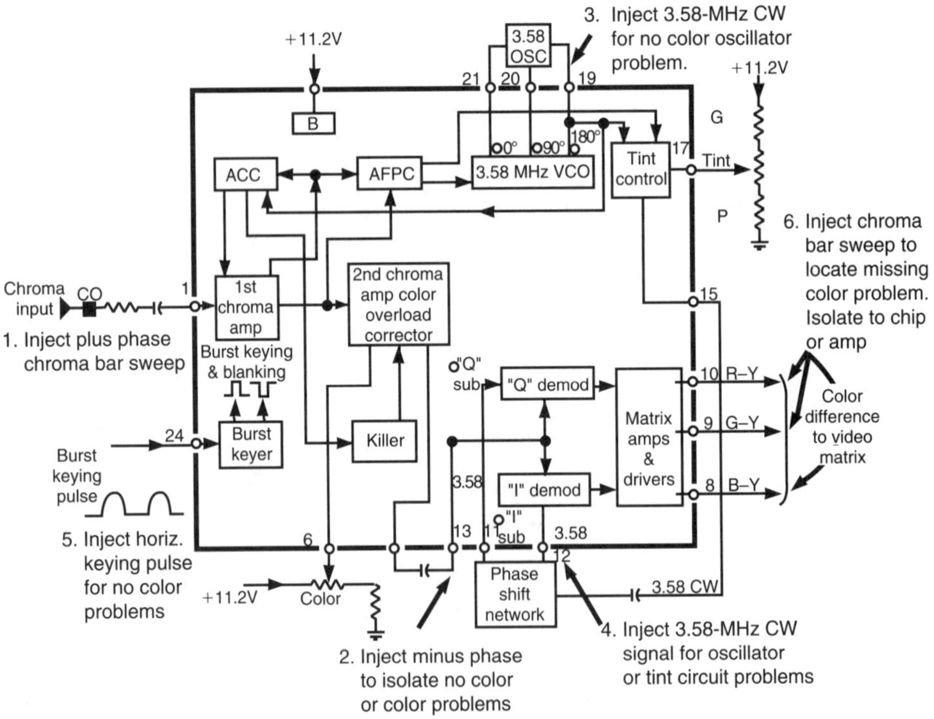

4-9 The IC chroma circuit showing test signal injection points.

If the color didn't return when the chroma bar sweep was injected in the demodulators, the 3.58-MHz oscillator or crystal is the suspect. Now use the 3.58-MHz drive signal from the analyzer to inject into the oscillator (pin 19) to replace the oscillator signal. If color returns, the oscillator in the IC or the crystal is defective. Now check the crystal. If the crystal checks out okay, then the IC is faulty.

If the color didn't return, inject the 3.58-MHz substitute oscillator signal into the demodulator input (pin 12). Color returning here indicates that the coupling capacitor or IC is faulty. By injecting a 3.58-MHz signal on both sides of the coupling capacitor, it can be checked right in the circuit. With the Analyzer, you can go through the complete circuit to determine where the defect is located.

In the case of a missing color, the signal can be injected at the output of the matrix amplifier to determine if the IC or following circuit is the cause. If color returns when the signal is injected at the IC output, the IC is faulty. If the missing color fails to return, the matrix amplifier or output stage is the problem.

Color circuit analysis and alignment

The VA62A Analyzer is designed to completely test and align all sections of a TV receiver that process the color information. The first pattern is a standard 10-bar rainbow color pattern. This pattern (unlike that produced by a standard color bar generator) provides an actual color burst that is phase-locked to the horizontal sync pulses. This provides two advantages compared to color bar generators:

1. The color pattern is more stable.
2. The color killer threshold adjustments can be performed as with an off-the-air TV station signal.

The second color pattern is the chroma bar sweep. This pattern provides a total test of the chroma bandpass amplifier and the color demodulators. The results of color circuit tests are monitored on the TV receiver's CRT, with the signals injected at the set's VHF antenna terminals. Also, drive signals are available to analyze any color problem with signal injection analyzing.

The chroma bar sweep pattern provides a reference for testing the total color system's operation. The systems tested with this one pattern are: the bandpass amplifier, the tint control, and the demodulator stages. For this general test, inject the signal at the VHF antenna terminals.

The chroma bar sweep pattern consists of three bars. These bars should be of equal amplitude at the output of the bandpass amplifier and at the output of the color driver stages. The balance between the upper and lower bar indicates the state of alignment of the bandpass amplifier. The center bar should be a bright blue color.

Additional tests are performed with the multiburst bar sweep and staircase signals. These tests include a check of the drive levels (grey-scale tracking) and the color killer circuits.

The signals for color circuit tests are injected at the antenna input, and the results are monitored on the set's screen. The tests for overall performance of the color system are as follows:

1. Connect the Analyzer's RF output to the VHF antenna terminals of the TV receiver.
2. Set both the VA62A and the TV receiver to the same channel.
3. Select the chroma bar sweep position with the video pattern switch.

4. Increase the RF-IF level control for a stable, snow-free picture. Fine tune the set's tuner, if required.
5. Depress all three of the chroma bar sweep selector buttons. The center bar of the pattern should be bright blue. If the bar is not blue, adjust the tint control to produce a blue bar. If a blue bar can't be obtained, this indicates a problem in the tint range control circuit, the demodulator, or the color drivers. If the colors are low in intensity with the color level control turned all the way up, this indicates a problem in the chroma amps, color drivers, or the CRT. Should there be no color in the center bar, even with the fine tuning properly set and the color level control adjusted for full output, this indicates a problem in the color killer circuit, the burst gate, the demodulators, or the 3.58-MHz color oscillator.
6. Reduce the setting of the brightness control until either or both of the outside bars turn black. If both bars turn black at the same time, this indicates proper operation of the bandpass amplifier and the demodulators. If the right-hand bar is the first to go black, the bandpass amplifier or IF amplifier is favoring the high-frequency portion of the response curve. If the left-hand bar is the first to go black, the bandpass amplifier is favoring the low-frequency portion of the response curve.
7. Now switch from the video pattern to the multiburst bar sweep pattern. Depress the 3.50-MHz bar sweep selector button. If the 3.50-MHz bar shows a color, the color killer adjustment is set too high. If the background is not black, one or more of the CRT bias controls is set too high. If the grey or white bars at the left of the pattern show a color, one of the CRT drive controls is set too high.

Chroma bandpass alignment

The bandpass amplifier alignment response is tuned while observing the screen on the TV receiver:

1. Connect the VA62A to the input of the first IF stage. Notice the control settings and signal injection connections shown in Fig. 4-10.
2. Select the chroma bar sweep position of the video pattern switch. Make sure all three of the chroma bar sweep selectors buttons are depressed.
3. Adjust the tint control until the middle bar is bright blue.
4. Reduce the brightness control until either the upper or lower sideband bar turns black.
5. Adjust the chroma bandpass amplifier coil(s) until both the upper and lower sideband bars have the same brightness level. The alignment of the bandpass amplifier is now complete.

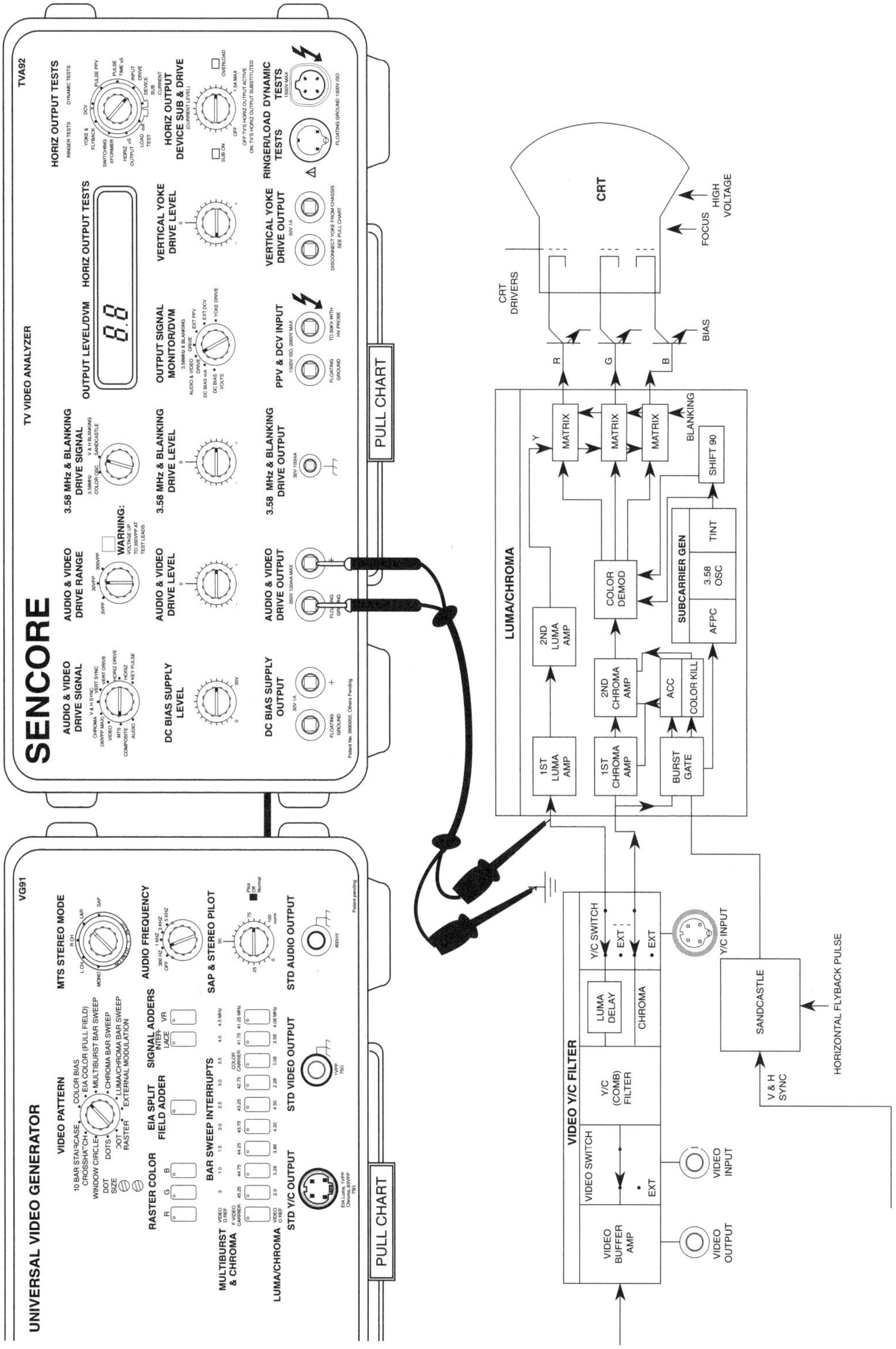

4-10 VA62A setup for chroma bandpass amplifier checks and alignment.

With an oscilloscope, use the following procedure for bandpass amplifier alignment.

1. Connect the VA62A Analyzer to the input of the first stage, as you would do for IF alignment.
2. Select the chroma bar sweep position of the video pattern switch. Make sure that all three of the chroma bar sweep selector buttons are depressed.
3. Connect the oscilloscope to the input of the chroma bandpass amplifier.
4. Adjust the chroma bandpass amplifier coil(s) until all three of the bars on the scope have the same amplitude. The alignment of the bandpass amplifier is now completed.

Setting the 3.58-MHz oscillator

The chroma bar sweep pattern provides a reference for setting the local 3.58-MHz color oscillator to the proper frequency with the automatic phase control (APC) disabled. Sets that don't have a switch for disabling the APC can be set with the bar sweep pattern by increasing the setting of the color killer until the 3.50-MHz bar shows color.

To set the color oscillator on sets with an APC switch:

1. Feed the RF-IF output from the VA62A to the TV set's VHF antenna input.
2. Set the VHF tuner and the Analyzer to the same channels and fine tune the set, if required.
3. Select the chroma bar sweep position to the video pattern switch.
4. Move the color APC switch to the service position.
5. Adjust the 3.58-MHz control or coil for the least amount of color change in the blue bar of the chroma bar sweep pattern. A barber-pole pattern of colors will be evident in the center bar when the 3.58-MHz phase is not correct. The color will be solid across the bar when the adjustment is properly set, although the color of the bar might change very slowly.
6. Return the color APC switch to the normal position. Should the TV receiver not have an APC defeat switch, the bar sweep pattern should be used. This pattern will in effect disable the APC circuit because the pattern doesn't have a color burst signal to reference the local color oscillator. The color killer must be set to produce a color in the 3.50-MHz bar before the adjustment to the local oscillator is performed.

To set the color oscillator is TV sets without an APC switch:

1. Feed the RF-IF signal output from the Analyzer to the set's VHF antenna input.
2. Set the VHF tuner on the set and the Analyzer to the same channels and adjust fine tuning, if required.
3. Select the bar sweep position of the video pattern switch. Be sure that the 3.50-MHz bar sweep selector button is depressed.

4. Set the color killer threshold control to produce color in the 3.50-MHz bar of the bar sweep pattern.
5. Adjust the 3.58-MHz oscillator adjustment for the least amount of color change in the 3.50-MHz bar. If there's a barber-pole effect in this bar, the adjustment should be changed until the 3.50-MHz bar produces a bar of the same color from top to bottom, with the slowest amount of color change.
6. Reset the color killer threshold control for proper color killer operation.

Color killer adjustment

The color killer threshold adjustment should be set to produce proper color information on both the color bars and the chroma bar sweep pattern, and no color information in the bar sweep pattern. This will result in a more accurate setting of this control than when a color bar generator is used because a color bar generator inserts only an extra color bar in the position normally occupied by the color burst. As the color level control of a color bar generator is reduced, the simulated burst is also reduced.

The procedure with the Analyzer will more closely simulate the operations of the color killer circuit when receiving TV station signals. These adjustments will compare the color killer operation with and without a color burst signal present.

Use the following procedure to set the color killer threshold adjustment:

1. Feed the RF-IF output from the Analyzer to the TV set's VHF antenna input.
2. Set the TV receiver's VHF tuner and the Analyzer to the same channels and adjust the fine-tuning control, if required.
3. Select the bar sweep position of the video pattern switch.
4. Depress the 3.50-MHz bar sweep selector button.
5. Adjust the color threshold control until the 3.50-MHz bar produces a color.
6. Reduce the setting of the color killer threshold control until the 3.50-MHz bar just loses color.
7. Switch the video pattern switch to the color bar position, and then to the chroma bar sweep position. In both of these positions, the TV receiver should display color information. If it doesn't, reset the color killer threshold control slightly to produce color. This concludes the color TV receiver chroma circuit checks and adjustments.

Introduction to comb filters

With the advent of comb filters, the resolution on the screens of color sets is very much improved. Before comb filters came into use, TV sets had less high-frequency resolution in the picture than what the TV station transmitted. The video frequency response had to be limited to minimize the beat interference (called color crosstalk) created by the high-frequency luminance beating against the 3.58-MHz chroma subcarrier and its AM sidebands. This beat interference created a grainy or noisy picture, especially in the larger areas of color.

The elimination of the color crosstalk or beat is accomplished by limiting the highest frequencies passed by the video amplifiers to about 3 MHz, even though TV stations transmit luminance signals to 4.2 MHz. This prevents most of the beat interference from being passed to the CRT, although there's still a small amount seen in large red or yellow color areas. Reducing the bandpass, however, also reduces the overall high-frequency video information, which represents the smallest picture detail that can be viewed.

A second problem with conventional color sets is that certain objects in the black-and-white (luminance) channel produce a video signal in the 3.08- to 4.08-MHz frequency range. Herringbone jackets or pin-striped suits, for example, often cause luminance signals in this frequency range. A conventional chroma bandpass amplifier can't distinguish between this high-frequency luminance information and the color information, which falls into the same frequency band. The result is rainbowing color on these objects. The color often appears and disappears at random as a person wearing such a pattern moves closer or farther away from the camera, changing the relative spacing between the small stripes and, thus, the frequency represented by the video signal.

The comb filter eliminates almost all of these problems. First, it allows the high-frequency luminance information to pass to the CRT without any of the color beat interference. Then, all but a very small percentage of the color rainbowing caused by closely spaced patterns is eliminated. Another minor advantage is that more of the color information is passed to the color demodulators, resulting in more color on smaller objects in the picture. This color response is restricted in a conventional chroma bandpass amplifier system.

Most set makers are now using comb filters in their high-end receivers. Comb filters are also used in VCRs to separate chroma and luminance signals and to eliminate certain types of color interference. Thus, comb filters provide superior separation of the chroma and luminance information as compared to a conventional color receiver.

I'll now cover the operation of the basic comb filter and then cover the differences between the Magnavox and RCA circuits. Later in this chapter, you'll find the alignment and checkout procedures for these two circuits so that you can align, check out, and troubleshoot the type of circuit you're working on.

Basic comb filter operation

The basic operation of the comb filter is based on four important characteristics of the transmitted NTSC composite video signal. These are as follows:

1. The luminance (black-and-white) information doesn't fill the entire energy spectrum between the video carrier and the upper 4.2-MHz limits allowed by the FCC for the luminance signal. The energy of the luminance information is actually found in small packages at each odd multiple of the horizontal sweep frequency. If you were to look at the energy spectrum of a composite video signal using a spectrum analyzer, you'd see a frequency response that looks like a picket fence with spaces between each energy package. (See Fig. 4-11.)

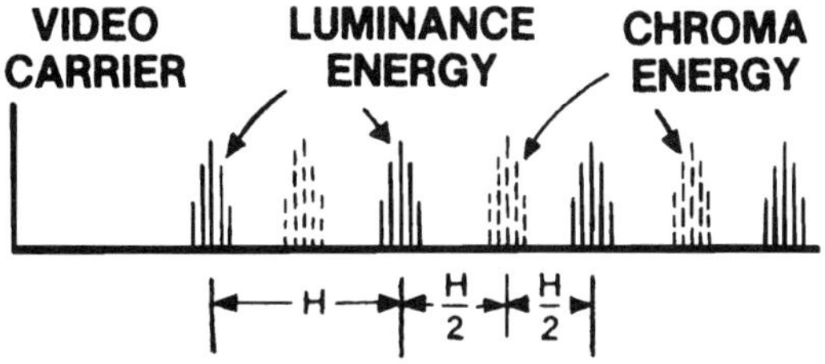

4-11 The energy of the luminance information is found in small bundles at multiples of the horizontal frequency, leaving room to "interleave" the chroma information between the energy bundles of the luminance information.

The luminance signal actually occupies less than one-half of the spectrum space between dc and 4.2 MHz. In the NTSC system, the chroma information is also in similar energy packages, except that each chroma package falls into the spaces between the luminance energy packages.

2. In order for the chroma information to fall into the unused spaces created by the luminance information, its frequency was made one-half of an odd multiple of the horizontal sweep frequency. The color subcarrier (3.579545 MHz) is the horizontal sweeping frequency (15734.264 Hz for color) multiplied by 455 (an odd multiple) and divided by 2 (to make the chroma fall into the spaces between the luminance signal). To understand the comb filter operation, just remember that the chroma signal is one-half a multiple of the horizontal frequency. The result is that the phase of the color signal changes 180 degrees from one horizontal line to the next. This is referred to as "phase alternation" of the color information.
3. Most of the transmitted luminance information is of a vertical format, meaning that there's usually very little difference between the information from one line to the next. This means that two successive horizontal lines can be added together without greatly changing the luminance information of either of the two added lines.
4. All signals transmitted from the TV station are phase-locked to each other. This means that the horizontal sync, color burst, luminance, and chroma information all have an identifiable phase relationship at the beginning of each horizontal line.

A drawing of the complete video spectrum showing the interleaving of the chroma and luminance information is shown in Fig. 4-12. The energy bursts appear similar to the teeth of a comb. Thus the circuit name "comb filter."

The basic operation of the comb filter can be broken into two sections. First, the luminance information is removed (or combed) from the composite video signal to leave only the chroma signal and its sidebands. Second, the "combed" chroma is used to remove the color from the original composite video, leaving only the luminance information. These two steps provide the clean chroma and luminance information required for a high-resolution TV picture. Let's start the circuit analysis with the removal or "combing" of the luminance from the composite video to obtain the chroma-only signal. Refer to the block diagram in Fig. 4-13.

The composite video is fed to the input of a delay device that delays the signal by one complete horizontal scanning line. The chroma information at the output of

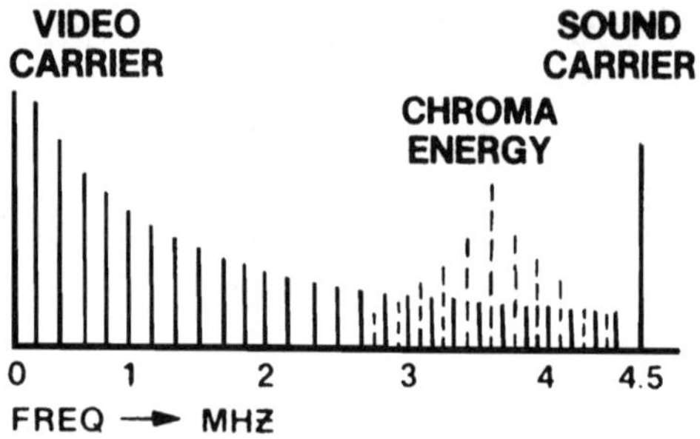

4-12 The full video frequency spectrum showing the interleaving of the chroma subcarrier, its sidebands around 3.58 MHz, and the luminance signal sidebands.

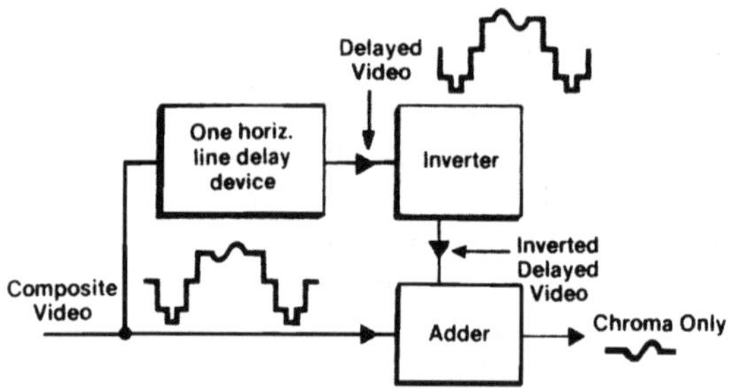

4-13 A block diagram of the basic "comb" filter operation to separate luminance from the composite video signal to leave the "combed" or chroma-only signal.

the delay device will be out of phase with the nondelayed chroma fed to the adder circuit because the color information is alternated and changes phase for each succeeding horizontal line. The first line is now delayed, and it's being compared with the second line of video information.

The delayed signal is then inverted so that the chroma information of the two signals will now be in phase. Because the luminance doesn't change phase from line to line, inverting the signal will now make the delayed luminance signal out of phase with the nondelayed signal fed to the adder circuit. When the inverted (and delayed) signal is mixed with the noninverted (and nondelayed) signal in the adder circuit, the chroma information will add. The luminance information is out of phase and will cancel. This leaves only the "combed" chroma signal at the output of the adder stage. Refer to Fig. 4-14 for a direct comparison of the waveforms involved.

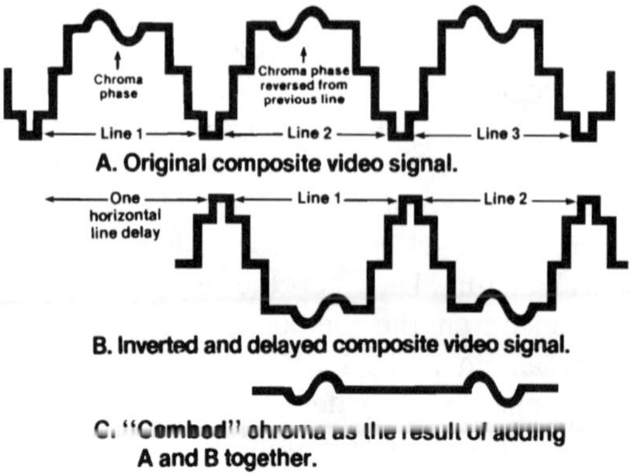

4-14 Part A represents the original composite video signal and B the delayed and inverted composite video signal. C is the result of adding A and B together. The luminance information in A and B will cancel, and the chroma information will add to produce only a chroma signal.

Now that you have the combed chroma signal (with the luminance information "combed" out), you can, in turn, use this signal to remove the color information from the composite video. You now invert the combed chroma signal so that it's out of phase with the chroma information in the composite video. The inverted, combed chroma signal is then added to the composite video so that the chroma information cancels, leaving only the luminance. In effect, you've used the combed chroma to "comb" the color from the composite video. (Refer to Figs. 4-15 and 4-16).

4-15 The basic "comb" filter operation is to separate chroma from the composite video to leave only a combed luminance signal.

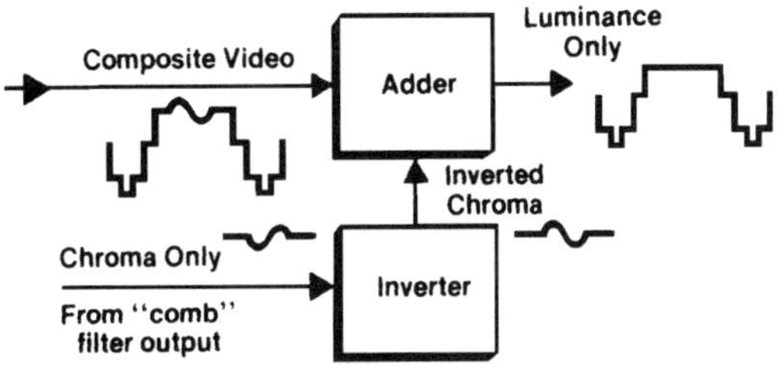

4-16 By adding the "combed" chroma signal back to the original composite video signal, the two chroma signals will cancel and "comb" the chroma from the composite video signal to leave luminance without chroma.

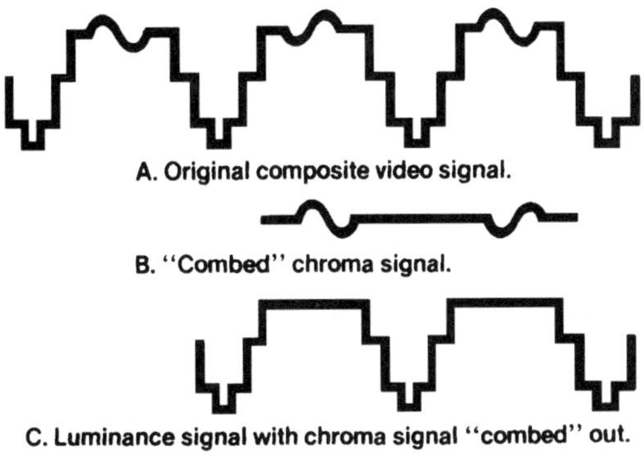

The overall results of the combing process is that you now have a high-resolution luminance signal (with all the chroma cancelled out), and this signal can be passed on to the video amplifiers. Note that the frequency response of the luminance signal has not been changed. This means that the video amplifiers can have a wider bandpass, resulting in more picture detail on the CRT without color-beat interference. You also have a clean chroma signal to be passed on to the demodulators for further separation into red, green, and blue. You've greatly reduced the tendency to produce rainbowing on pin-striped suits (and other similar objects) because the only signals passed to the chroma circuits are the ones with an exact phase-reversal between horizontal lines. All other luminance signals in the 3.08- to 4.08-MHz band are cancelled.

There's one very important fact about the operation of the comb filter when considering alignment and troubleshooting of the circuit. The comb filter circuit will only operate properly if the signals being applied are properly phase locked to each other like those from the TV station. The chroma information must be phase alternating and phase locked to the horizontal sync so that the signals will add and subtract properly. Nonphase-locked signals, such as from a color bar generator, will not allow you to properly align the circuit because the phases are at random. The results will be poor separation of the chroma and luminance information.

Not only will nonphase-locked signals result in improper operation and alignment of the comb filter, but the resulting waveforms will not tell you how the circuit's operating. You won't be able to determine if the circuit's functioning correctly or if there's a circuit defect. The VA62A provides the properly phase-locked signals that allow you to align and troubleshoot the comb filter circuits.

RCA and Magnavox comb filter comparisons

Although the RCA and Magnavox comb filter circuits accomplish the same end results, there are several differences between them. The main difference is in the way in which the composite video is delayed for one horizontal line. Let's compare the block diagrams of the two systems to see how they're similar and how they differ.

The Magnavox system (see Fig. 4-17) uses discrete transistors and has four adjustments for proper phasing: two in the chroma and two in the luminance channels. This circuit's very similar to the comb filter used in VCRs. The delay device used is a glass acoustical delay line that passes signals between 3 and 4 MHz. This means that combing only takes place between 3 and 4 MHz, while signals below 3 MHz are passed virtually unchanged. The lower-frequency luminance must be removed in a bandpass filter to complete the signal separation process. The combed and filtered signal is then used to "comb" the color from the composite video, leaving high-resolution combed luminance to be passed to the video amplifiers. The Magnavox comb filter inverts the output of the delay line before adding the delayed signal to the composite video and again before the "combed" chroma is added back to the composite video.

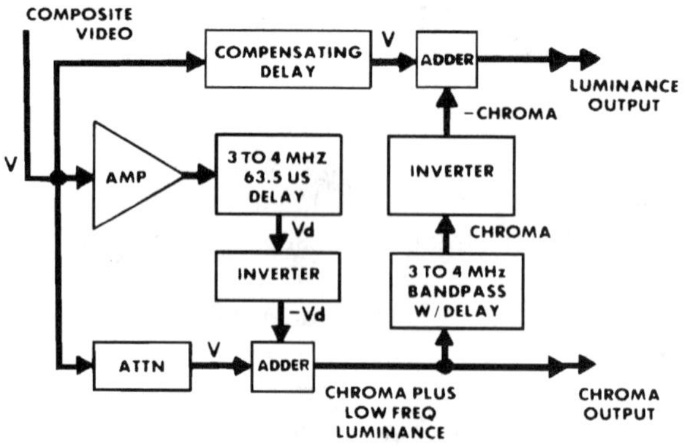

4-17 A block diagram of the Magnavox comb filter system.

The RCA comb filter (see Fig. 4-18) uses a special integrated circuit called a "CCD" (charged coupled device) to delay the composite video. The CCD is not frequency sensitive and delays the entire video spectrum so that a bandpass filter is not required. The CCD requires a clock signal of 10.7 MHz, which is obtained by tripling the frequency of the 3.58-MHz color oscillator. The internal switching of the CCD is thus phase locked to the composite video signal applied to its input. All of the circuitry,

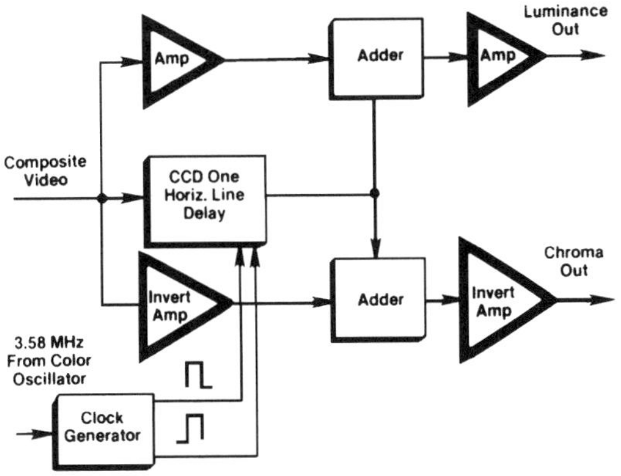

4-18 The RCA comb filter circuit.

with the exception of several output and buffer transistors, is contained in one integrated circuit chip. There are no phase adjustments required in the RCA comb filter.

Magnavox comb filter alignment

The Magnavox comb filter has four adjustments. The service data doesn't provide alignment procedures for these adjustments because they shouldn't drift enough to cause reception problems. However, from past experience, some technician or set owner will probably sooner or later attempt to set the adjustment for improved operation. Adjusting the comb filter without the proper test signals will result in poor operation. This might require the comb filter module to be replaced when it only requires alignment.

Having an alignment procedure allows you to confirm if the comb filter is operating properly. An adjustment that has no effect (or the wrong effect) indicates a circuit problem other than simple alignment.

The VA62A has the correct phase-locked signals required to align the comb filter. Phase locking can be quickly confirmed by observing the edges of the color bars on any TV receiver. The edges should look like tiny gear teeth and be standing perfectly still. If the teeth are moving along the edge, the color generator doesn't have phase-locked color signals.

The following alignment procedure results in the correct setting of the comb filter adjustments. The adjustment locations for the Magnavox comb filter are shown in Fig. 4-19.

1. Inject the chroma bar sweep pattern into the UHF input of the VHF tuner. Connect the scope to pin 7 of the comb filter module.
2. From the rear of the set, adjust R16 fully counterclockwise. Adjust L3 to bring the slug to the outer edge of the coil form. Slowly adjust the slug back into the form until the 3.50-MHz bar in the chroma bar sweep pattern is reduced to its lowest amplitude. You might have to increase the scope gain to observe the action of coil L3.
3. Adjust R16 to null the 3.50-MHz bar to minimum.

4. Move the scope to TP4 on the comb filter module and set the VA62A to the bar sweep pattern.
5. Adjust R10 fully counterclockwise (CCW). Adjust L1 for maximum amplitude of the 3.00-MHz bar.
6. Adjust R10 to null the 3.00-MHz bar to minimum. You might need to increase the scope gain to observe the null of L3 for the best setting. Refer to scope waveforms shown in Fig. 4-20A and B).

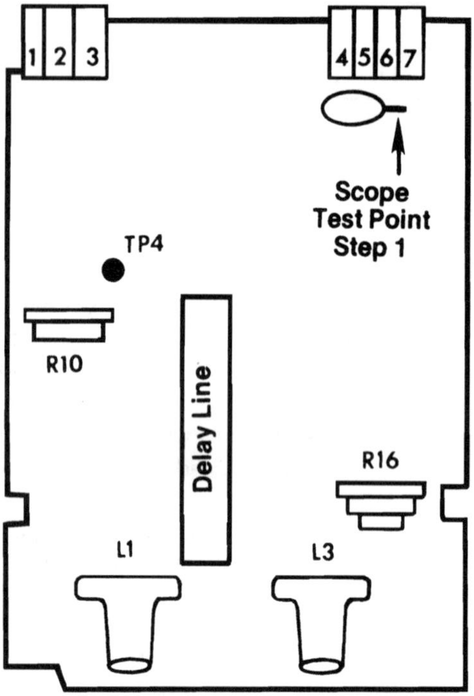

4-19 Location of adjustments on the Magnavox comb filter module.

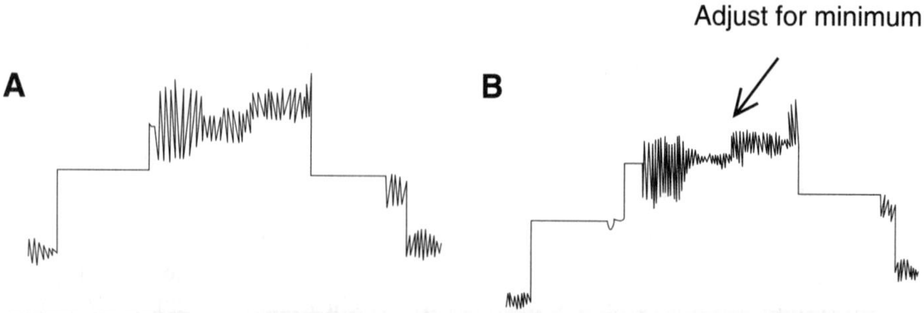

4-20 Part A shows R16 and L3 misadjusted. B shows when the two controls are properly set to null the 3.58 MHz bar in the pattern. The waveforms have been expanded to see the detail of the 3 bars in the chroma bar sweep pattern.

This completes the alignment of the Magnavox comb filter circuit. To check the operation of the comb filter, tune in a TV station signal and observe the large color areas for noise. If any noise is seen, recheck the circuit adjustment settings. The adjustments are crucial and must be done properly for correct separation of the chroma and luminance information.

The alignment procedure can be used to aid in the troubleshooting of the Magnavox comb filter circuit. The lack of a signal at the test point or the inaction of a control should lead you to the area of the circuit that's faulty. The chroma section of the filter must be working before you begin the alignment or troubleshooting of the luminance section. A poor separation of the chroma will cause poor separation of the luminance, even though the luminance section might be working perfectly. The scope waveforms illustrated in Fig. 4-21A and B show proper and improper comb filter adjustments.

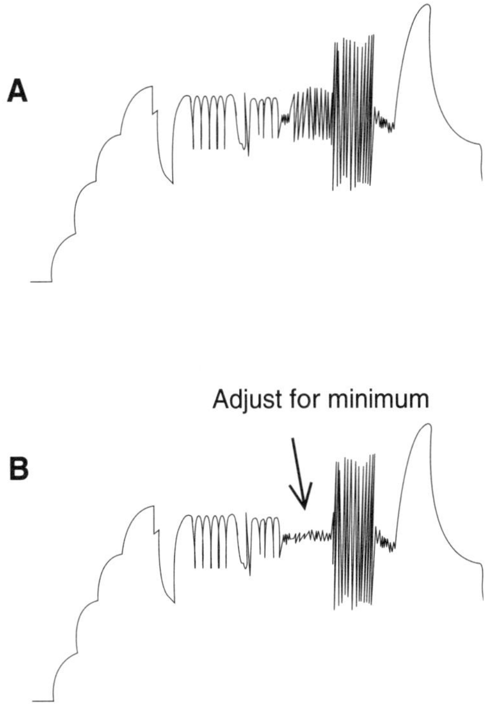

4-21 Waveform A shows R10 and L1 misadjusted. Waveform B shows R10 and L3 properly set for a null of the 3.02-MHz bar in the bar sweep pattern. The waveforms have been expanded to show the detail of the bars in the bar sweep pattern.

RCA comb filter alignment

The RCA comb filter circuit has three adjustments: A 10.7-MHz tripler, a combed chroma ("C") adjust, and a vertical detail output adjust. These adjustments can be made quite easily using the VA62A phase-locked signals and a dual-trace scope. The adjustment should be made in the proper sequence to ensure the best operation of the comb filter.

Adjust the 10.7-MHz tripler coils as follows:

1. Connect the VA62A to the set's antenna terminals, set the Analyzer and TV receiver's tuner to the same channel, and switch to the bar sweep pattern.

2a. Make a loop of wire using four turns about the size of a dime and connect one end to the scope probe and the other end to the scope ground lead. Place this over L6 on the comb filter board or over C23. Don't make direct contact to the 10.7-MHz tripler circuit, as the scope probe might load down the circuit. An alternative method is as follows:

2b. Connect an RF pick-up loop (part #PL207) to the input of the scope. Place the loop near the comb filter PC board on either side of L6. Note: Be sure that neither pick-up loop is placed in a location that will allow it to remain stationary. It must not move around during the adjustment because it could produce a misleading indication of the adjustment.

3. Adjust coil L10 and L6 on the comb filter for maximum amplitude of the observed signal. Note that the actual amplitude is not important because the pick-up loop can't be calibrated. Chroma comb filter ("C") adjust:
4. Connect the scope probe to the VDO (vertical detail output) pin 13 on the comb filter PC board.
5. Connect the second scope channel (or external trigger input) to the collector of the horizontal driver transistor, Q407.
6. Set the timebase switch to 10 microseconds/division and lock in the signal.
7. Adjust R62, the chroma comb ("C"); adjust for a minimum horizontal sync amplitude in the waveform. See the scope waveforms in Figs. 4-22 and 4-23. IF (comb filter) preset:

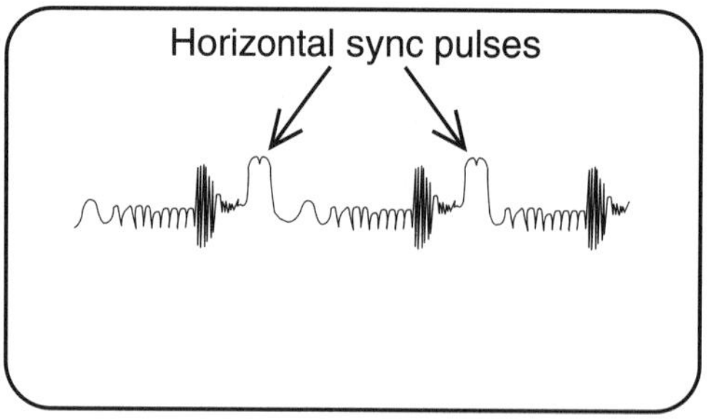

4-22 Control R652 "C" adjust improperly set, as shown by the presence of a horizontal sync pulse.

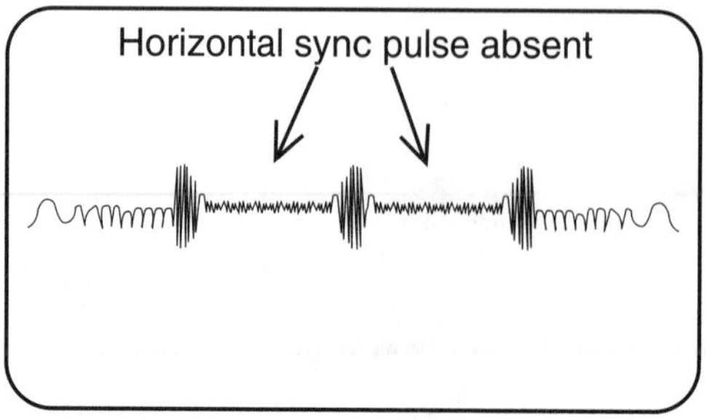

4-23 Control R652 will be properly set when you find no horizontal sync present in the waveform.

8. Move the second scope channel (or the external trigger input) to the collector of the vertical output transistor, Q504.
9. Set the timebase switch to the .2 msec/division setting and lock the vertical signal in on the scope.
10. Adjust the IF preset control, R344, located on the main PC board between the IF shield and the comb filter PC board, to obtain a 520-millivolt peak-to-peak signal amplitude of the VDO signal, as shown in Fig. 4-24. See Fig. 4-25 for the location of the comb filter in the RCA chassis.

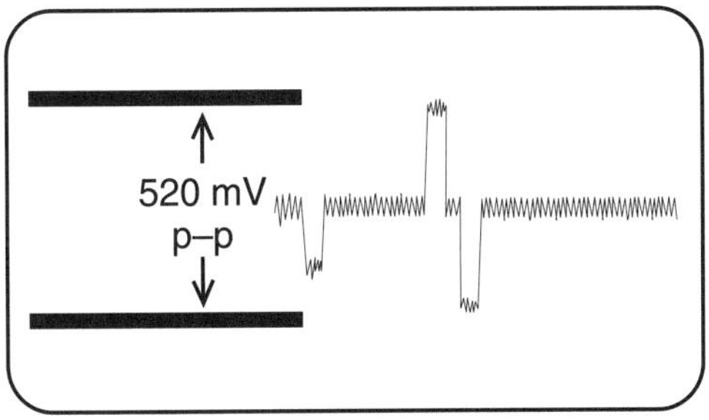

4-24 The IF preset, R344, set to provide a 520-mV peak-to-peak signal at VDO (vertical delay output)

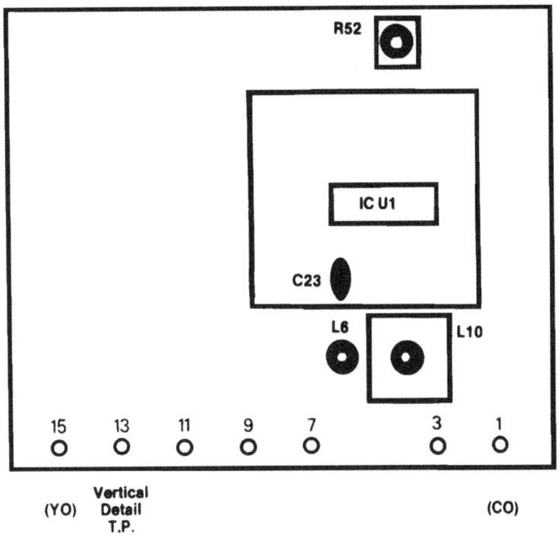

4-25 The RCA comb filter PC board as viewed from the front of the chassis.

Checking out RCA comb filter operation

The operation of the comb filter can be quickly checked out by using the VA62A multiburst bar sweep and chroma bar sweep patterns and the scope to confirm the alignment of the filter. The following quick check can help in troubleshooting as you're trying to locate a defect in an improperly operating comb filter. A check of the pattern will quickly tell you if the filter is properly aligned or if there's a circuit problem and where the problem might be located.

1. Check the bar sweep pattern at the video detector. It should appear similar to that shown in Fig. 4-26. Loss of video or weak video might be caused by an inoperative filter circuit.

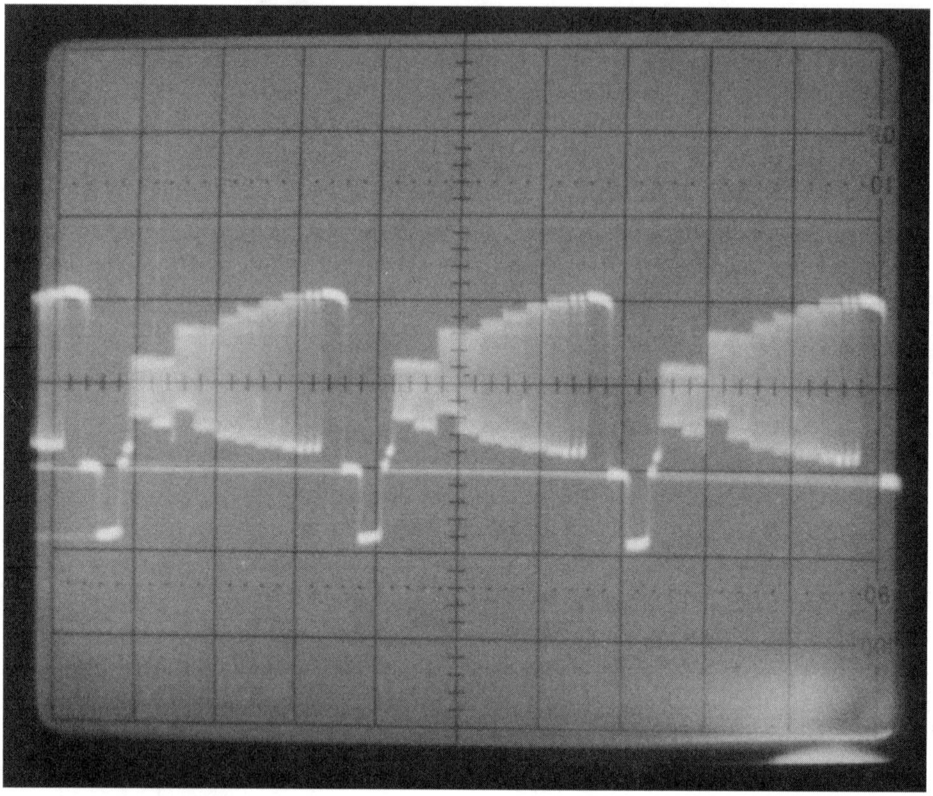

4-26 Multiburst bar sweep pattern waveform seen at the video detector output.

2. Check the multiburst bar sweep and chroma bar sweep patterns to the comb filter input at terminal 3 on the comb filter PC board. Note that the 3.00- and 3.50-MHz bars in the bar sweep pattern have been peaked, and that's a normal condition at this test point. Refer to the scope traces in Figs. 4-27A and B.

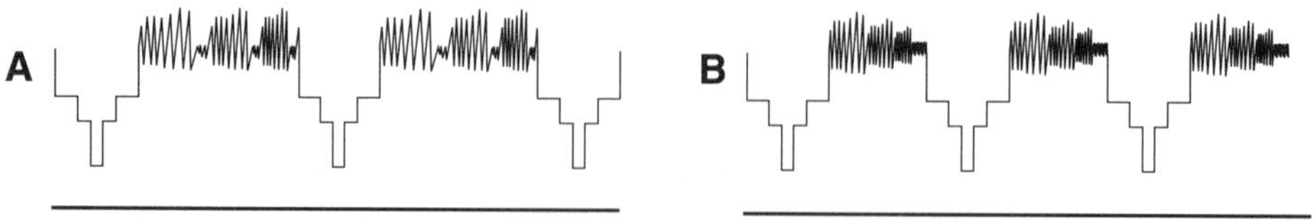

4-27 Waveform A shows the bar sweep pattern, and waveform B shows the chroma bar sweep pattern at the input (pin 3) to the comb filter.

3. Check the 3.58-MHz sine wave input at pin 7. For proper operation, the signal should be at least 1 volt peak-to-peak in amplitude.
4. Check the chroma only signal at pin 1 (CO) on the comb filter PC board. The bar sweep pattern should appear like that shown in Fig. 4-28, with only the

3.50-MHz bar being present. The chroma bar sweep pattern should appear with all bars present, as shown in the Fig. 4-29 pattern.

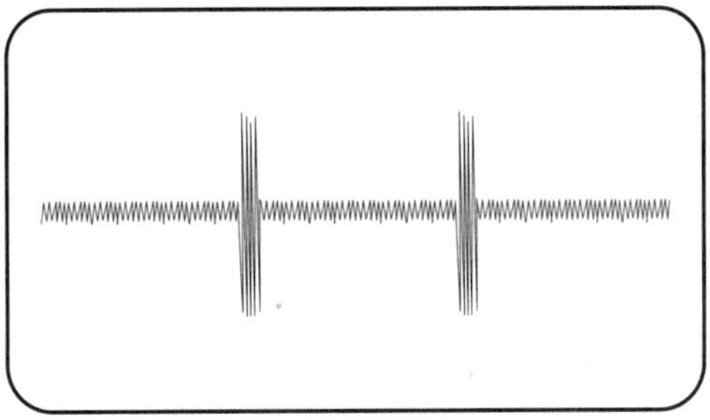

4-28 The bar sweep pattern waveform seen at CO (chroma output) of a properly operating comb filter. Note that only the 3.50-MHz bar (chroma frequency) is present.

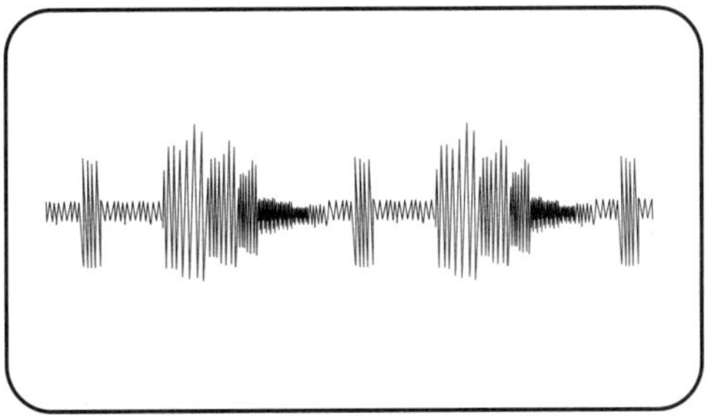

4-29 The chroma bar sweep pattern waveform seen at CO (chroma output) of a properly operating comb filter.

5. Check the luminance signal at pin 15 (YO) on the comb filter PC board. The multiburst bar sweep pattern should appear as shown in Fig. 4-30. The 3.50-MHz bar should show almost no signal information. The chroma bar sweep pattern, as shown in Fig. 4-31, should show almost no information.

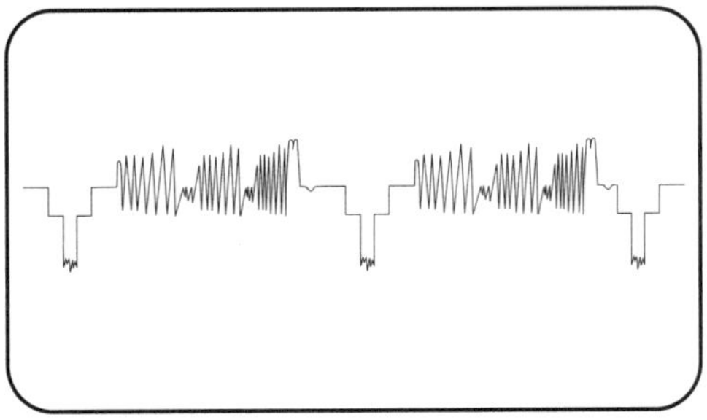

4-30 Bar sweep pattern waveform scoped at YO (luminance output) of a properly operating comb filter.

If one of the waveforms is incorrect, that section of the comb filter should be checked for proper alignment or a circuit defect. Always be sure to check the chroma-only signal first. The lack of "combed" chroma might appear as chroma in the luminance-only signal. This might cause you to look in the luminance section instead of the chroma section where the actual problem might be located. If the IC is replaced, be sure to perform the alignment procedure and check the operation of the comb filter by observing the waveforms as indicated.

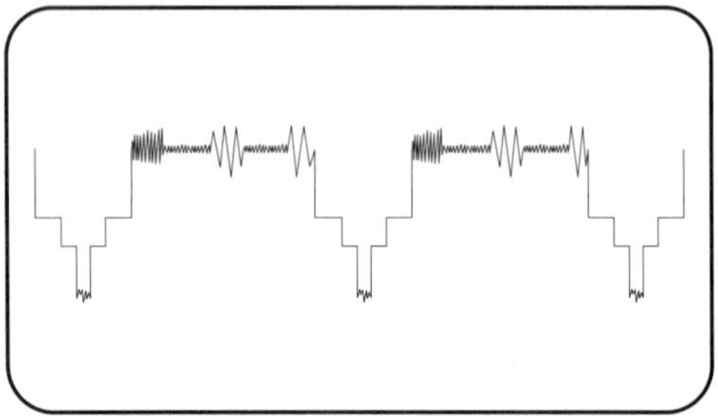

4-31 The chroma bar sweep pattern waveform as scoped at YO (luminance output) of a properly operating comb filter circuit.

Zenith comb filter circuitry

The Zenith comb filter (found on the 9-152-01 module) is a 5-transistor circuit that contains a precision-ground piece of glass as the delay element. The comb filter circuit is shown in Fig. 4-32. This comb filter makes use of acoustic waves transmitted through glass. At the output of the filter, the acoustic signal is changed back into an electrical signal 63.5 microseconds later.

The composite video signal is applied to Q2404, which inverts the signal and feeds it to the glass delay line. At the output it's added back to the original signal and chroma that's free of any luminance found at the base of Q2401. This signal is fed to the chroma channel and also to one input of a differential amplifier formed by Q2401 and Q2404. A differential amplifier subtracts one input from the other, in this case the luminance-free chroma from the composite video. The result is luminance that's chroma-free at the collector of Q2401. Transistor Q2405 is a stage to equalize the preshoot and overshoot of the luminance, and Q2403 is a driver for the remaining luminance processing circuitry.

There are four adjustments in this comb filter—two for nulling the luminance out of the chroma and the remaining two for nulling the chroma out of the luminance. On field service calls, only the latter two should be adjusted, R2401 and L2426. The other two controls adjust the nulling of luminance in the chroma, which is not as important as nulling the chroma out of the luminance.

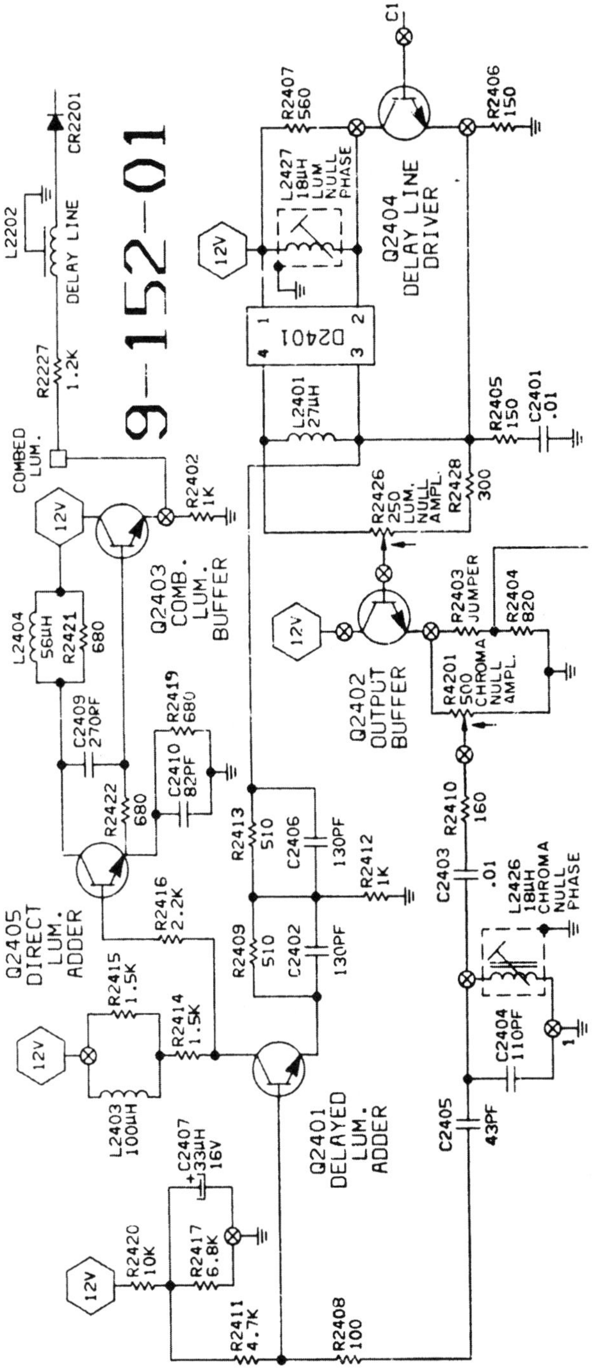

4-32 Circuit diagram of the Zenith comb filter circuit used in the 9-152-01 module.

5

Video IF/video detector systems

IN THIS CHAPTER I DEMONSTRATE THE VARIOUS WAYS YOU CAN USE THE VA62A Analyzer for troubleshooting video IF stages, checking stage gain, performing video IF alignment, and setting the IF traps. Then I show how to do some IF alignment quick-checks, and I discuss sweep and marker generator alignment.

You'll also find detailed instructions for aligning synchronous video detectors and some tips on SAW filter alignment. These surface acoustic wave (SAW) filters are being used in the newer TV-receiver IF systems. I also cover techniques for "spot aligning" IF stages.

Video IF alignment

The Analyzer provides a multiburst bar sweep pattern that's designed for both IF alignment and IF stage troubleshooting. This pattern provides reference frequencies that relate directly to the key marker frequencies provided by a sweep-and-marker alignment generator.

The bar sweep pattern is part of a composite TV waveform. This technique lets you perform a complete IF alignment without biasing the AGC circuit. The bar sweep pattern is available with an RF or IF carrier. This allows the signal to be fed into the antenna terminals or into the IF input of the receiver under test. The results of injecting the bar sweep signal through the TV receiver can then be evaluated on the CRT of the set under test, or with an oscilloscope connected to the video detector output.

The video IF bar sweep signal is ideal for the latest circuit designs such as the synchronous video detector. This circuit requires a 45.75-MHz carrier for video detection. If a standard sweep and marker generator is used for the alignment of the IF stages of this type of circuit, a second generator signal must be mixed with the sweep output to produce detection. The bar sweep pattern will work equally as well with the synchronous detector or a standard diode detector.

Sweep and marker comparison

Before I cover the applications of the bar sweep pattern, let's see how it relates to the standard sweep and marker system. The video IF amplifiers process five key frequencies for proper operation. The crystal-controlled markers of a sweep and marker generator monitor the amplification of these frequencies. In addition to the five frequencies that the IF amplifiers are designed to amplify, there are three frequencies that would potentially cause picture interference. The IF traps are designed to reduce the amplitude of these frequencies. The trap setting signals and the bar sweep pattern combine to adjust the IF coils and traps for alignment or troubleshooting. The five desired frequencies are:

1. Video carrier.
2. Low-frequency video information.
3. Medium-frequency video information.
4. High-frequency video information.
5. Color subcarrier.

The three frequencies that are trapped out of the video signal are:

1. Video carrier of the next higher channel.
2. Sound information from the next lower channel.
3. Sound carrier from the selected TV channel.

In practical terms, an IF amplification system provides three levels of amplification. The three undesired frequencies should be as close to zero amplification as possible. The video carrier and the color carrier should be at 50 percent of the peak amplification, and the remaining three frequencies should be given maximum amplification. A sweep and marker generator provides all eight of these frequencies at about the same amplitude. The output waveform then indicates the amount of gain at each of the frequencies.

The multiburst bar sweep pattern of the VA62A approaches IF alignment from the opposite point of view. The bar sweep starts with signals of different levels. After passing through the IF amplifiers, each key frequency should have the same amplitude. This special signal is part of a composite video pattern to eliminate the need for biasing the AGC circuits during alignment. One of the many advantages with the bar sweep concept is that every type of IF amplifier system provides the same output pattern when properly aligned. Notice the pattern drawing illustrated in Fig. 5-1.

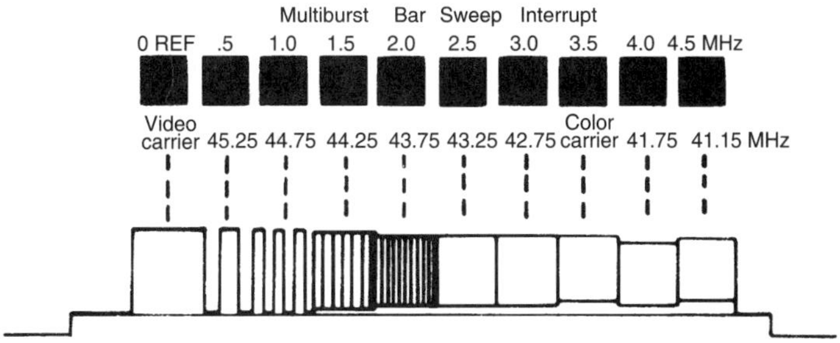

5-1 Multiburst bar sweep waveform, produced at the video detector, indicating correct IF response.

The trap frequencies are treated separately. Each of the trap signals are microprocessor-controlled, modulated signals. Each trap is then adjusted for minimum modulation. The modulation of a misadjusted trap can be monitored directly on the TV screen or by using an oscilloscope connected to the output of the video detector.

The VA62A provides a dynamic system to align the IF stages of any TV receiver. The dynamic nature of the multiburst bar sweep works equally as well with tube, transistor, IC, or SAW filter IF stages. The VA62A provides the proper signals for use with conventional diode detectors or synchronous video detectors.

In addition, the Analyzer provides a method to dynamically set the IF stages. The three standard trap frequencies (for the sound, adjacent sound, and adjacent video carriers) are generated with a unique combination of a modulated video carrier and an interfering trap signal, both generated to crystal accuracy. This combination offers several advantages over any other trap-setting method:

1. The combination of signals exactly duplicates the video carrier and the interference that the traps are designed to eliminate.
2. The modulated video carrier holds the keyed AGC circuits at a fixed value so they don't "fight" you as the traps are being adjusted.
3. The crystal-controlled video carrier causes synchronous video detectors to operate in a dynamic fashion so the interfering carrier is properly detected.

Another feature is the programmable IF generator, allowing any carrier frequency from 35 to 50 MHz to be generated with crystal accuracy. The carrier can be generated as an unmodulated signal or as an amplitude-modulated signal, depending on your application. This feature allows "spot alignment," where a badly misaligned system can be roughed-in before a final adjustment is made. It also provides a method of setting "odd-ball" trap frequencies that aren't covered by the standard trap-setting signals.

Bar sweep alignment procedures

I recommend that you refrain from trying to learn bar sweep alignment on an IF stage that you know has problems or on a chassis with an "odd" IF design, such as an RCA CTC38 or a system with only two IF amplifier stages. Bar sweep alignment is best learned on a TV receiver that you know is operating properly. This prevents the problems of trying to align a defective IF. Usually, poor alignment is not the only problem with a faulty receiver, and, too many times, a technician tunes and turns the coil slugs, thinking that there must be a way to set them properly. Generally, the technician will end up frustrated, and might even think that the VA62A can't be used for alignment.

The VA62A multiburst bar sweep will show imperfections in IF performance that can't be seen with any other IF testing or alignment methods. Many of these imperfections are inherent in the IF stages and can't be corrected with any amount of alignment. Thus, you should work for the best possible results and then stop.

As a general rule, a particular bar of the multiburst pattern can vary in amplitude by as much as 50 percent from the ideal waveform before there's an objectionable change in picture quality. This is especially true of the higher video frequencies, which contribute the small detail to a picture and aren't noticeable at average viewing distances.

TV alignment comments

Many technicians claim that modern IF circuits have eliminated the need for alignment. These claims go back to the first solid-state IF circuits because (supposedly) the interelement capacity of a transistor doesn't change like it does in tubes. While solid-state and SAW filter designs result in smaller amounts of alignment change than found in tube stages, circuits still change for various reasons. These changes are due to aging, component replacement, or unsuccessful attempts to align the circuits. There are fewer adjustments in IF stages now than a few years ago. However, the adjustments that remain are crucial and must be set correctly for the best picture quality.

A good question to ask is, "When should I align?" The answer to this is easy. The IF stages should be aligned whenever a test of the video performance shows a noticeable degradation of picture quality. The VA62A alignment test is so fast and easy to perform that it can be done on every TV receiver. Then, use test results to make your decision about aligning a receiver.

Trap comments

More and more receivers are being used with cable TV. The IF traps must be set right on the nose any time a TV is used for cable reception. This is because cable produces the signal condition where there's a channel immediately above and immediately below the one tuned in by the viewer. Off-the-air signals are designated so as not to produce these adjacent channels.

Therefore, you should learn to set traps, even if you never plan to align a single IF stage. Some technicians think of IF "alignment" as a nasty word, so it might help to think of trap setting as a procedure separate from IF alignment. Regardless of how you deal with the idea of alignment, remember that the traps must be properly adjusted for any receiver that will be used with a cable system.

Checking alignment without a scope

Note: Sencore doesn't recommend attempting to align IF stages without using an oscilloscope. Problems in the video amplifiers might cause you to severely misalign the IF stages if you attempt to do alignment from the screen. Use the following procedure only to test the overall performance of the video system.

The dynamic nature of the multiburst bar sweep provides a method of testing the performance of the IF and video amplifier stages without removing the back from the receiver under test. You'll know the IF stages and video amplifiers work correctly if the receiver passes this test. If it doesn't, you know there's a problem and you should use functional analyzing to determine whether the problem is before detection (the tuner or IF amplifiers) or after detection (the video amplifiers).

There are two things to watch for when you observe the multiburst bar sweep pattern on the CRT. First, the bars through 3.0 MHz should have sufficient amplitude, as explained in the next paragraph. Second, the lower frequency bars shouldn't show signs of ringing or overshoot.

There are two things to watch for as you test for the correct bar amplitude. First, a bar that has made it all the way through the video system should show the individual vertical stripes of video. A bar that has been restricted in amplitude becomes a uniform shade of grey. A restricted bar also has a lower brightness level, allowing the relative amplitudes to be checked with the following procedure.

Watch the multiburst bar sweep bars as you slowly turn down the receiver's brightness and contrast controls. The bars will disappear (turn black) at different times, depending on each bar's peak-to-peak amplitude at the CRT. Restricted frequency response causes the bars to disappear in the wrong sequence.

The 3.5-MHz bar ("color carrier") should always be the first bar to turn black during the test because the video amplifiers have a trap to eliminate the color signal. The 4.5-MHz bar should turn black at nearly the same time as the 3.5-MHz bar when feeding the signal through an IF strip. (The 4.5-MHz bar should contain detail and brightness when injected via the direct-video input of a high-resolution system (high-tech) because there's no 4.5-MHz sound trap to restrict the higher frequency response.)

The 4.0-MHz bar will usually not show any detail, although its brightness should be higher than the 3.5-MHz bar. The 4.0-MHz bar should disappear after the 3.5- and 4.5-MHz bars. After that, the bars should begin to disappear in sequence, starting from the 3.0-MHz bar. The 1.5-, 2.0-, 2.5-, and 3.0-MHz bars should all turn black at very close to the same brightness/contrast setting. The "0 Ref.," 0.5-, and 1.0-MHz bars should be the last to disappear.

You should also watch for signs of ringing or overshoot in the "0 Ref.," 0.5-, and 1.0-MHz bars as you reduce the picture controls. In a properly operating set, the bars should have a uniform intensity from the left to the right side, in both the white and black areas. If there's a bright or dark line, you know a circuit's causing video distortion.

Note: The video peaking control (sometimes called the "sharpness" control) might affect the ringing on the low-frequency bars. Set this control for maximum detail when testing the video response. If there's some ringing, set the control to midscale to determine if the ringing will be a problem when the control is adjusted for normal viewing.

IF signal injection checks

The VA62A can be used for signal injection into any IF video stage. Two features of the VA62A's IF generator make it ideal for IF analyzing:

1. The carrier frequency is microprocessor controlled to crystal accuracy.
2. The levels of the output signals are set for a normal voltage gain of 10 per stage. This means that testing for stage gain is as easy as moving the RF-IF

signals switch to the position corresponding to the injection point and injecting the IF signal into each individual stage.

Refer to the drawing in Fig. 5-2 for the control settings and the injection test points.

The following steps can be used as a guide to isolate a fault in the IF stages:

1. Set the RF-IF signals switch to the third IF (Hi) position.
2. Inject the IF signal (using the balun coil) at the third IF input and adjust the RF-IF level control for a snow-free picture.
3. Move the injection cable to the input of the second IF amplifier and change the RF-IF signals to the second IF (Med) position.
4. If the signal is still being passed at this point, repeat the test at the input to the first IF stage using the (Lo) position of the RF-IF signals switch.

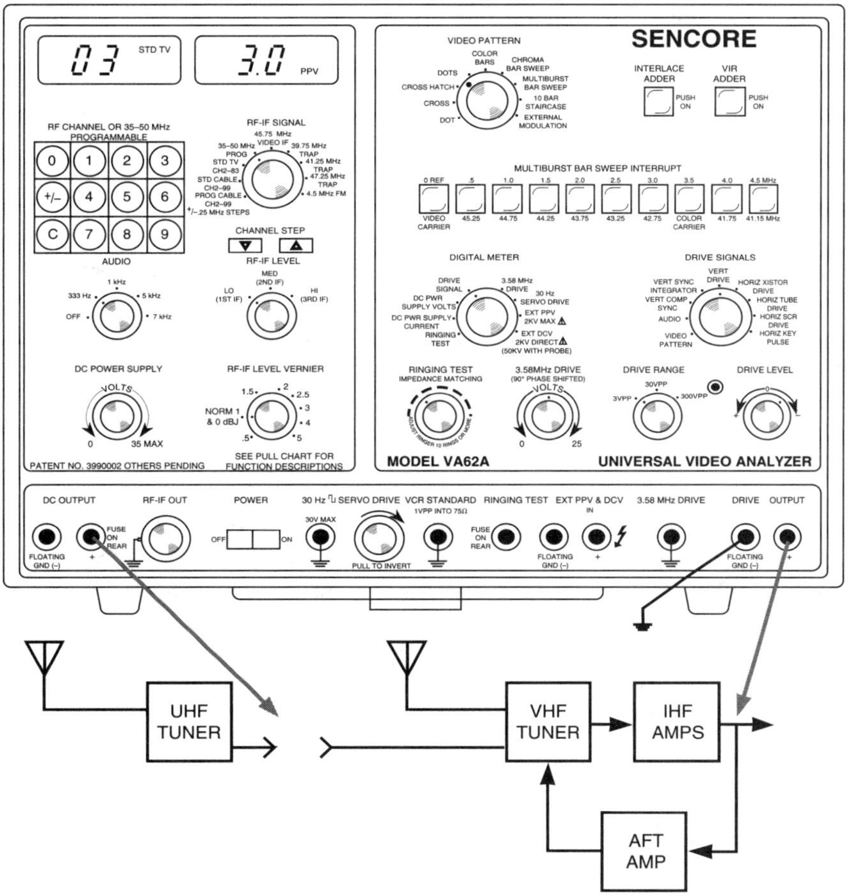

5-2 VA62A control positions for injecting IF signals into the IF stages.

Stage gain checks

The amount of gain is easily determined with the calibrated RF-IF level control. The signal is injected at the output of the stage and adjusted until a snowy picture is produced. The connection is then moved to the stage input and the control again set for a snowy picture. The voltage gain of the stage is then found by dividing the output signal by the input signal.

Remember that the signal level with the RF-IF signals switch in the third IF (HI) position is ten times that in the second IF (MED) position, which in turn is ten times that signal strength that will be produced in the first IF (LO) switch position.

Setting the IF traps

Always keep in mind that the traps must be set properly before making any other IF adjustments. Often, the traps are the only adjustments required to complete the required IF alignment.

The three special VA62A trap-setting signals provide a fixed-level carrier, modulated with a video signal. This signal mixes with an interfering signal that can be adjusted to a wide range of amplitudes. The video signal dynamically clamps the AGC to a fixed level so you can adjust the trap-setting signal for best contrast. The trap-setting signal can be mixed with the video signal as an unmodulated carrier, or modulated with any of the tones selected by the audio switch, depending on which provides the best contrast. The amplitude of the interfering signal is adjusted until good contrast is obtained. The receiver trap is then adjusted for the least amount of the interfering signal.

Note: Use the smallest level of interfering signal that provides noticeable interference. If the trap adjustment completely eliminates the interference before reaching a solid null, increase the signal level slightly to confirm a proper null.

With a few exceptions, the results of the trap setting are usually easier to see on the CRT screen than on a scope. This allows you to set the traps without hooking up the scope. The trap-setting signal is generally easiest to use with the multiburst bar sweep. You might find it easier to turn off the 3.5-, 4.0-, and 4.5-MHz bars to provide a large black area on part of the screen for maximum contrast. The 333-Hz or 1-kHz position of the audio switch usually provides the best contrast for the interference signal.

If you want to use a scope to set traps, the best results are obtained by using the "10-bar staircase" pattern and no audio modulation. The trap is simply adjusted for minimum "thickness" of each trace pattern step.

Setting the traps

1. Connect the VA62A to the input of the IF amplifier.
2. Set the RF-IF SIGNAL switch to the desired trap frequency.
3. Set the video pattern switch and the audio switch to the desired positions for best contrast.

4. Adjust the RF-IF level switch and the RF-IF level vernier until the interfering signal provides a noticeable amount of contrast to the selected video pattern.
5. Adjust the trap coil(s) until the interfering signal has the least effect on the video pattern. (If necessary, increase the setting of the RF-IF level vernier.)
6. Repeat steps two through five for the remaining trap frequencies.

Special notes on traps

1. It's not necessary to use an external attenuator when using the VA62A, as the step attenuator reduces the signal by 40 dB compared to the Sencore VA48. Use the "Lo" position of the RF-IF level switch for all the trap settings.
2. Some older TV sets don't have a 39.75-MHz trap. Simply ignore any reference to setting this trap. You will, however, need to reduce the amplitude of the 3.5- and 3.0-MHz bars when aligning the bandpass coils if the receiver will be used on cable, or the set might receive some strong adjacent channel interference.
3. Some TV sets have a 41.25-MHz trap in the first IF stage and a second one in the last IF stage. If so, misadjust the one in the last IF stage and set the one in the first IF stage first. Then, increase the level of the VA62A trap signal to force a signal through the first IF stage while adjusting the trap in the last IF stage for minimum interference.
4. Some traps consist of a coil and a separate potentiometer. The coil sets the tuned frequency of the trap and the pot sets its depth (Q). Always adjust the coil first, before the pot.
5. You might find that some traps will have two slugs in one coil. Normally, the slug closest to the PC board (bottom slug) is the frequency adjustment and should be set first. The top slug is usually the depth adjustment and should be adjusted after the bottom slug is set.
6. A few older chassis have three slugs in one IF can; one was a trap and the other two were bandpass coils. There's no way to eliminate the interaction between these slugs. You'll have to set the traps, go through the bandpass adjustments, and then set the traps again. If the bandpass is affected, you might have to return to the bandpass slugs a second time.
7. The 39.75-MHz trap is sometimes difficult to set because the response of the lower frequencies is already established by the 41.25-MHz trap. If so, detune the 41.25-MHz trap (with the 39.75-MHz trap signal selected on the VA62A) until the trap interference is at maximum. Set the 39.75-MHz trap, and then switch the VA62A to 41.25-MHz to return the 41.25-MHz trap to its correct setting.
8. If you aren't sure the traps are being set to the correct point, select the multiburst bar sweep pattern with all the bars turned on and watch the bars as you adjust the trap. The trap will affect one bar at a time as it moves through the pattern. The trap should just be seen in the pattern just before

it reaches its correct setting. The 41.25-MHz and 39.75-MHz bars should move off the pattern to the right, and the 47.25-MHz bar to the left. The video pattern will lose sync any time a trap is set on top of the video carrier. Turning the trap adjustment further returns the carrier as the trap moves past the carrier frequency.

9. A few traps might not appear to null when using the trap-setting signals because the IF response is already reduced considerably at the frequency where the trap comes into the picture. These traps can usually be set using the Programmable IF Generator, which only feeds one signal to the IF stages at a time.
10. The amplitude of the interfering trap-setting signal should always be as low as possible. If the trap signal overpowers the video pattern signal, check that the RF-IF Level switch is set to the "Lo" position.

Setting traps with the programmable generator

The VA62A provides a second method to set IF traps. The 35–50-MHz Programmable IF Generator can be used to adjust traps with nonstandard frequencies. Plus, it can be used to adjust traps that might not show a proper null when using the special trap-setting signals. The Programmable IF Generator provides either an unmodulated carrier or a carrier that's amplitude modulated with an audio tone. The tone modulation is generally used for setting traps.

When using the Programmable IF Generator, you must be careful to use the least amount of signal that will give a modulation pattern on the CRT. If you use a larger amount of signal, the keyed AGC might latch to the audio modulation and flatten out the null-point of the trap adjustments. Using a small signal level, however, forces the AGC circuits to run wide open, giving a much better null.

The Programmable IF Generator can be modulated with any of the tones selected by the audio switch. The 1-kHz position generally gives the best contrast. Note the trap bar pattern shown in Fig. 5-3.

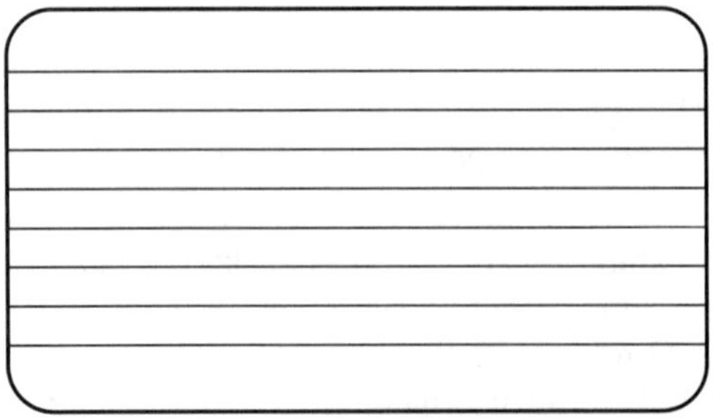

5-3 The programmable IF generator can be used to set traps with oddball frequencies. The VA62A is adjusted until the modulation pattern is barely viewable on the CRT, and the trap is then adjusted for the least amount of picture signal.

Using the programmable IF generator

1. Connect the VA62A Analyzer to the IF input of the TV receiver to be checked.

2. Set the RF-IF signal switch to the "35–50 MHz PROG" position.
3. Modulate the signal by setting the AUDIO switch to the desired frequency.
4. Select the desired frequency by entering four digits into the keyboard. For example, enter 4-0-5-0 for a 40.50-MHz signal. If you make a mistake while entering the number, either enter a total of 4 digits to complete the program or press the "C" (clear) button to start over.
5. Adjust the RF-IF level switch and the RF-IF level vernier until there's a noticeable modulation pattern on the CRT screen. Use the least amount of signal needed to obtain a modulation pattern.
6. Adjust the trap for the least amount of modulation on the CRT. Increase the setting of the RF-IF vernier control, if necessary, to locate the true null point.

Setting the bandpass coils

After setting the traps, you need to adjust the bandpass coils for the best balance between two parameters: the best sensitivity and the best overall video response. Because of this balancing, you're always setting the coils for a compromise. It's normal for the setting of one coil to affect the setting of another, thus you should repeat the adjustment two or three times for best results. Each time you go through the coils, you'll be setting the coils closer and closer to their ideal alignment. This will hold true for any IF alignment technique.

Using the correct sequence

You must make the adjustments in the correct sequence because each stage of an IF system is designed to operate at a slightly different frequency. Should you adjust the coils in the wrong order, you might exchange the tuning of two stages and cause unstable operation, "critical" fine tuning, excessive noise, or reduced sensitivity.

The correct sequence (after setting the traps) is:

1. Set the first IF.
2. Set the remaining coils.

The block diagram in Fig. 5-4 shows the correct sequence to follow for consistent results.

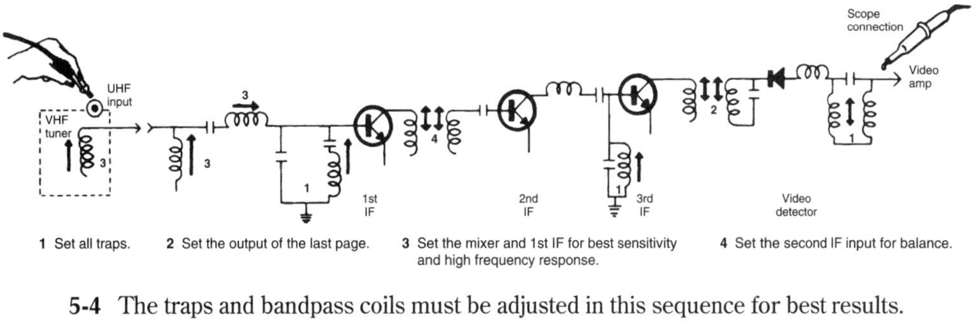

5-4 The traps and bandpass coils must be adjusted in this sequence for best results.

Set the detector first

You first tune the video detector to the 45.75-MHz IF carrier. The coils between the last IF amplifier and the detector affect detector tuning. (Traps in the last IF stage aren't detector tuning adjustments.) Some schematics might label these coils "detector," but most just label them "3rd IF" or "4th IF." However, they have more effect on the detector than on the last IF amplifier.

The "10-bar staircase" video pattern provides the best reference for tuning the detector. The detector adjustments affect the linearity of the detected signal between the sync pulse tip and peak white. The "10-bar staircase" signal lets you adjust the detector tuning for best overall linearity.

Adjust the detector coils for the largest amplitude signal without compression of the sync pulse or any of the stair steps. Stages other than the detector can cause sync or stair-step compression. Therefore, set the detector coils for the best linearity, even though the linearity might not be ideal. Then move on through the alignment of the rest of the IF stages. You'll then eliminate small linearity errors when you run through the detector tuning a second or third time.

Never adjust an earlier IF stage for so much gain that the video begins to compress. You don't need to switch to the "10-bar staircase" pattern when adjusting the earlier IF stages because compression will affect the sync pulses before any other part of the waveform. All you need to do is avoid compressing the sync pulses as you move through the earlier bandpass adjustments.

Set the first IF input next

After tuning the detector, go back to the input of the first IF stage. The first IF input includes the adjustments before the first IF amplifier (except for traps) and the mixer coil in the VHF tuner. You must consider the mixer part of the first IF stage because it works together with the tuned stages of the first amplifier to match the impedance of the tuner to the first IF stage.

The alignment of the first IF stage affects sensitivity and high-frequency video response. You can peak the tuning of the first IF adjustments for maximum sensitivity, but the IF stages will then have restricted video frequency response. This results in poor video detail, weak color, and critical fine tuning. In order to avoid this, do the adjustments in two steps.

First, adjust the stages for maximum sensitivity without worrying about bandwidth. Then, readjust the same coils for good bandwidth at a normal (1000-microvolts) signal level.

Adjusting for maximum sensitivity involves reducing the output of the VA62A to the lowest setting that produces an output at the video detector. If the receiver produces an acceptable output with the VA62A set to minimum, you don't need to worry about sensitivity. If, however, the picture is lost before you set the output control to minimum, you adjust the first IF coils for maximum gain.

In the case of severe misalignment, this can be corrected in steps. First, you reduce the setting of the RF-IF level vernier control until the signal on the scope becomes noisy. Then, tune the first IF adjustments for a clean waveform. Continue to alternate between reducing the signal level and tuning the IF adjustments until no further improvement is possible.

Note: Remember that a defective IF amplifier could be the cause of poor sensitivity. If you must set the RF-IF level switch to the "Med" or "Hi" positions to force a signal through the first IF stage, or if you can't adjust the first IF stages for enough gain to feed a signal through with the VA62A's RF-IF attenuators set to "Lo" and ".5," you should suspect a circuit defect that must be isolated before completing the alignment.

After you adjust the input stages for best sensitivity, increase the setting of the RF-IF level vernier to the "norm" position. This feeds 1000 microvolts to the input, which is the standard level used for the rest of the alignment. You then readjust the same coils you just adjusted for the best amplitude on the 3.0- and 3.5-MHz frequency bars. Don't worry if these bars seem low in amplitude after you adjust the first IF controls, as the other IF stages also affect the high-frequency bars.

At this point, you should begin to watch for symptoms of overpeaking an adjustment. For example, you might be able to increase the amplitude of the 3.0- or 3.5-MHz bars, but you might compress the sync pulses in the process. Or increasing the high-frequency gain might cause severe overshoot or ringing on the low-frequency bars ("0 REF," and 1.0 MHz). Adjust a control only as long as the high-frequency bars improve, but stop adjusting it if any other part of the waveform is affected. Don't worry about balancing problems, as the final balance comes in the next step and is then improved further during the second or third time through the complete set of adjustments.

Adjusting remaining stages for balance

The final step is to adjust the remaining coils for best overall balance. Most AGC-controlled IF stages reduce the high video frequencies when receiving low-level signals to keep "snow" to a minimum. To avoid this, leave the VA62A set to 1000 microvolts so the AGC stage holds the IF amplifier gain at a fixed level that's not restricting frequency response.

During the final step, you must watch for several conditions in the multiburst bar sweep waveforms:

1. The bars must roll off at the correct rate.
2. The lower frequency square waves must show the least amount of overshoot and ringing.
3. None of the bars must drop below the imaginary line representing the black level.
4. None of the bars should be larger in amplitude than the one to its left.
5. None of the bars should be missing or greatly reduced in amplitude.

Black-and-white TV alignment

Most black-and-white TV receivers don't have as much IF bandwidth as color receivers. The IF response normally extends only to 2 or 3 MHz, and then rolls off quickly before reaching the 3.5-MHz color carrier. This allows extra sensitivity, eliminates the need for color traps in the video amplifiers, and reduces high-frequency "snow" on weak signals.

You can determine the normal rolloff of the IF stages by comparing the multiburst bar sweep to the sweep generator curve in the set's data alignment manual. In the next section, I'll cover details on comparing the VA62A signal to a sweep response curve.

Comparing the bar sweep to sweep curves

There might be times when you want to use the manufacturer's data to determine how quickly the IF response should roll off at the high video frequencies. This can be especially helpful when aligning the IF stages of a color receiver that does not have a 39.75-MHz trap, or when aligning the IF stages of a black-and-white TV receiver.

You can determine the normal rolloff of the IF stages by comparing the multiburst bar sweep to the sweep generator curve in the manufacturer's alignment instructions. Ignore the area on the sweep and marker curve around the IF carrier (45.75-MHz marker). Use the VA62A multiburst bar sweep and the "10-bar staircase" video patterns to align for the correct response on the low video frequencies. The dynamic nature of the VA62A patterns provides the best possible alignment of the IF stages at these frequencies.

Use the sweep comparison to determine the correct video response for the multiburst bar sweep frequencies of 2 MHz and higher. The VA62A produces all of these bars at the same amplitude to help you in making this comparison.

The easiest way to compare the rolloffs is to use a dual-trace scope. Connect one of the scope's vertical input channels to the video detector output. Ground the second scope channel input. Select the dual-trace function and position the straight line from the grounded channel at the middle of the "0 Ref." bar. (If you don't have a dual-trace scope, use one of the graticule markings on a single-trace scope. Adjust the position control until the "0 Ref." bar is centered on a major graticule.)

Now look at the amount of bar rolloff for the 2.0-, 2.5-, 3.0-, 3.5-, and 4.0-MHz bars (the 4.5-MHz bar should always be a straight line near the center line). The rolloff of the bars (referenced to the center line established earlier) should agree closely with the rolloff of the sweep curve near the 42.17-MHz color carrier marker.

Notice that the IF frequency of each of the VA62A multiburst bar sweep bars is marked below each of the multiburst bar sweep interrupt buttons. Use these frequencies when you want to compare directly to a sweep and marker pattern. Remember that higher video frequencies produce lower IF frequencies because the sideband frequencies are reversed when the incoming RF signal is mixed with the local oscillator frequency in the tuner. You usually don't have to compare the exact frequencies if you remember which bars agree with which areas of the sweep curve in general terms. Do this by thinking of a sweep curve with the 45.75-MHz carrier on the left and the color carrier on the right. The VA62A multiburst bar sweep bars then follow in the same direction.

Tips for "rough-in" IF alignment

The multiburst bar sweep alignment pattern is designed to be injected into the first IF amplifier (usually via the VHF tuner mixer stage) and a full alignment performed by connecting an oscilloscope to the output of the video detector.

There might be times when the initial alignment of the IF amplifiers is so poor that no signal can be pushed through the IF stages at all. In these cases, the alignment signal can be fed into any later stage that will allow some portion of the signal to pass through. If, for example, the only test point that will pass the signal is the input to the third IF amplifier (and you're sure that the problem is in the alignment of the amplifier and not a defective stage), you can set the Analyzer to the third IF input function and align the third IF stage by itself.

In the case of the single-stage alignment, the even-amplitude bars normally found at the output of a properly aligned IF section will probably be very difficult to obtain. The main point is to align the various stages just to the point that the signal can be passed through the entire IF section. This "rough-in" alignment should always be followed with a complete alignment.

IF alignment comments

The IF stages perform two important functions in any TV receiver:

1. They determine the receiver amplification (gain) or sensitivity.
2. They limit the bandwidth of the received signal to eliminate interference from other signals.

The traps establish the upper and lower frequency limits of the video portion of one TV channel. The IF bandpass coils, on the other hand, allow you to tune the individual IF amplifiers to make sure that the set can detect picture detail from dc (pure black) to around 3 MHz. The upper frequency portion of the picture represents the smallest portions of the picture detail. There must be even amplification over the range of picture information so that different-sized objects have the correct brightness relationships.

Makeup of the IF response curve

The IF stages must take the very small signal from the output of the tuner and amplify it until there's enough signal to operate the video detector. The combined gain of the IF amplifiers is typically around 1000 times or 60 dB. Each stage is designed to provide gain at a slightly different frequency (stagger-tuned) to prevent feedback (regeneration) and to make sure that all portions of the video signal are amplified at about the same level. Notice the typical IF response curve shown in Fig. 5-5.

The IF stages in a color receiver must also pass the 3.58-MHz color subcarrier along with sidebands extending 500 kHz on either side of 3.58 MHz. Remember, however, that the 4.5-MHz information doesn't want to be passed on to the video or chroma circuits because this signal will cause interference in the picture. We can't hold the frequency response flat to 4.08 MHz (the top limit of the upper color sideband) and then drop off suddenly to eliminate the 4.5-MHz sound. A steep drop-off of this type would cause severe ringing in the fast rise-time square-wave information that makes up the picture. I compromise by rolling the IF frequency response off more gradually. This means that the 3.58-MHz color subcarrier is amplified half as much as the lower-frequency picture information. This gives enough color informa-

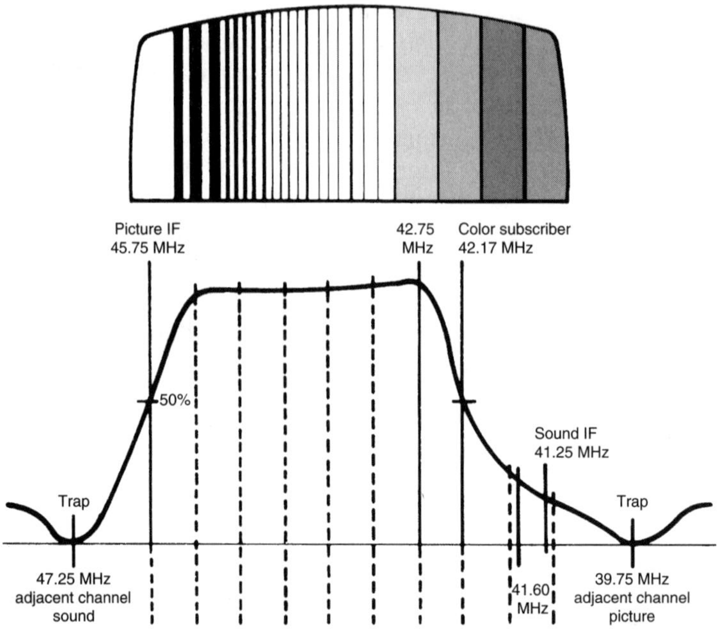

5-5 Typical IF response curve with frequency points.

tion to operate the color detector stages, plus enough ringing on square-wave information, and allows for trapping out the sound information. This is illustrated in the waveform drawing shown in Fig. 5-6.

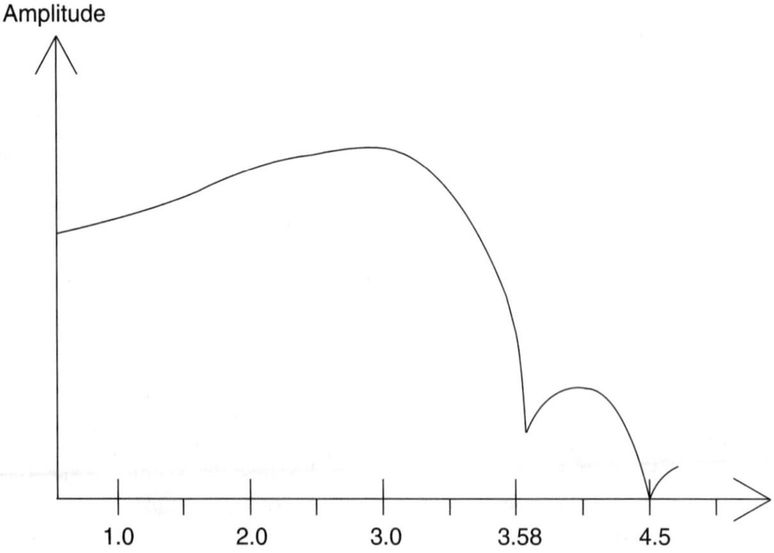

5-6 How bar sweep alignment prevents tuning problems at the color end of the IF curve.

Alignment information

Actually, all alignment should do is let you adjust the IF stages to give the best possible picture over a wide range of input signal levels, and to set the traps to reduce interfering signals. The sweep and marker generator is designed to test the actual frequency bandwidth of the IF stages by feeding a constantly changing frequency into the input and looking at this same signal after it has passed through the IF stages. The input frequency is changing at a 60-Hz rate, which is totally unnatural to the keyed AGC circuits that are tuned to the horizontal sync pulses. The time constants in the AGC circuit react to the 60-Hz sweep signal and produce a constantly changing gain, thus the AGC action must be blocked with a dc bias voltage. The AGC bias voltage, applied to the tuner and first IF amplifier, sets the gain of these stages to a single level so the output is closer to the actual frequency response of the total IF system.

Clamped AGC is one key limitation to the sweep and marker alignment system. Many IF stages are designed to have different frequency response curves for different input signal levels. For example, if you use a large bias voltage and a strong input signal, you usually get more high-frequency detail than with a small bias and a weak input signal. This design provides better noise immunity on a weak signal, such as in fringe areas or when using an indoor antenna. The actual amount of bias required for a given signal level varies from one chassis to another because of the tolerance of the components. You never really know the correct AGC bias voltage because the circuits are tweaked in for each chassis to compensate for component tolerances.

A second limitation is that there's no test for video ringing because the sweep and marker system produces only sine waves. It takes a square wave to cause a stage to ring. The transmitted TV signal is actually made up of squares, thus the sweep and marker signal will not duplicate the signals the TV set would normally receive. You might have attempted to tune the ringing out by resetting the IF stages with an off-the-air signal. The results are always questionable because the off-the-air signal is constantly changing. This "eyeball technique" of alignment might also sacrifice picture or color detail because the effects on the other portions of the picture aren't readily seen.

Overall, the one big problem with the sweep and marker method is that you have no way to tell if the alignment has accomplished what was intended: to provide a good picture at different input signal levels.

The "spot align" alignment technique

Let's now review the "spot align" technique because the term might be new to some technicians. *Spot alignment* allows you to set each of the stagger-tuned stages to the correct frequency before you balance the alignment adjustments. The advantage of spot alignment is that you know that each stage is tuned to the correct frequency. The procedure is usually done with a generator with an adjustable (nonsweep) frequency. You set the generator to a given frequency and adjust the coils that correspond to that frequency for the maximum reading on a dc meter connected to the video detector.

More alignment comments

Any time a set is totally out of alignment, you should try to find out the reason for the poor response. Thus, you'll find that IF alignment and IF circuit troubleshooting will go hand in glove.

Most sets should require only a touch-up alignment to correct for slight component drift or to compensate for the fact that the set was aligned with a sweep and marker that left some ringing in the picture. (For tube-type IFs, there might be differences in the internal capacitances of a replacement tube.) Also, of course, the two most obvious causes for poor alignment are circuit defects and "screwdriver drift."

"Screwdriver or diddle stick drift" is caused by someone who has attempted to align the set and has botched the job. You should ask the set owner if the set has been worked on recently. In a few cases, a do-it-yourselfer might have tried to improve the IF alignment. Symptoms like missing or broken slugs, or slugs that are turned completely into or out of the coils, point to screwdriver drift. In these cases you might be able to correct the condition by doing a complete alignment job.

When you encounter a set that has very poor alignment (but no signs of screwdriver drift), it's best to leave the slugs where they are until you locate the cause of the alignment problem. This might save you time because you'll only need to do touch-up alignment once you've located the defect.

Always keep in mind that the VA62A is designed to inject an IF signal into the first, second, or third IF stages. The multiburst bar sweep pattern gives you a reference that allows you to interpret the results as you inject into each stage. Connect the scope to the input of the first video amplifier to see the results of each step of injection.

Troubleshooting a typical IF defect

The VA62A Analyzer lets you quickly narrow down a defect to a particular stage. Each IF stage simply represents a block in the total TV receiver system. The VA62A Analyzer will be more effective for troubleshooting when you understand the function of each block in the TV receiver system. In Fig. 5-7, you'll find a schematic for an IF stage that uses three transistors for amplification. It doesn't matter whether each stage contains a tube, transistor, or part of an integrated circuit, because the function will be the same in each case. The application of the VA62A IF generator is also the same for each type of IF stage. Begin by injecting the substitute signal at the input to each stage until you find out that the known-good signal from the VA62A doesn't give you an adequate output. The defect must then be between the last point that gave a good output and the point that doesn't give an output.

Let's use an example of an IF strip that will not give an output when you inject at the input to the first IF stage. You know that a defect is somewhere between the first IF input and the video detector, but now the fault will have to be isolated to the stage or component.

Let's start by setting the VA62A RF-IF level control to the third IF position and injecting at the base of Q103, the third IF amplifier transistor. An output signal tells you that both the third IF amplifier and the video detector are working. The next log-

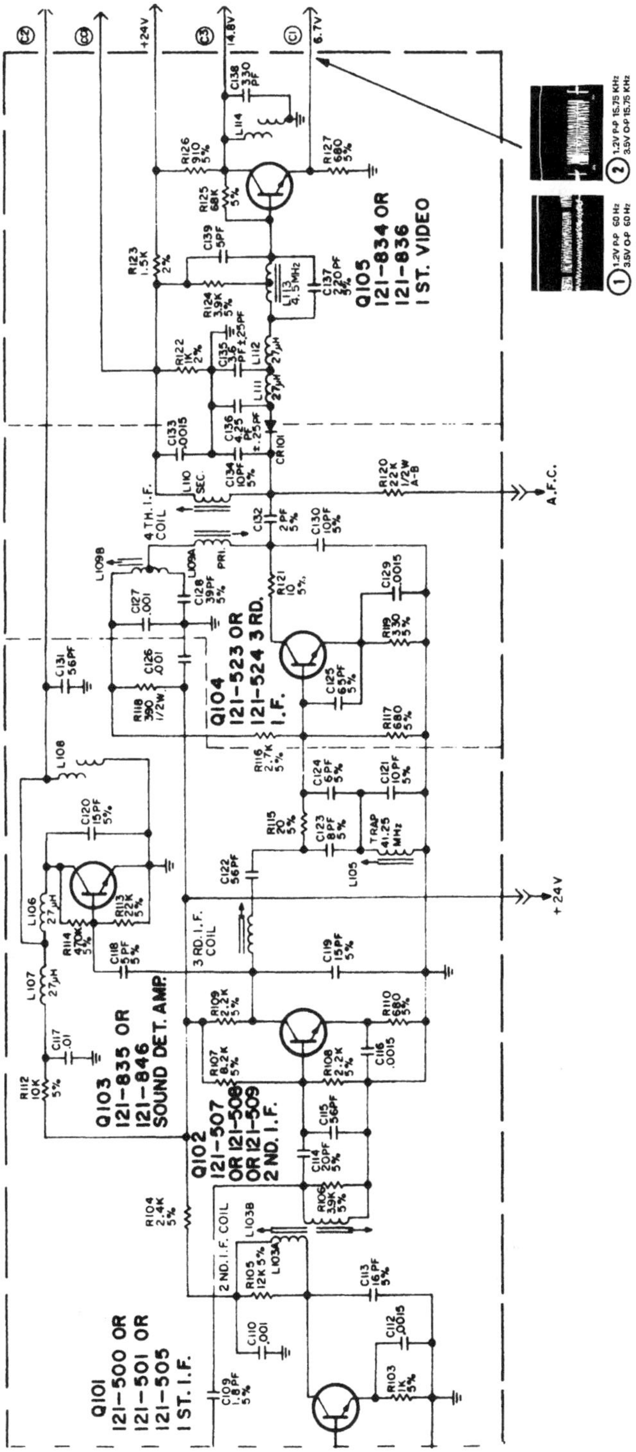

5-7 Zenith IF system schematic

ical step is to move back to the second IF amplifier, Q102. At this point, no output signal is found.

This, then, isolates the defect to the second IF amplifier. The problem could be a defective second IF transistor, missing B+ voltage to this stage, or a fault in the coupling components between the second and the third IF transistor. You can narrow the defect down even more by moving the injection point to the collector of the second IF transistor.

Remember to move the RF-IF selector switch to the third IF position on the VA62A, as the signal level at the output of the second stage is the same as the input to the third IF stage. If an output is now obtained, you know the amplifier stage itself is defective. Use a transistor checker to test the transistor itself. The built-in DVM on the VA62A will let you confirm if the transistor stage is being fed the correct B+ voltage and bias voltages.

If the amplifier is not the problem, injecting the test signal at the collector will not give a satisfactory output. You then know that the problem is somewhere between the output of the second IF stage and the input of the third IF stage. All that's now required is to move the injection probe to the input and output of each interconnecting component.

Let's say, for example, that you have an output when the signal is injected at the output of the coupling transformer, but not when you inject at the primary side of the same transformer. This tells you that you have a defective transformer or a shorted component on the primary side of the transformer.

At this point you're down to just a few components that can be individually tested or substituted. You can isolate the components on the input side of the transformer, as an example, by unsoldering the primary of the transformer and again injecting the test signal. A defective transformer will still not produce an output. A shorted component in the primary circuit, on the other hand, will now be isolated from the transformer, and the signal will now pass on through the transformer.

In summation, the VA62A Analyzer lets you troubleshoot an IF defect down to a single stage and then trace the signal through each component in series with the signal path. You can then use conventional troubleshooting techniques to locate the actual cause of the circuit failure.

SAW filter and synchronous detector comments

The SAW (surface acoustic wave) filter is a device that's tuned to pass a certain set of frequencies by the way it's made, not by tuning after it's installed in a circuit. The SAW filter has been used in audio RF systems for many years because of its sharp tuning ability. With careful manufacturing, a SAW filter capable of producing a perfect TV IF waveform is now almost standard in most makes of color TV receivers.

The SAW device is made the same way a transistor or other semiconductor device is made—by slow build-up of layers. However, the big difference in the SAW device, compared to a semiconductor, is that when it's finished, it has the structure of a crystalline device. While it's being manufactured, a thin metal film is deposited on the surface of the device, which forms a transducer with interleave fingers. The 45.75-MHz IF signal is applied to the SAW device's input. When the crystal is stressed by the incoming signal, a mechanical wave moves along the surface of the transducer. This wave was named a "surface acoustic wave," and the device was given the same name.

The main reason for using a SAW filter is the reduction, or elimination, of the IF tuning coils. The SAW filter eliminates all interstage coupling transformers and usually reduces the number of trap coils to one. This reduction of tunable stages makes the SAW filter IF less susceptible to the drift problems that stagger-tuned IF strips develop with age. However, tuning has not been eliminated completely. Instead of the familiar IF alignment, you'll now use a procedure called "IF matching." This is because the IF strip doesn't need to be aligned. The need is to match all of the parts of the SAW filter IF strip to each other. This is required so as to obtain the maximum amount of signal to the video detector. The theory behind IF matching is exactly the same as the theory behind maximum power transfer in an audio system; the output of one stage must match the input of the next stage.

Maximum signal transfer becomes quite important in SAW filter IF strips because, as a trade-off for a perfect waveform, the SAW filter has a loss of close to 20 dB. By adding a linear, high-gain amplifier and setting the matching coil(s) properly, this loss has no effect on the signal processed at the video detector. Refer to the SAW circuit in Fig. 5-8 and positions of the remaining matching coils. Thus, IF tuning has been replaced by IF matching.

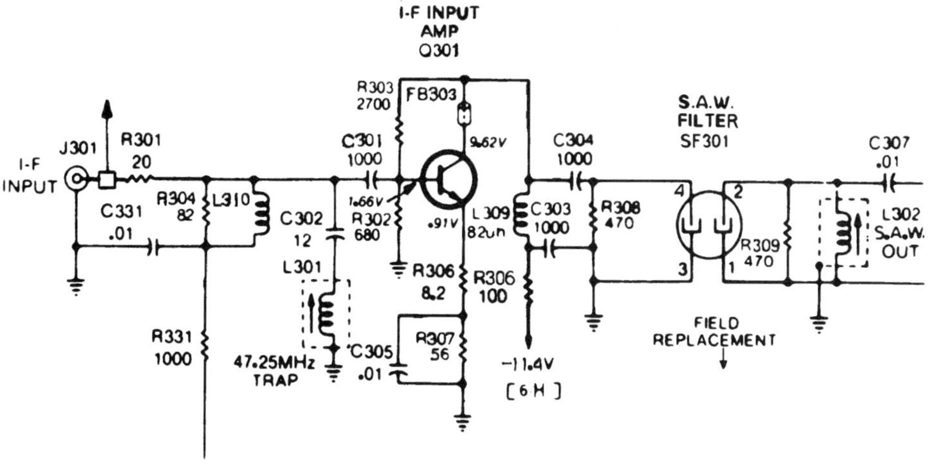

5-8 Circuit diagram of a SAW matching coil.

The *synchronous video detector* is a marvelous circuit that takes a simple diode detector circuit and makes it into a very complex circuit built within the chip package. The added complexity, however, provides several major benefits over the diode detector.

The diode video detector has always had problems because of the nonlinear action of any rectifier diode just at the point where it turns on. This nonlinear action causes beat frequencies that radiate back to earlier stages and must be trapped out. If not, the beat will show up on the TV screen.

Also, the diode detector will turn on any time the forward bias voltage is reached. This means noise spikes can turn the detector on. Extra noise elimination circuits are thus required to hold the IF strip amps off if any noise spikes are present. The final drawback with the video detector diode is that it has a signal loss, not a signal gain. This is because the diode is a passive type of device.

The synchronous video detector eliminates these problems. It has a carrier generator section that only keys on when the 45.75-MHz picture carrier is present, which eliminates any turn-on by noise spikes. It has an *active bridge rectifier* made up of four transistors that are operating in their linear range any time the carrier generator keys on, so any beat interference is eliminated. It also has a differential amplifier section that provides gain instead of losses. These synchronous video detector benefits far outweigh the size and cost disadvantages, and these detectors are being used by most TV receiver manufacturers.

Aligning synchronous video detectors

The synchronous video detector is an active video detector using a circuit tuned to the video IF carrier to drive a transistorized bridge demodulator. The synchronous detector is within an IC. Not all IC detectors, however, are synchronous detectors. The way to determine if the IC is a synchronous detector is to look for a coil marked "oscillator," "45.75-MHz reference," or "detector tuning." A conventional detector inside an IC doesn't have a similar adjustment. The IC might include one, two, or three adjustments.

One of the alignment adjustments always affects the linearity of the detector. This is the only adjustment in a one-coil system. Some multicontrol detectors call this coil the *limiter coil*. This adjustment is the most crucial one because it causes a major shift in the brightness or contrast of the detected video signal. In addition, minor misalignment of this coil often causes severe overshoot on the leading edge of square waves.

The other adjustments generally determine the lock-in range of the detector's tuned circuits or the amount of dc bias applied to the first video amplifier stages. These controls aren't as crucial as the linearity adjustment, but they still must be properly adjusted for best performance.

Whenever you're aligning the detector, you should inject the VA62A's 45.75-MHz video IF signal into the IF input, as you would for alignment of a conventional IF stage. A scope must also be used, as with conventional alignment, and connected in the same way.

To align a one-coil system:

1. Adjust all traps before adjusting the detector.
2. Set the VA62A to "Lo" and "Norm" output levels.
3. Set the VA62A video pattern switch to the "10-bar staircase" pattern.

4. Trigger the scope at the horizontal video rate.
5. Adjust the detector tuning control until the steps on the CRT scope display are as straight as possible. Avoid compression of the top or bottom steps, or bowing of the steps between the two extremes. Also, watch for signs of overshoot or ringing on the edges of the steps.

To align Zenith systems with three adjustments:

1. Adjust all traps before adjusting the detector.
2. Adjust the "zero carrier" control for the correct dc voltage at the composite video test point, as found in the Zenith service data.
3. Feed the VA62A "45.75-MHz video IF" signal to the alignment test point with the output level controls set to "Lo" and "Norm."
4. Adjust the "oscillator" coil clockwise until the pattern on the scope drops out to noise.
5. Count the number of times you must turn the same slug in the counterclockwise direction to run the coil to the second point that the pattern drops out to noise.
6. Readjust the slug to the clockwise direction ½ the number of turns from step 5 to center the slug between the two points where the detector loses its reference.
7. Adjust the "limiter" coil for the best linearity of the stair-step signal on the scope.

SAW filter alignment

Note: IF stages using SAW devices almost always use synchronous video detectors. Always confirm that the detector is aligned properly before working on the SAW adjustments.

IF stages using SAW filters don't require alignment in the normal sense. The frequency response of the SAW system is determined almost entirely by the design of the SAW device. The one or two adjustable coils that remain match the input and/or output of the SAW device to the amplifiers. These adjustments have much less effect on the detector output than the coils found in conventional bandpass systems because they provide impedance matching instead of frequency response shaping.

Most SAW systems also include at least one IF trap. As with any IF system, the trap must be adjusted before adjusting any other matching coils. Traps in SAW systems are less responsive than those of conventional systems because the SAW device already provides a great deal of attenuation of the unwanted signal. Thus, you need to watch the results of the trap-setting signal very closely when adjusting a SAW trap. Use the least amount of trap signal that will show interference on the video pattern. Switch off several bars from the multiburst bar sweep in order to produce a black area on the TV screen. Then increase the brightness and contrast controls until you can see the modulation of the trap pattern in the black area of the TV set's screen that's open where the bars are turned off.

You might find it easier to use an oscilloscope connected to the video detector test point in some chassis if you can't see the trap null on the screen. When using a scope to set traps, use the "10-bar staircase" pattern and set the audio switch to "off." Then, adjust the trap for the least amount of thickness on each of the steps.

You might prefer to use the programmable IF generator in a chassis that shows little trap action. If so, use 1000-Hz audio modulation. Always use the least amount of IF signal that will show a trap pattern on the set's screen.

After setting the trap(s), you're ready to adjust the other matching coils. The coils usually affect low-end sensitivity rather than frequency response. The easiest way to adjust for best sensitivity is to monitor the AGC voltage with the VA62A Analyzer's digital meter. The procedures will vary slightly, depending on whether there's a test point available to monitor the IF AGC voltage. Some chassis don't provide a test point for this purpose because the AGC detector and the IF amplifier are inside the same IC chip. In such cases, you need to monitor the tuner AGC voltage, which might only provide dynamic action over a limited range of signal levels. Let's now look at the procedures for adjusting the coils when an IF test point is available, and then check out another procedure if you must use the tuner AGC voltage.

To use the IF AGC voltage as a reference:

1. Connect the VA62A external dc meter to the IF AGC test point.
2. Feed the VA62A's "45.75-MHz video IF" to the IF input using the IF input jack or tuner test point. If neither is available, you might feed to the antenna input, but make sure that the AFT circuits are properly adjusted.
3. Set the RF-IF level switch to "Lo" and the RF-IF level vernier to .5.
4. Adjust the SAW matching coils for the highest possible reading on the digital meter. This provides the best IF sensitivity.

When using the tuner AGC voltage as your reference, you must determine two things: the correct signal level, and whether the voltage must be peaked or nulled for correct response. First, connect the dc meter to the AGC line running to the tuner and monitor the voltage as you change the settings of the RF-IF level switch and the RF-IF level vernier controls. You should notice that there are two extremes of voltage. One occurs at high signal levels, and one at low signal levels. Generally, there's only a limited range over which the voltage will change with changing signal levels.

You want to adjust the voltage to duplicate high signal levels. For example, if the voltage is at its minimum value with the high signal levels, you'll adjust the SAW coils for the least amount of voltage. Similarly, if the voltage is at its maximum value with the highest signal level, you want to adjust the SAW matching coils for the highest voltage.

After you've determined whether you need to peak or null the SAW adjustments, use the following procedures to adjust the SAW matching controls:

1. Adjust the VA62A output through the range of levels to determine where the AGC circuits provide dynamic action (the range of signals that cause the AGC voltage to change in step with changing signal level).
2. Set the VA62A output to the lowest level that's part of the dynamic range determined in step one.

3. Adjust the SAW matching coil until the AGC voltage is either peaked or nulled (as determined earlier) with the setting of the VA62A output level.
4. If the SAW adjustment causes the AGC voltage to reach a maximum or minimum value, reduce the setting of the VA62A and repeat the SAW adjustment until no further improvement is noted.
5. Repeat steps one through four for any additional SAW matching coils.

Analysis of the synchronous detector

The simple germanium diode has been used for years as the video detector in TV receivers, even though much better systems were known. The germanium diode has many shortcomings, but due to the complexity and cost of the other systems, it was the only economical choice. Modern IC technology has now changed all of this. The synchronous video detector system is now taking the diode detector's place in many TV receivers and offers many advantages when compared to the simple diode detector. To better understand the advantages of the synchronous detector, let's review the simple diode detector, how it operates, and its shortcomings as a video detector.

The simple diode detector acts just like the rectifier in a power supply. The diode actually half-wave rectifies the incoming IF signal to form half-cycle pulses of IF carrier signal. Figure 5-9 shows that the sine wave signal is cut in half and only the positive portions are present at the diode output. The valleys between the peaks are smoothed out by the resistor and capacitor filter network so that the resultant output of the circuit is an "envelope" of the amplitude of the modulated RF carrier or the video information. The output "envelope" is an exact replica of the signal that modulated the TV transmitter.

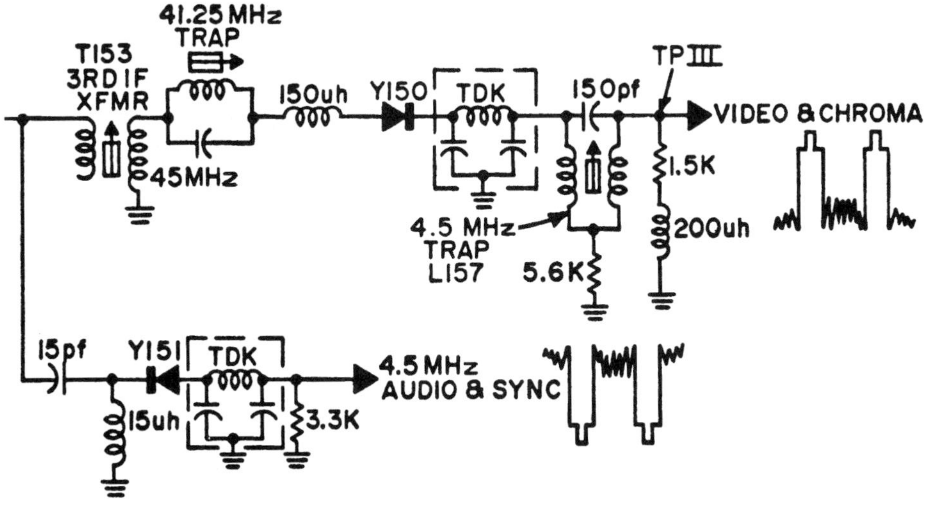

5-9 The simple diode detector rectifies the incoming RF signal and converts it to half-wave pulses that are smoothed out by the resistor and capacitor filter network.

The diode is a very simple detector, requiring very limited circuitry. However, it has four basic shortcomings that must be overcome or compensated for to obtain an acceptable picture on the set's screen. These are as follows:

1. A threshold voltage that must be overcome before the diode begins to conduct.
2. A nonlinear region or knee.
3. Beat interference caused by the mixing action of the nonlinear portion of the knee.
4. Detection of anything that's applied to its input, including adjacent channel sound, adjacent channel video, and noise.

Two of the shortcomings, the beat interference and the ability to detect whatever is applied to its input, can be reduced to an acceptable level by the addition of traps in the IF stages to reduce the adjacent channel sound and adjacent channel video signals. The traps reduce these two interfering signals to such a small level that they don't usually reach the diode to be detected or create interference beats.

The 920-kHz beat interference created by the mixing of the 42.17-MHz color carrier and the 41.25-MHz sound carrier is more difficult to reduce in the detected video, as this frequency is in the middle of the video spectrum. This 920-kHz beat is reduced by trapping the sound signal to a very low level and removing it before it reaches the video detector. The sound then must be detected with another diode to change it from 41.25-MHz to the 4.5-MHz IF signal that's amplified in the sound IF circuits.

The effects of the diode on the linearity of the detected signal can be reduced by increasing the applied signal level. The nonlinear portion, which includes the threshold voltage and knee of the diode, will then be small parts of the overall signal. This will reduce the effect, but not eliminate it completely. This means that the IF stages will have to provide more gain to the IF signal. The high-gain stages in the IF often add more distortion and noise to the detected signal.

Technically, the simple diode is not the best choice for a video detector, but until recently, it was the only economical answer for the design engineer. The synchronous detector required about 50 or more parts, including closely matched transistors and resistors. IC technology has made the synchronous detector cost effective for the design engineer to use in TV receivers. It is, for example, very easy to make the matched transistors that are required for the synchronous detector on an IC chip. A block diagram of an IC synchronous detector is shown in Fig. 5-10. As shown in Fig. 5-11, the synchronous detector alignment is simplified by using the 10-bar staircase pattern.

Cable-shifted channels

Cable TV systems often shift the frequency of the video carrier of some or all of the channels to prevent interference. This interference sometimes comes from an over-the-air station. At other times, the interference comes from the carrier of one channel mixing with the harmonics of another channel within the cable system itself. The

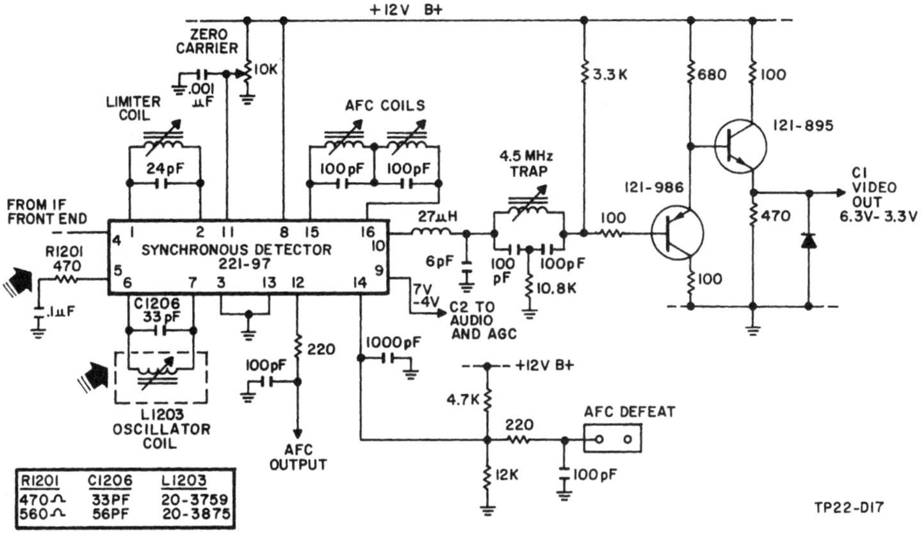

5-10 Simplified diagram of the synchronous detector circuit.

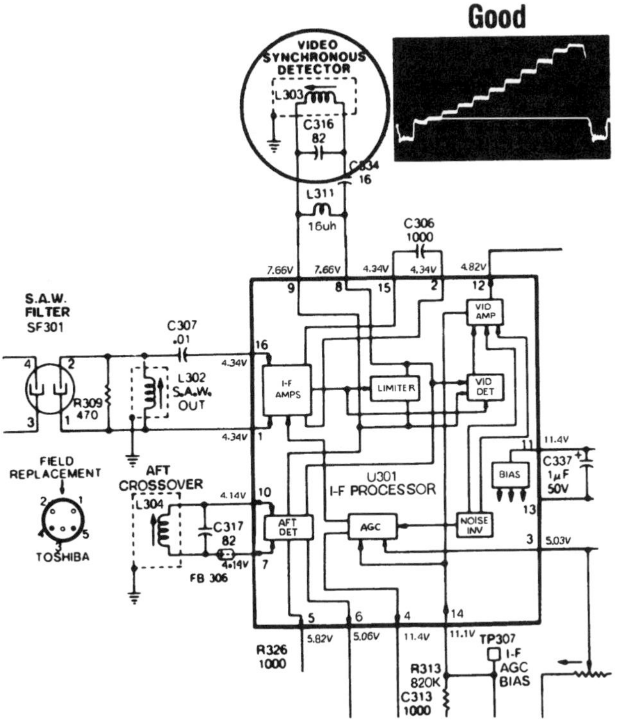

5-11 Synchronous detector alignment is simplified by using the 10-bar staircase pattern.

two most common schemes for shifting the frequencies are called the *harmonic related carrier* (HRC) and the *incremental coherent carrier* (ICC).

The HRC scheme shifts all of the video carriers so that the video carrier frequencies of higher cable channels land at exact multiples of the video carrier frequencies of low VHF channels. Thus, the upper channels are harmonically related to the lower channel carriers. All carriers that are harmonically related are then phase-locked together so the harmonics of a lower channel can't cause interfering beats in the carrier of a higher channel.

All of the video carrier frequencies must be shifted from the FCC-assigned frequencies in order to get all of the harmonics to fall in line with each other. This is done by shifting all of the carriers (except for VHF channels 5 and 6) down to 1.25 MHz from the FCC channel frequencies. The FCC frequency assignment includes a gap between channels 4 and 5, which causes channels 5 and 6 to be out of the normal 6-MHz step sequence compared to all other channels. Channels 5 and 6 are, therefore, shifted up by 0.75 MHz to make them fall into line with the other channels. This results in a full 6-MHz gap between channels 4 and 5, which is filled with an extra channel on some cable systems. (See Fig. 5-12.)

Channel	HRC Frequency Shift	ICC Frequency Shift
	(MHz)	(MHz)
2	− 1.25	0
3	− 1.25	0
4	− 1.25	0
4 +	4 (+ 4.75) or 5 (− 5.25)	4 (+ 6.00) or 5 (− 4.00)
5	+ 0.75	+ 2.00
6	+ 0.75	+ 2.00
7	− 1.25	0

5-12 Both cable shifting methods provide space for an extra channel between 4 and 5.

The ICC system leaves all of the channels (except for channels 5 and 6) at the standard FCC-assigned frequencies. Channels 5 and 6 are moved up 2 MHz to make them fall into normal 6-MHz steps. Now, the harmonics of each carrier never fall close enough to the carrier of higher frequency channels to cause interference. Like the HRC system, moving channels 5 and 6 results in a full 6-MHz gap between channels 4 and 5, which can be filled with an extra cable channel if desired.

The extra cable channel between channels 4 and 5 is called "4+," and "5A," or "A-8." I'll call this extra channel "4+" in this book. The "4+" and "5A" names are pretty obvious, but not "A-8."

Originally, cable channels were given letter designations, starting at 121.25 MHz. This kept the frequencies occupied by the FM band (88.1 through 107.9 MHz) open. Some cable systems didn't carry FM stations and found that TV channels could be placed in the FM band. These extra channels were numbered backwards from "A": A-1, A-2, etc. If you continue to count backwards until you get to the gap between channels 4 and 5, you'll find that you're eight 6-MHz channels down from channel "A," leading to the designation "A-8."

The VA62A can provide a signal at channel "4+" for either the HRC or IIC frequency, as needed. Like most digital tuners, however, 4+ is not part of the normal numbering sequence because it's so seldom used in cable systems. Some tuners, for example, assign "4+" channel number "00." Others assign it some other number out of the normal cable tuning sequence. Still others leave 4+ off altogether.

You can use the programmable cable channel feature to move channel 4 up or channel 5 down in frequency to produce the correct channel 4+ carrier. The amount of shift will depend on whether you're making the adjustment for an ICC or an HRC system. The HRC system needs a carrier of 72.00 MHz for channel 4+. This can be obtained by shifting channel 4 up by 4.75 MHz, or channel 5 down by 5.25 MHz. The ICC system assigns channel 4+ to 73.25 MHz (6 MHz higher than FCC channel 4). This can be reached on the VA62A by shifting channel 4 up to 6.00 MHz or channel 5 down by 400 MHz.

You might want to program in all of the other carrier shifts, except for channel 4+, and only shift the VA62A when it's necessary to test an occasional tuner. The instructions for shifting the carrier are found in the pull chart under the unit.

Some local cable systems might shift the carriers of some cable channels using a scheme that's different from the HRC or ICC methods. If so, the VA62A programmable cable channel feature can be programmed for any frequency shift up to ±9.75 MHz. Once the memory has been programmed, the shift instructions are kept in nonvolatile memory (until reprogrammed), even if the power to the unit is removed. Thus, no battery is needed to maintain the memory. Note that the frequency allocations are shown in Figs. 5-13 and 5-14.

VHF And Cable Frequency Allocations

BAND	DIGITAL TUNER NUMBER	ALPHA NUMERIC DESIGNATION	VIDEO CARRIER FREQUENCY
Low-Band VHF	2	2	55.25
	3	3	61.25
	4	4	67.25
	5	5	77.25
	6	6	83.25
Mid-Band Cable	95	A 5	91.25
	96	A 4	97.25
	97	A 3	103.25
	98	A 2	109.25
	99	A 1	115.25
	14	A	121.25
	15	B	127.25
	16	C	133.25
	17	D	139.25
	18	E	145.25
	19	F	151.25
	20	G	157.25

5-13 Standard (non-shifted) frequencies for VHF and cable channels.

BAND	DIGITAL TUNER NUMBER	ALPHA NUMERIC DESIGNATION	VIDEO CARRIER FREQUENCY
	21	H	163.25
	22	I	169.25
High–Band VHF	7	7	175.25
	8	8	181.25
	9	9	187.25
	10	10	193.25
	11	11	199.25
	12	12	205.25
	13	13	211.25
Super–Band Cable	23	J	217.25
	24	K	223.25
	25	L	229.25
	26	M	235.25
	27	N	241.25
	28	O	247.25
	29	P	253.25
	30	Q	259.25
	31	R	265.25
	32	S	271.25
	33	T	277.25
	34	U	283.25
	35	V	289.25
	36	W	295.25
Hyper–Band Cable	37	AA	301.25
	38	BB	307.25
	39	CC	313.25
	40	DD	319.25
	41	EE	325.25
	42	FF	331.25
	43	GG	337.25
	44	HH	343.25
	45	II	349.25
	46	JJ	355.25
	47	KK	361.25
	48	LL	367.25
	49	MM	373.25
	50	NN	379.25
	51	OO	385.25
	52	PP	391.25
	53	QQ	397.25
	54	RR	403.25
	55	SS	409.25
	56	TT	415.25
	57	UU	421.25
	58	VV	427.25
	59	WW	433.25
	60	XX	439.25
	61	YY	445.25
	62		451.25
	63		457.25
	64		463.25
	65		469.25

5-13 Continued

BAND	DIGITAL TUNER NUMBER	ALPHA NUMERIC DESIGNATION	VIDEO CARRIER FREQUENCY
	66		475.25
	67		481.25
	68		487.25
	69		493.25
	70		499.25
	71		505.25
	72		511.25
	73		517.25
	74		523.25
	75		529.25
	76		535.25
	77		541.25
	78		547.25
	79		553.25
	80		559.25
	81		565.25
	82		571.25
	83		577.25
	84		583.25
	85		589.25
	86		595.25
	87		601.25
	88		607.25
	89		613.25
	90		619.25
	91		625.25
	92		631.25
	93		637.25
	94		643.25

5-13 Continued

UHF Frequency Allocation

DIGITAL TUNER NUMBER	VIDEO CARRIER FREQUENCY	DIGITAL TUNER NUMBER	VIDEO CARRIER FREQUENCY
14	471.25	49	681.25
15	477.25	50	687.25
16	483.25	51	693.25
17	489.25	52	699.25
18	495.25	53	705.25
19	501.25	54	711.25
20	507.25	55	717.25
21	513.25	56	723.25
22	519.25	57	729.25
23	525.25	58	735.25
24	531.25	59	741.25
25	537.25	60	747.25
26	543.25	61	753.25
27	549.25	62	759.25
28	555.25	63	765.25
29	561.25	64	771.25
30	567.25	65	777.25
31	573.25	66	783.25
32	579.25	67	789.25
33	585.25	68	795.25
34	591.25	69	801.25
35	597.25	70	807.25
36	603.25	71	813.25
37	609.25	72	819.25
38	615.25	73	825.25
39	621.25	74	831.25
40	627.25	75	837.25
41	633.25	76	843.25
42	639.25	77	849.25
43	645.25	78	855.25
44	651.25	79	861.25
45	657.25	80	867.25
46	663.25	81	873.25
47	669.25	82	879.25
48	675.25	83	885.25

5-14 UHF frequency allocation channels.

6

Troubleshooting TV sync and AGC systems

THIS CHAPTER SHOWS YOU HOW TO USE THE DELUXE VA62A VIDEO ANALYZER TO troubleshoot different sync and AGC systems used in various color TV receivers. You'll also find out how to check out PLL (phase-locked loops) and closed-loop-feedback circuits with the Analyzer, and how to use the interlaced adder signal to test the vertical countdown IC chips now used in most modern TV receivers.

Sync circuit testing

Sync circuits don't seem complicated on the surface, but they can be big troublemakers if they aren't understood and checked out thoroughly. Moreover, the sync stage in a TV receiver seems to have been designed last in many TV sets.

Technicians are currently confronted with tube-operated sets, transistor-designed TVs, countdown divider chips, plus interlaced and noninterlaced sync systems and integrated circuits. Also, you never know when you'll run into a tricky, gated sync separator that adds complexity beyond belief because of the various feedback circuits. The VA62A Analyzer will let you service all of these types of TV receivers, simulate the TV station signal, and make all circuits look alike for standardized testing.

The sync stages actually begin in the video amplifier in most receivers (except some Zenith models) because video amplifier limiting is used to clean up the signal for better operation. This is all well and good if the AGC system is operating properly. If it's not, excessive IF gain will drive the video output stages to cutoff, and a compressed (but very clean) sync pulse will be produced. The sync separator doesn't "know" the sync is compressed, and the result is that lower black portions of the picture pass through the sync separator, producing the "horizontal bends" or wiggles.

Once they've eliminated the video amplifier as a troublemaker, most technicians attack a sync problem from the sync takeoff point. For this reason, the VA62 must have adequate amplitude and correct polarity to feed that point with a composite video signal to check out the entire sync system. The sync pulse should be as good as the pulse from the TV station in order to perform correct troubleshooting circuit analysis. The sync pulse from the VA62A is a genuine FCC spec pulse with a rise-time standard horizontal sync pulse exactly like that transmitted by a TV station.

At this point, let's talk about amplitude and polarity for a moment. It's very obvious that you'll require an adjustable sync output drive signal, with both positive and negative polarity, if you're going to substitute for both input and output of various stages. Amplitude can be a very different question, as you must cover all current and future sets and must not damage any components with the test signal pulses. For this reason the VA62 has three variable output pulse ranges of 3 volts, 30 volts and 300 volts. Use the 300-volt range for tube sets, 30-volt range for transistor sets and the 3-volt range for sets with IC chips.

A good question is: What happens if I connect the VA62 Analyzer test sync pulse to the input of a sync separator, for example, and the normal TV sync pulse is already present? Of course, you don't want the TV signal to be present at the injected signal test point while you're testing. Also, you want to use the test signals to tie down the input circuits with known-good signals for these checks. This is accomplished by using an impedance that's much lower in the Analyzer drive-level output than in the circuits being tested. Thus, the substitute signal "swamps" out any other signal that's present. These impedances must be chosen so that they swamp the TV signal but don't load down the TV circuits. This is how the impedances were selected for the VA62 driver-test output signals.

TV sync circuit troubleshooting would in many cases be thought of as one of the simpler circuits to troubleshoot if it weren't for the gated sync separators. These gated systems can complicate sync circuit troubleshooting because the feedback gating pulses from the horizontal sweep output transformer travel in a closed-loop system. You have to find a way to tie down troubles in this system, where any trouble in one section of the loop reflects the same trouble throughout the loop. (Refer to the block diagram of a gated system loop shown in Fig. 6-1.) Let's now put the Analyzer to work in these more complex circuits.

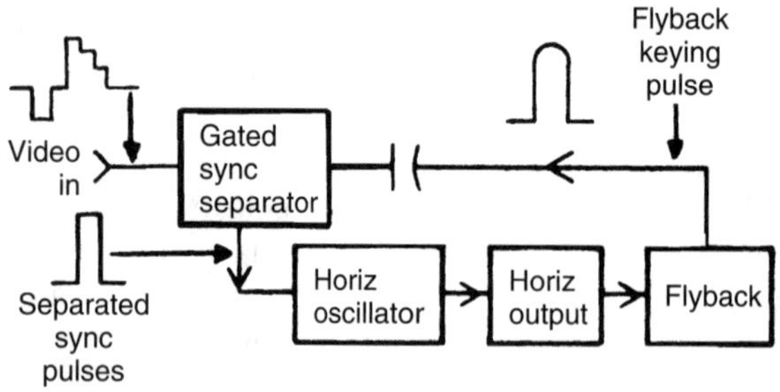

6-1 Block diagram of a gated, closed-loop sync system.

The input of the gated sync separator is checked just like a conventional separator, by feeding the composite signal into the input stage. If the input signal is missing or distorted, the set should now return to normal operation with the injection of the test signal. Refer to Fig. 6-2 for the Analyzer control settings and the sync pulse-test injection points.

A substitute signal for the sync separator output is just as easily available. Simply reverse the phase, increase the drive-level amplitude, and substitute for the sync pulses in the output stage. The plus and minus vertical synchronization integrated pulses are shown in the Fig. 6-3 scope waveform. The Fig. 6-4 scope trace shows the plus and minus vertical and horizontal synchronization pulses.

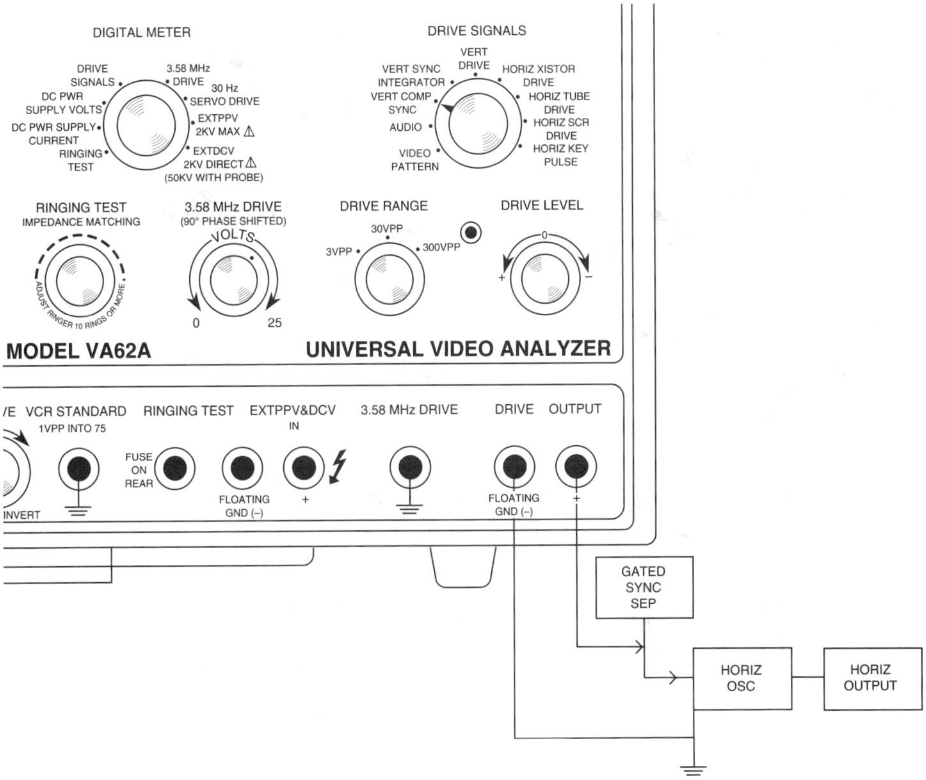

6-2 Control settings and test signal injection points for sync system troubleshooting with the VA62.

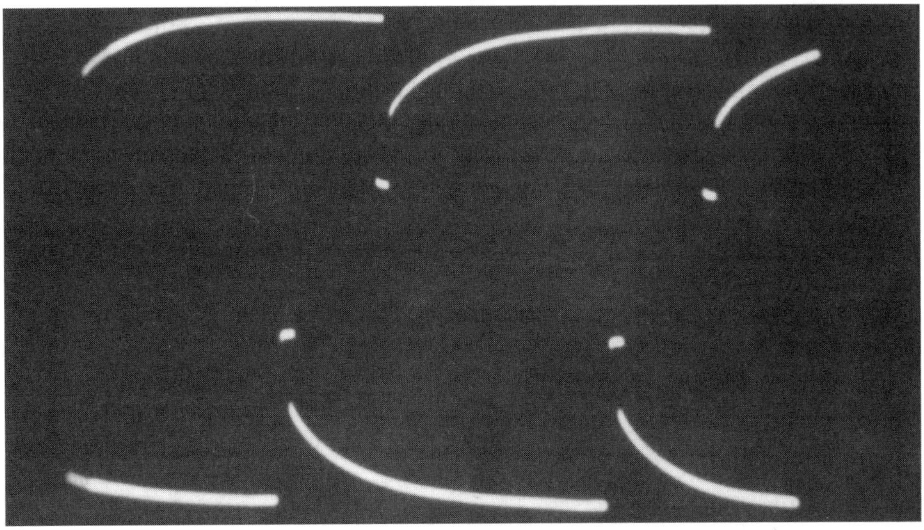

6-3 Wave form shows the horizontal and vertical sync integrated pulses ±.

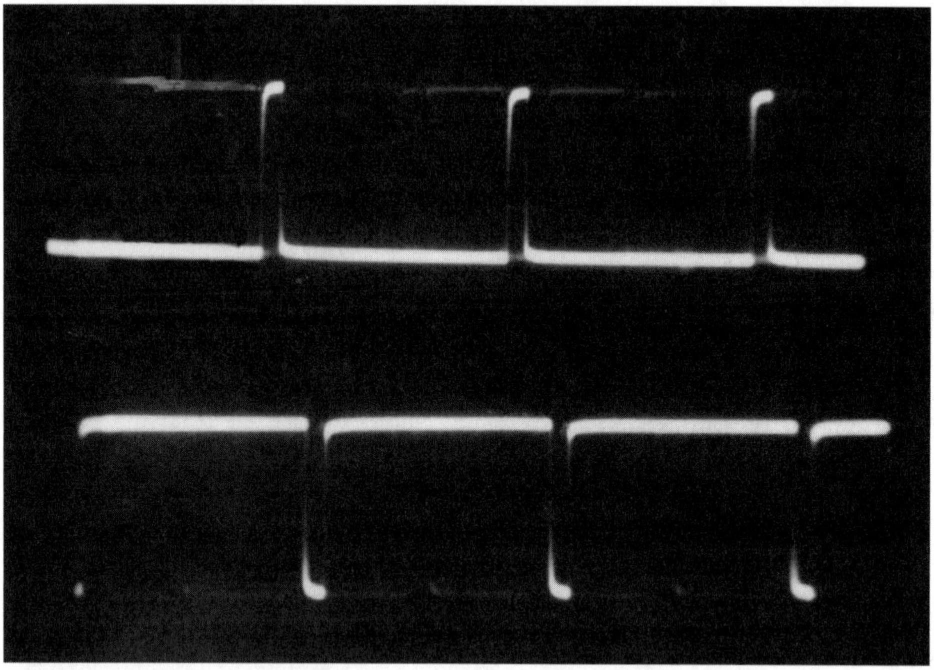

6-4 V and H Compositor Sync pulses, ±.

Now, if the TV set's picture returns to normal with the test signal applied, the problem might be due to a faulty gating pulse. To substitute for this pulse, just switch to the horizontal key pulse and feed the test signal directly into the sync separator output.

All of these signals are phase-locked together in the Analyzer, and there's no concern about the signals not arriving at the same time. The same pulse can be used to drive the horizontal sync discriminator by just lowering the amplitude and feeding it directly in before any waveshaping takes place. This takes the place of the feedback pulse from the horizontal sweep transformer.

The sync discriminator itself can be checked the same way you check other automatic circuits, by connecting the bias supply and feeding the dc control voltage directly to the horizontal oscillator. Check a block diagram of the horizontal AFC and locate the dc control voltage point. A small variation in the control voltage should vary the horizontal oscillator frequency until the picture falls into sync.

The amplitude of the test pulses are very important for proper sync operation. You can easily check these amplitudes with the built-in digital peak-to-peak meter that automatically switches as you make these pulse drive selections. This occurs as the drive signal digital meter monitors your action to tell you how much peak-to-peak signal you're driving each stage with. This meter monitor becomes very important when troubleshooting stages that you suspect are being fed too much signal to make up for low stage gain, etc.

It's possible to brute-force a signal through a defective stage by overpowering it. This can be avoided by using the drive signal monitor because it will tell you that you're somewhere in the ballpark or out of the ballpark and the stage has low gain. There are those times when you locate the problem, repair the stage, and just want to know the pulse level that you're driving the circuits with, so you can compare it with the recommended peak-to-peak amplitude on the service data. The digital drive signal monitor does this for you automatically.

There might also be times when you'd like to check the sync section (for example) by feeding the input of the sync separator and checking signal amplitudes at subsequent stages to speed up your analyzing time (especially when you have no sync at all). This is particularly true in those kinds of circuits where amplitude is very important because you're basically separating the signal amplitudes. The Analyzer enables you to signal trace as well by switching the digital meter to the EXT/PPV range and using the external test prods to trace throughout the circuit. This peak-to-peak signal-tracing technique can be used in any stage after signal detection because of the high-input bandwidth of the digital meter.

Specified NTSC sync

If the TV set checks with the signals supplied by the Analyzer, you can be sure that the set will operate properly when tuned to a broadcast station signal. The sync pulses (see Fig. 6-5) generated by the Analyzer meet EIA specifications, RS-240 for broadcast stations that pertain to the sync pulse rise time and width. The Analyzer gives you this big plus of operating the TV set just like an off-the-air signal for more precise troubleshooting.

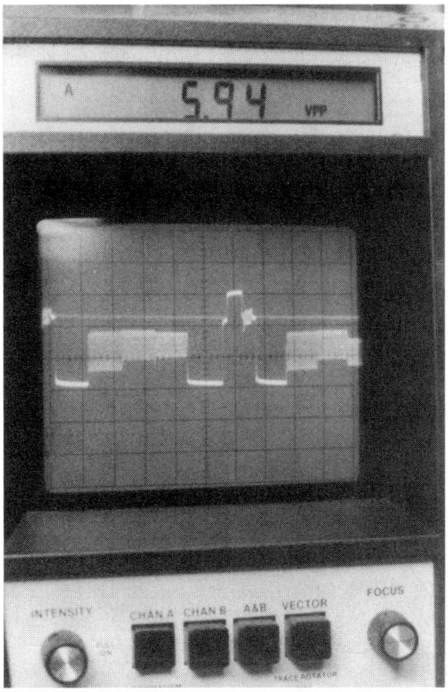

6-5 Broadcast-quality sync pulses are produced by the VA62 Analyzer.

Sync circuit applications

The loss of horizontal or vertical synchronization can be due to the loss of composite sync pulses. The composite video signal from the Analyzer can be injected into the sync separators to test their operation. If this doesn't correct the defect, composite sync pulses can be injected at the output of the sync separator, or at the input to the horizontal or vertical oscillators. Refer to Fig. 6-6 for the control settings and injection test points for sync trouble location.

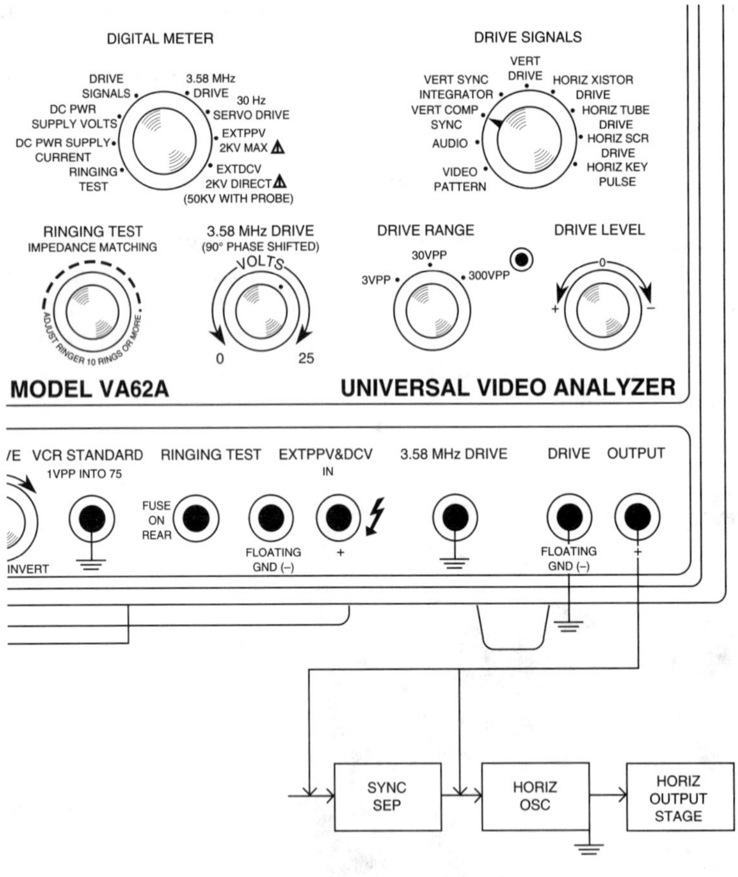

6-6 Control positions and test signal injection points for sync-circuit troubleshooting.

When using these substitute signals, you should apply the RF output of the Analyzer to the antenna input to provide video information. Since the substitute sync pulses are phase-locked to the video modulation on the RF signal, the set will begin operating normally if the test signal is injected beyond the defective stage.

Analyzing sync defects

Use the following steps for sync circuit analysis:

1. Connect the RF signal output of the Analyzer to the VHF antenna input terminals.
2. Select the same channel on the Analyzer and the TV receiver under test.
3. Select the desired video pattern.
4. Determine the proper signal level and polarity of the substitute sync signal.
5. Select the drive signals monitor position of the digital meter switch.
6. Select the desired sync test pulse.
7. Adjust the drive level control for the proper polarity signal amplitude. Warning: To prevent possible damage to solid-state devices, make sure the test signal polarity and level are within limits before injecting into the circuit under test.
8. Inject the substitute test signal at the input of the sync separator. If the set operates normally, the defect is at the output of the video detector.
9. If there's no change, set the drive signal switch to the composite vertical/horizontal sync position. Make sure the amplitude is not too high for solid-state sets.
10. If there's still no change, inject the signal at the input of the vertical or horizontal oscillator. If synchronization still doesn't return, the problem is in the vertical or horizontal sweep circuits.

Integrated vertical pulse

The VA62 has an additional drive signal that simplifies sync troubleshooting. The Analyzer produces an integrated vertical sync pulse to feed directly into the vertical oscillator. This integrated vertical pulse (+) is shown in the Fig. 6-7 scope trace. This pulse works much better than a normal vertical sync pulse, which can often be forced through a defective oscillator. Because the analyzer actually integrates the composite sync to form this pulse, the shape and timing exactly duplicate the output of the vertical integrating filter inside the TV receiver.

Horizontal keying pulse applications

The Analyzer provides a substitute horizontal keying pulse for use in place of signals derived from the horizontal sweep transformer. These pulses are used to trigger circuits such as keyed AGC, sync separators, or color burst gates. If the keying pulse is either missing, low in amplitude or mistimed (phase shifted), the keyed circuit will not function properly.

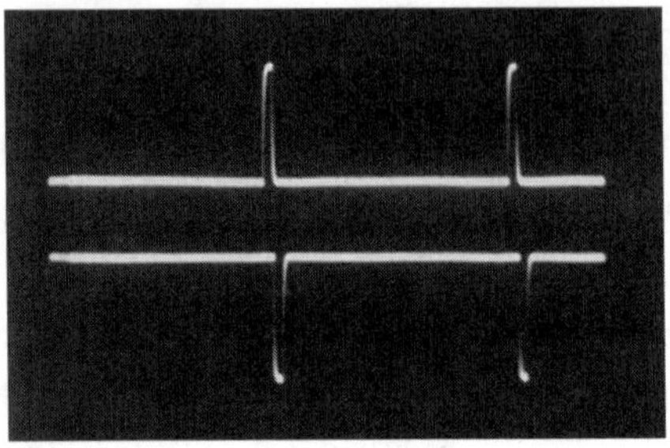

6-7 Vertical sync integrated pulses, ±.

The VA62 Analyzer provides a horizontal keying pulse that's properly timed for substitution of the missing pulse in the circuit under consideration. This signal will operate most circuits of this type. In some cases, the SCR horizontal drive pulse can also be used.

The block diagram in Fig. 6-8 shows the control setup positions and test points for injecting the horizontal keying pulse to troubleshoot AGC, sync, or burst-gate circuits.

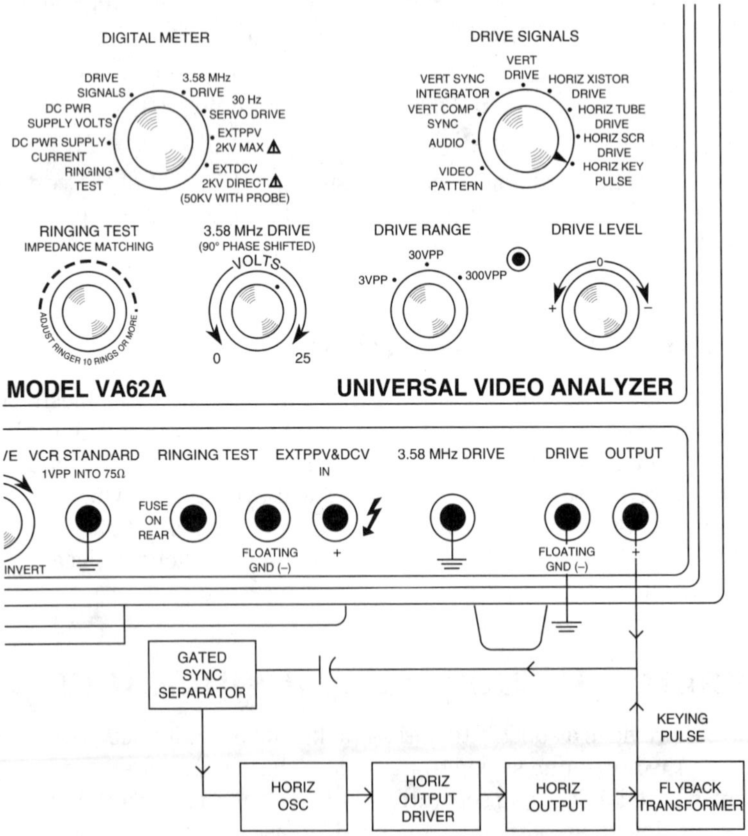

6-8 Analyzer setup and test points for horizontal keying pulse injection into AGC, sync, or burst-gate circuits.

Use the following procedure for keying pulse injection:

1. Connect the RF output of the Analyzer to the VHF antenna input terminals.
2. Select the same channel on the Analyzer and the TV receiver under test.
3. Select any pattern with the video pattern switch.
4. Determine the proper signal level and polarity for the substitute signal. Note: Observe proper polarity and signal-level recommendations for the drive pulses before injecting to prevent possible damage to solid-state circuits.
5. Select the drive signal monitor position of the digital meter switch.
6. Select the horizontal key position of the drive signals switch.
7. Substitute the horizontal sweep keying pulse into the circuit under test.
8. If proper operation returns to the circuit, the defect is a missing (or low in amplitude) keying pulse from the sweep transformer. Use the ringer test to test the keying-pulse winding on the sweep transformer for shorted or open windings. Also, check any coupling capacitors.

The dual-trace scope waveform photo in Fig. 6-9 shows the correct timing for an AGC keying pulse, as found in a solid-state TV receiver.

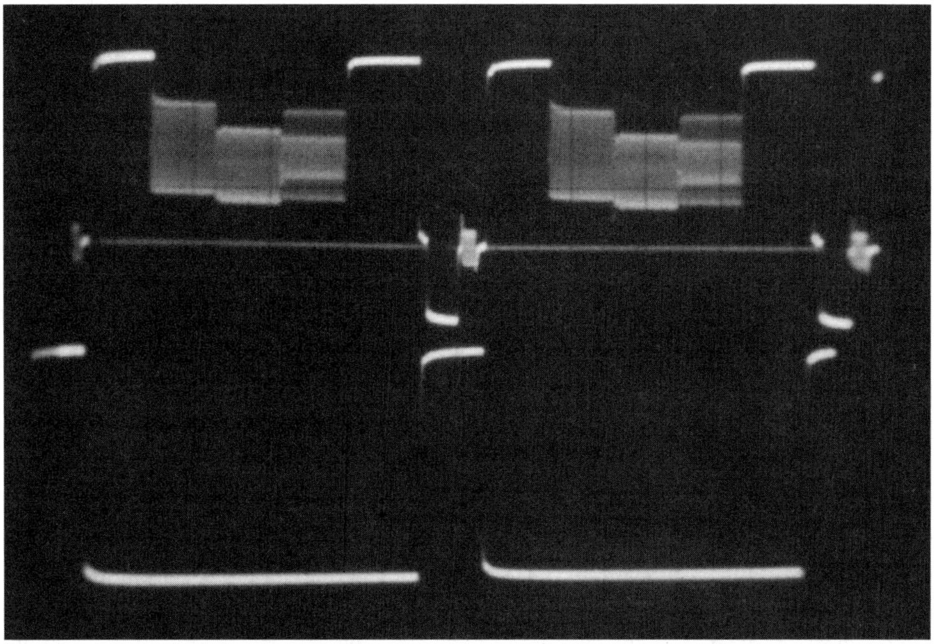

6-9 Correct timing for an AGC keying pulse as found in a solid-state TV receiver.

Interlace adder feature

You can select any of the VA62A video patterns as an interlaced or noninterlaced signal. The noninterlaced signal provides the most stable patterns for convergence.

The selectable interlace feature also allows a complete test of the digital vertical countdown ICs used in modern TV receivers. These ICs detect whether or not the incoming signal has equalizing pulses (found only in interlaced signals) and switches to one of two internal modes. By having both types of these test signals, the automatic detector circuits and the digital counting stages can be dynamically checked out.

Vertical countdown circuitry

Let's now look at this countdown system found in late model Zenith TV receivers. The vertical section, which contains IC 221-103, accepts the 503 kHz signal from a master scan oscillator, composite sync, and various dc voltages and thus provides vertical deflection, horizontal drive to the predriver, and a blanking output to the low-level luminance circuits. Refer to block diagram in Fig. 6-10.

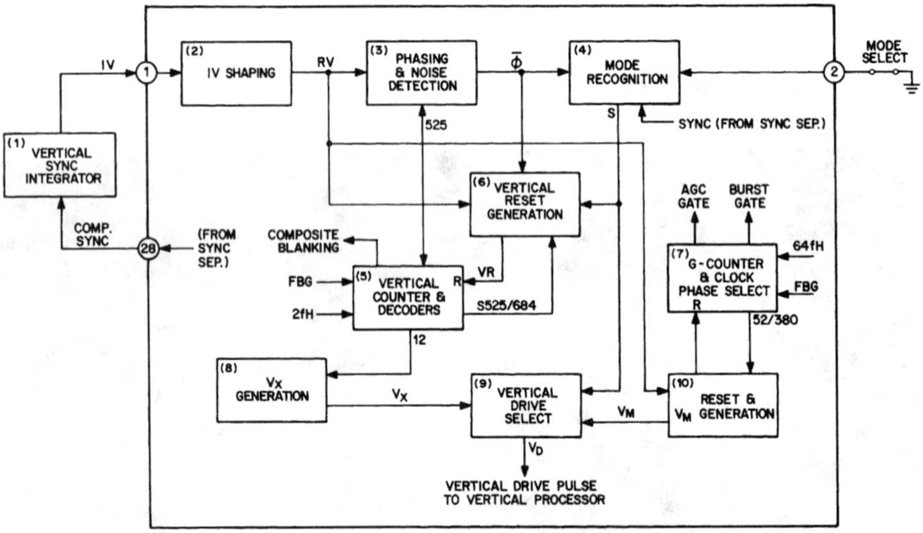

6-10 Block diagram of vertical countdown chip and its associated circuitry.

Although the vertical system provides horizontal drive, it doesn't contain the master scan oscillator or phase-lock loop. These circuits are located in another module. The vertical system accepts a clock signal that operates at 32 times the horizontal frequency (32 H is approximately 503 kHz). (See the block diagram in Fig. 6-11.) The IC divides this by 32 (down to H) before buffer-stage amplifying to provide drive to the horizontal sweep predriver. The resulting flyback pulse is compared to the incoming horizontal sync pulses in an external phase-lock loop, and the output of the phase-locked loop regulates the 32-H oscillator. These circuits are all located on another module.

The vertical system consists of an IC (part no. 221-103), two power transistors, and the peripheral circuit elements. Of course, you'll find no vertical or horizontal controls on this type of system.

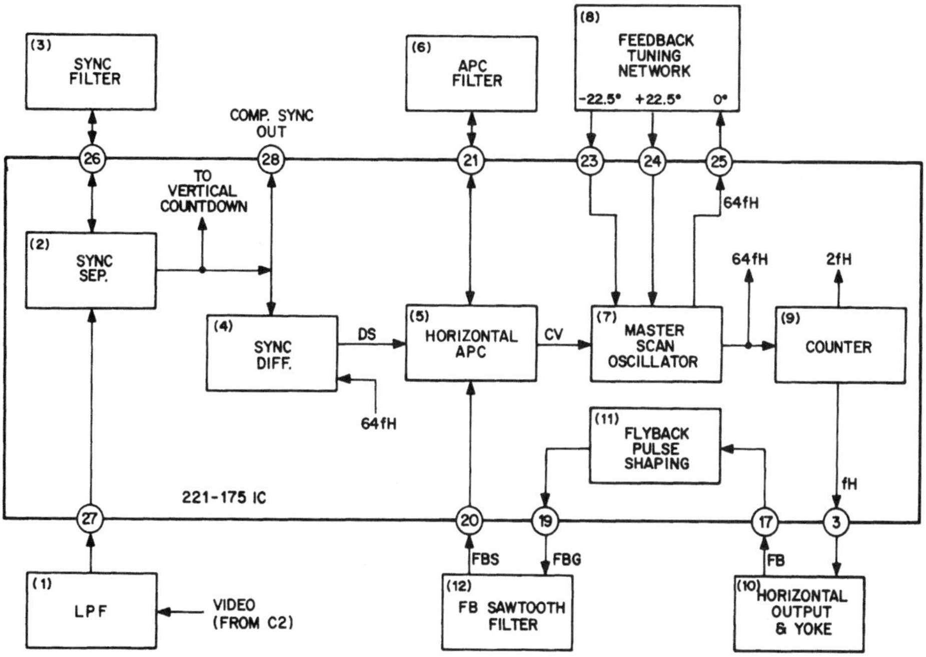

6-11 Expanded vertical countdown IC block diagram.

To divide by 32, the frequency is first divided by 16 (down to 2 H) for use in the logic circuitry of the IC. The signal is then further divided by 2 (down to H) before being buffered out as horizontal drive to the sweep output stages.

The logic circuitry provides a pulse to trigger a ramp generator. Associated with the ramp generator is a size control and a ramp-forming capacitor. The voltage ramp from the ramp generator is dc coupled inside the IC to one side of a differential amplifier. The differential amplifier drives a transistor output stage, and the output stage drives the vertical windings of the yoke. The yoke current is sensed and wave-shaped in the feedback network, and a sample of that yoke current is coupled to the feedback side of the differential amplifier. Thus, the differential amplifier and output stages function to make the sample of yoke current look just like the voltage ramp at its input.

The logic section also provides a blanking pulse that's amplified by a buffer stage and coupled to the 221-96 low-level luminance IC. The logic circuitry within the IC also has other inputs. It receives integrated vertical pulses to make sure that the vertical retrace interval coincides with the vertical interval of composite sync. Also, the logic circuit receives composite sync because it must decide whether or not the signal being viewed has 525 lines per frame. If the signal has 525 lines per frame, then the ramp generator is triggered by the countdown chain in the IC, and all of the advantages that come with a countdown system result, such as good interlace and good noise immunity. If the signal doesn't have 525 lines per frame, then the ramp generator is triggered by the integrated vertical pulses, which still drive the yoke, but interlace and noise immunity might be affected. Thus, the logic circuit must know if the incoming signal is standard or nonstandard. The logic circuit "looks" at the composite sync signal during the vertical interval for a period of 6 lines. If the signal is a standard NTSC signal, the composite sync will have proper serrations and the logic circuit should count 12 pulse edges during the 6-line time period. If the logic circuit doesn't count at least 9 edges during the 6-line period, it decides that the signal is non-NTSC (or nonstandard) and, therefore, doesn't necessarily have 525 lines.

Examples of nonstandard signals (no equalizing pulses) are those from some cable systems, some pattern generators, and possibly some closed circuit TV camera systems. Examples of standard TV signals are produced in broadcast television programs and high-quality pattern generators.

However, not all signals with enough serrations are standard NTSC signals with 525 lines per frame. Unfortunately, there are signals that exist that can confuse the decision-making process in the logic circuit. Thus, to force the correct response, a mode override jumper is provided on the module board at pin 8 of the IC. If this jumper is left plugged in, the logic circuit makes its own decision about whether the signal is standard or nonstandard. If the jumper is disconnected and left open-circuited, the logic is forced into a nonstandard mode.

Normally, the jumper should be left plugged in. However, in some reception areas, a signal might exist that this system will not lock onto. In this unusual case, the jumper should be disconnected. The circuit should then lock onto the signal, but the advantages of countdown (which are good picture interlace and noise immunity) will be lost.

With the signal injected from the VA62A Analyzer, push the interlace adder button on and see if the circuit goes into the standard mode. Push it off to check out nonstandard signal operation.

Closed-loop system checks

Troubleshooting automatic closed-loop systems is usually complicated by the fact that either end of the feedback loop can be the cause of various circuit faults. Probably the quickest technique for troubleshooting such a system is to provide a substitute feedback voltage and observe whether the operation of the stage returns to normal. If varying the external bias voltage causes a change in operation of the automatic circuit, the defect is the feedback voltage, "sending" end. If the operation is the same with or without the substitute voltage, the defect is most likely in the "receiving" end of the feedback loop. The block diagram in Fig. 6-12 shows a typical closed-loop system and injection voltage point.

Closed-loop testing procedure

1. Connect a 1000-ohm current-limiting resistor in series with the dc voltage from the analyzer and connect this resistor to the feedback loop.
2. Ground the other end of the dc supply.
3. Connect the digital meter to the feedback loop control point.

4. Adjust the dc voltage control to vary the substitute voltage through the range of voltages indicated on the service data for normal operation of the feedback loop system.
5. Now, should varying the voltage cause the circuit operation to change, you should test the sending end of the feedback loop. If the varying voltage doesn't correct the circuit symptoms, you should closely check out the receiving end.

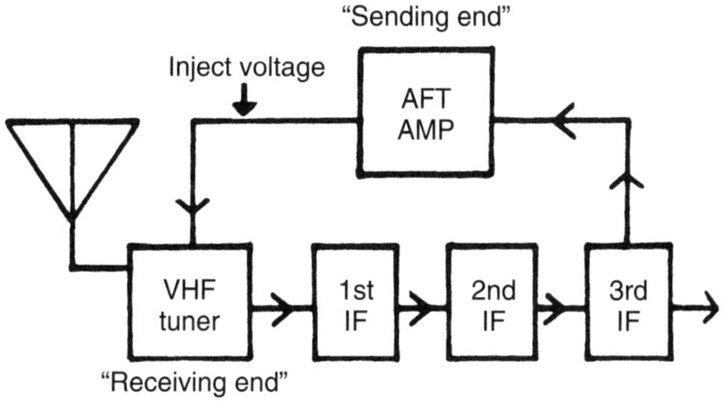

6-12 Block diagram of closed-loop feedback system.

The AGC (automatic gain control) system

Because the TV signal varies from station to station and from time to time, an AGC system is used to minimize these variations so that the viewer doesn't notice any of these changes. AGC systems vary from the simple, which is not very satisfactory because the AGC uses the negative dc output from the detector to bias the IF and RF amplifiers. As the signal increases, the negative dc voltage increases, putting more negative bias on the IF and RF stages. This bias reduces the gain and maintains a relatively constant detector output. This system works like the automatic volume control (AVC) used in radios. In some of the simple AGC systems, a diode is used to achieve some AGC delay for the RF amplifier stage. This is used to ensure a high RF gain for very weak TV-station signals.

Keyed AGC circuits

All modern-day TV receivers use the keyed AGC system, which overcomes all of the faults found in the simple and delayed AGC circuits. In the keyed system, the AGC voltage is developed only during the sync pulse interval. The keyed system eliminates the long-time constant of the peak detector, and the noise between sync pulses will have no effect on circuit operation and picture quality.

The block diagram in Fig. 6-13 is for a typical keyed AGC system. In a tube-type set, the controlled tubes are the RF amplifier and the first and second IF amplifier stages. The control grids of these tubes are connected also, and thus receive bias

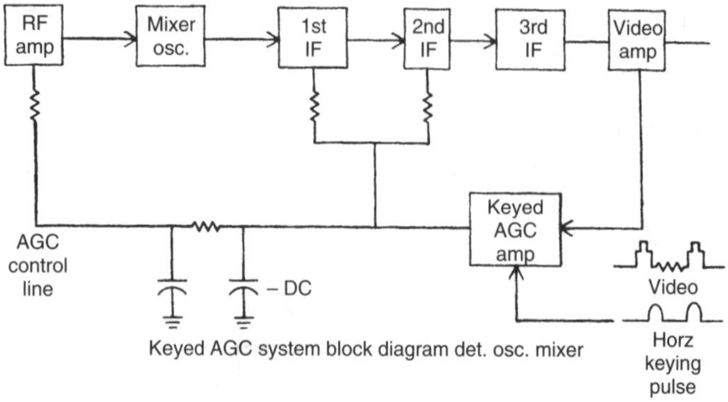

6-13 Block diagram of a keyed AGC system.

from the AGC control line. For this reason the gain of these stages is controlled by the AGC voltage that's developed by the keyer stage.

For the keyed automatic gain control system, the control voltage is developed by the AGC amplifier stage. The control grid of this amplifier receives a positive-going composite video signal and a dc voltage variation, which is dependent on the level of the RF signal. You'll notice that no B+ voltage is applied to the plate of the AGC amplifier tube. However, a narrow positive-going pulse from the sweep output system is fed to the plate at a 15.750-kHz horizontal rate.

During normal operation, the keying pulse and horizontal sync pulse should appear at the AGC amplifier at exactly the same time. Waveforms of these pulses are shown in the Fig. 6-14 dual-trace scope photo. The amplifier conducts only when the horizontal keying pulse appears at the plate. The AGC voltage level developed is pro-

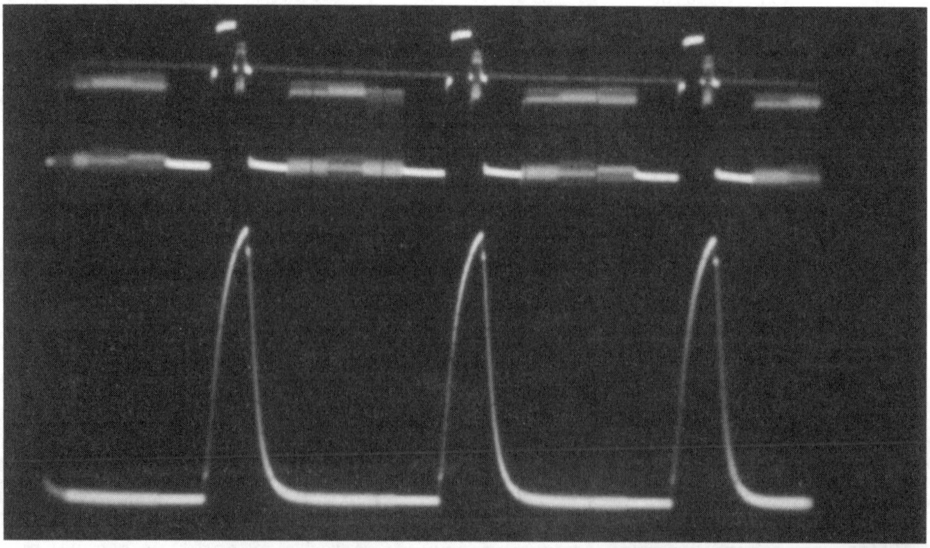

6-14 Dual trace scope traces for the horizontal keyer pulse and sync pulse found at the plate and grid, respectively, of a normally operating, keyed AGC stage.

portional to the sync pulse amplitude of the composite video signal and the dc bias on the control grid. Thus, any noise pulses in the video signal between sync pulses cannot cause the AGC amplifier to operate because there's no keying pulse at its plate at that time. Hence, noise pulses don't affect the level of the AGC output voltage. The circuit is immune to noise that might occur between horizontal sync pulses.

A fault in the AGC keying system can cause picture overload or bending, as shown in the Fig. 6-15 symptom photo. A problem in the sync circuits will cause the picture to lose vertical or horizontal sync (no picture lock-in). The picture will tend to roll up, down, and around sideways, as shown in Fig. 6-16.

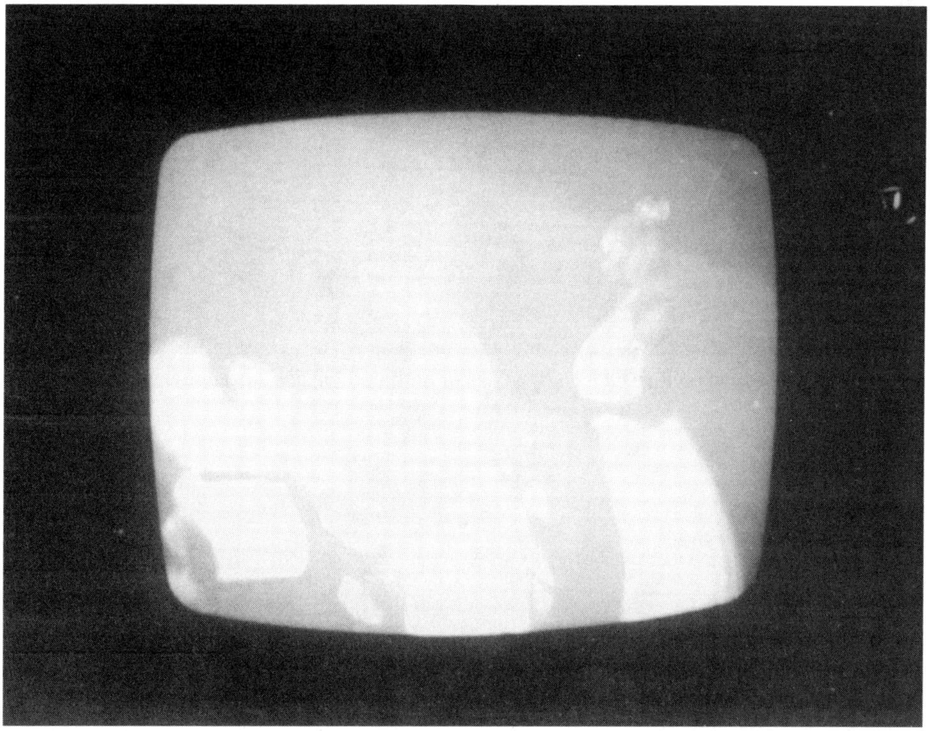

6-15 Picture bending caused by a defect in the set's AGC circuits.

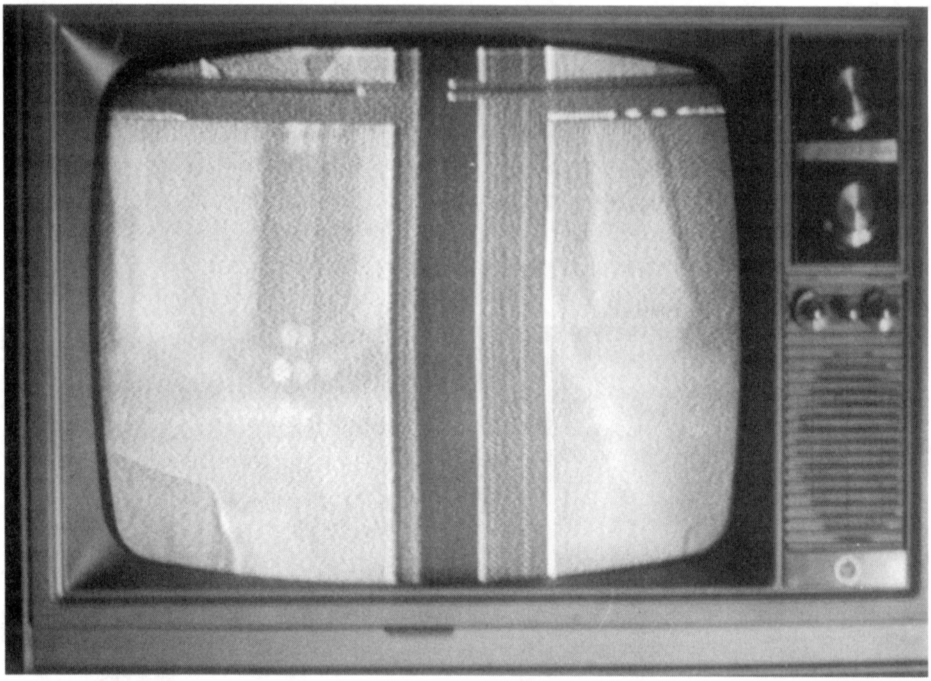

6-16 Out-of-sync picture caused by a loss of both vertical and horizontal sync lock-in.

7

Analyzing vertical sweep systems and sandcastle circuits

IN THIS CHAPTER I EXPLAIN HOW TO USE THE VA62A ANALYZER TO TROUBLESHOOT vertical sweep circuits. I also cover various types of vertical sweep circuit operation, including conventional sweep circuit and modern countdown and divider systems.

Zenith vertical countdown system

Let's do a brief review of how the Zenith vertical countdown system operates. Referring to the block diagram in Fig. 7-1, you'll note that the vertical section, which uses an IC 221-103 chip, accepts the 503-kHz signal from another module. Other signals and voltages include composite sync, 12 volt, 62 volt, and 130 volt as inputs providing vertical deflection, horizontal drive to the predriver, and a blanking output to the low-level luminance circuits.

Although the vertical system provides horizontal drive, it doesn't contain the master scan oscillator or phase-locked loop. The vertical system accepts a clock signal that operates at 32 times the horizontal frequency (32 H is approximately 503 kHz). Now refer to the block diagram in Fig. 7-2. The IC divides this by 32 (down to H) before buffer amplifying to provide drive to the horizontal predriver stage. The resulting flyback pulse is compared to the incoming horizontal sync pulses in an external phase-locked loop, and the outputs of the phase-locked loop regulate the 32-H oscillator.

The vertical system consists of IC (221-103), two power transistors, and the peripheral circuit elements. Of course, these systems have no vertical or horizontal hold controls or adjustments.

To divide by 32, the frequency is first divided by 16 (down to 2 H) for use in the logic circuitry of the IC. The signal is then further divided by 2 (down to H) before being buffered and going out of the module as horizontal drive. The logic circuitry provides a pulse to trigger a ramp generator. Associated with the ramp generator is a size control and a ramp-forming capacitor.

The voltage ramp from the ramp generator is dc coupled inside the IC to one side of a differential amplifier. The differential amplifier drives a transistor output stage, and the output stage drives the vertical windings of the yoke. The yoke current is sensed and waveshaped in the feedback network, and a sample of that yoke current is coupled to the feedback side of the differential amplifier. Thus, the differ-

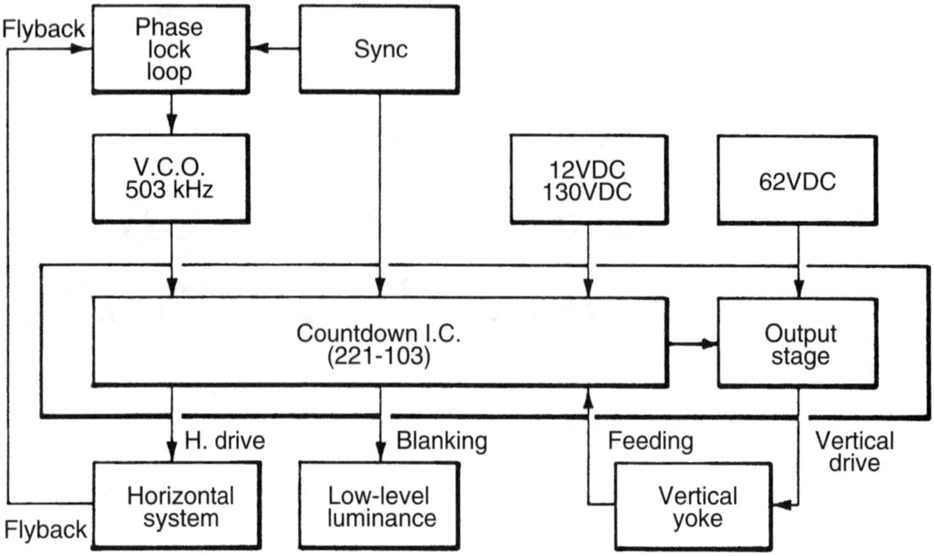

7-1 Block diagram of a vertical countdown system.

ential amplifier and output stages function to make the sample of yoke current look just like the voltage ramp at its input.

The logic section also provides a blanking pulse that's amplified by a buffer stage and coupled to the low-level video chip. The logic circuitry within the IC also has other inputs. It receives integrated vertical pulses to make sure that the vertical retrace interval coincides with the vertical interval of composite sync. Also, the logic circuit receives composite sync because it must decide whether or not the signal being viewed has 525 lines per frame. If the signal has 525 lines per frame, then the ramp generator is triggered by the countdown chain in the IC and all of the advantages that come with a countdown (such as good interlace and good noise immunity) are the result. If the signal doesn't have 525 lines per frame, then the ramp generator is triggered by the integrated vertical pulses that still drive the yoke, but interlace and noise immunity might be affected. Thus, the logic circuit must know if the incoming signal is standard or nonstandard. The logic circuitry "looks" at the composite sync signal during the vertical interval for a period of 6 lines. Should the signal be a standard NTSC signal, the composite sync will have proper serrations and the logic circuit should count 12 pulse edges during the 6-line time period. If the logic circuit doesn't count at least 9 edges during the 6-line time period, it decides that the signal is non-NTSC (or nonstandard) and, thus, might not have a total of 525 lines.

Examples of nonstandard signals are those from some cable systems, some pattern generators, and some closed-circuit TV camera systems. Examples of standard signals are broadcast TV station programs and good pattern generators.

However, not all signals with enough serrations are standard NTSC signals with 525 lines per frame. Sometimes there are signals that exist that can confuse the decision-making process in the logic circuit. Therefore, to force the correct response,

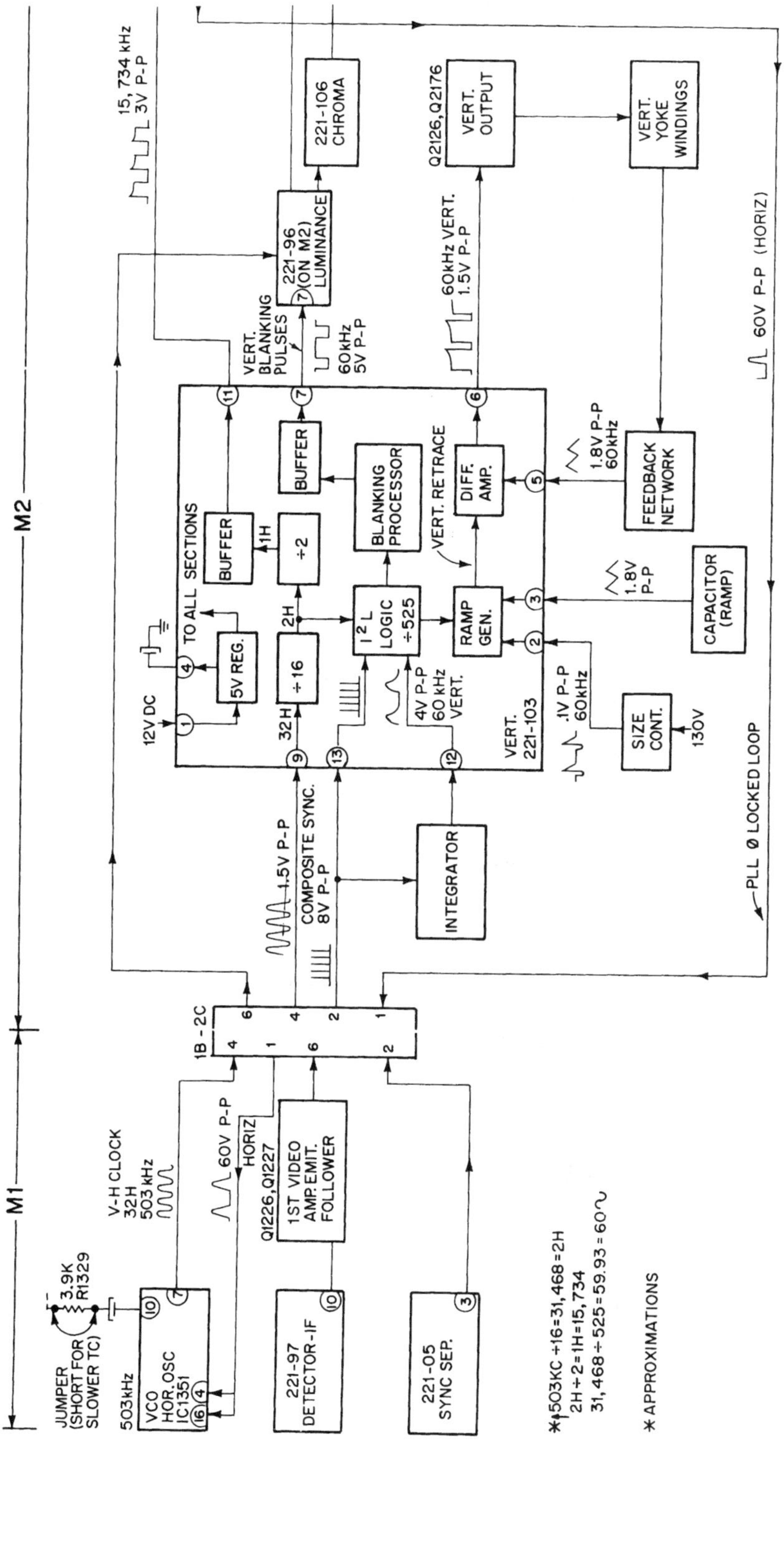

7-2 Complete block diagram of a V/H countdown system.

a mode override jumper is provided on the circuit board at pin 8 of the IC. If this jumper is left plugged in, the logic circuit makes its own decision about whether the signal is standard or nonstandard. If the jumper is disconnected and left open-circuited, the logic is forced into a nonstandard mode.

Normally, the jumper should be left plugged in. However, in some areas, a signal might exist that this system will not lock onto. In this rare instance, the jumper should be disconnected. The circuit should then lock onto the signal, but the advantages of countdown will be lost. This includes good picture interlace and noise immunity. Switch the interlace adder on the VA62 off and on to see if the set goes in and out of the standard and nonstandard mode properly.

A 5-volt regulator is also included inside the countdown chip. Most of the circuitry inside the chip uses 5 volts, which is generated from the 12-volt supply. The 5 volts is filtered at pin 4 and is used for bias networks at pins 12 and 9.

The yoke current passes through a resistor, and the resultant voltage across the resistor is directly proportional to the yoke current. This voltage is shaped by the feedback network to correct vertical linearity. The feedback waveform is applied to pin 5 of the countdown chip. Use the vertical drive signal from the VA62 Analyzer to inject test signals throughout the vertical output stage.

Vertical sweep analysis

In this vertical sweep troubleshooting section, I'll explain how to use the VA62 to inject test signals into the vertical sweep stages in order to isolate faulty components. In a conventional vertical sweep circuit, you can start at the output of the sync circuits that feed the vertical oscillator.

The composite sync is fed through a vertical integrator filter to the vertical oscillator. There had never been a good way to check out the integrator filter until the VA48 and VA62 Analyzer, even with a good scope. Many technicians have assumed that good horizontal sync and poor vertical sync indicate a defective integrator filter. You might wish that this is the case, but you'd like to know for sure if it's the integrator filter before disconnecting all those lead wires.

Now, use the Analyzer to feed a signal into the integrator filter, as has explained in the past tests, or jumper the filter and connect it directly to the input of the vertical oscillator. The Analyzer forms its own integration ramp to trigger the oscillator directly. If vertical sync is possible after the integrator, you know that you have a defective integrator filter.

The vertical oscillator itself is checked by driving the input of the vertical output stage. Note that two vertical pulses are available from the VA62. The labeled VERT SYNC (integrated) is used for solid-state stages and the one labeled VERT DRIVE is for tube-type vertical stages. The vertical drive signal (note waveforms in Fig. 7-3) provides a vertical signal with a linear drive ramp for tube-type sets. Note that the polarity of the signal can be changed with the drive level control to let you inject into the input or output of any vertical stage.

Every design attempt with the Analyzer was made to generate a waveform that will deliver nearly full height with good linearity. Linearity is close enough to detect

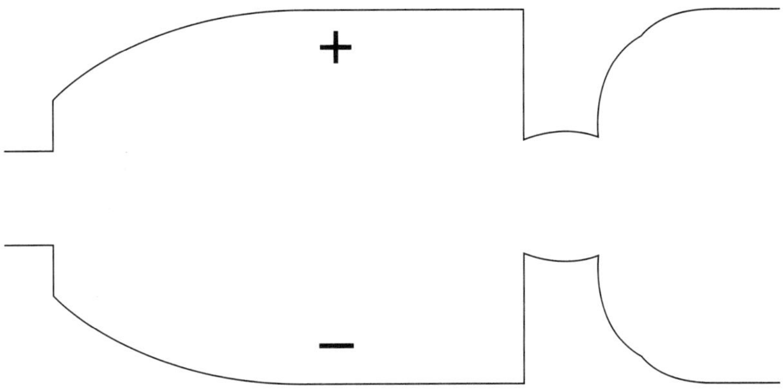

7-3 Vertical drive signal from the Analyzer (±).

faults in the waveform-shaping circuits at that point, thus serving another troubleshooting function.

After you've performed these signal injection checks, if you've not located the problem, you should suspect the vertical deflection yoke. The vertical yoke winding can be checked on the Analyzer ringing test mode. You have now isolated the problem to the vertical output stage and transformer. A quick check of the amplifier itself (tube or transistor type) will pinpoint the trouble to a defective output transformer.

Vertical drive test signal

The vertical drive signal is a modified ramp signal that's synchronized with the vertical sync pulses of the Analyzer's RF-IF signals. Since the signals are synchronized, the signal can be directly substituted into any vertical stage with proper picture lock-in.

Note: The exact shape of the vertical drive signals varies from one brand of TV set to another. This means that the linearity of the vertical drive will also change from set to set, or possibly from stage to stage in the same set. The substitute signal will give a direct indication of the operation of the stages in question, although the sweep might not be totally linear.

Testing the vertical sweep stages

A defect in the vertical oscillator, vertical sweep amplifier, or vertical output stages can be located by substituting the vertical drive signal from the Analyzer in place of the signals normally found in these stages. The RF output of the Analyzer should be applied to the antenna input of the receiver under test to provide video information and horizontal synchronization. The drive levels available from the VA62 Analyzer are usually high enough to feed directly into the yoke. This allows a dynamic test of the yoke itself. If the yoke signal is substituted, expect only half-screen vertical deflection, since most vertical yokes have two sets of windings. Of course, the yoke can be tested with the ringing test described in another section of this book.

Note the following when checking out the various vertical sweep stages:

1. The waveshape of the vertical drive signal might vary from one set to another, or even from stage to stage within a given TV set. The result might be a nonlinear sweep when the vertical drive signal is substituted.
2. Solid-state vertical circuits are usually direct-coupled. These dc stages can be checked by using the variable B+ sub-voltage from the Analyzer. Direct-coupled circuit checks are covered in another section of this book.

VA62A control settings for vertical sweep testing

1. Connect the RF output of the VA62A to the VHF input of the TV set. Refer to the Analyzer control setup shown in Fig. 7-4.

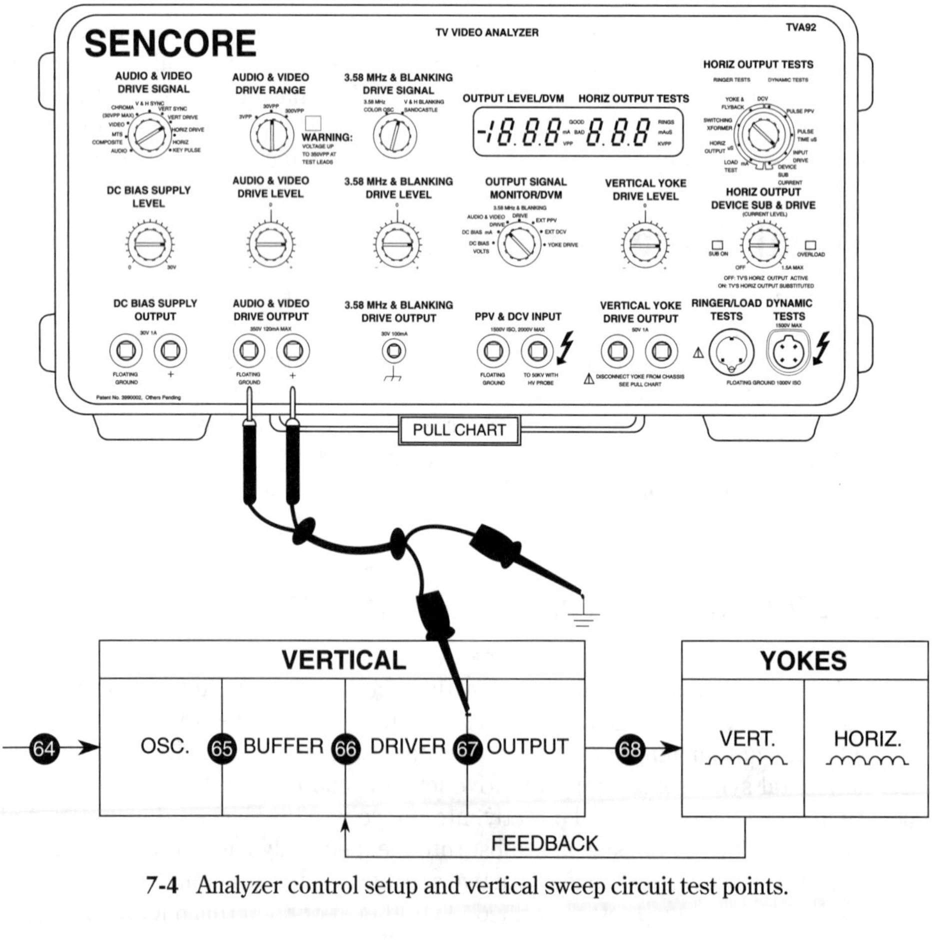

7-4 Analyzer control setup and vertical sweep circuit test points.

2. Select the same channel on the VA62 and the TV receiver under test.
3. Select any test pattern with the video pattern switch.
4. From service data, determine the proper signal level and polarity for the substitute signal. Note: To prevent possible damage to solid-state circuits, observe proper polarity and the specified level of the drive signals before injecting.
5. Select the drive output position of the digital meter switch.
6. Select the vertical position of the drive signal's switch. Now select the proper peak-to-peak signal output range.
7. Inject the signal at the output of the vertical oscillator. If the operation returns, the defect is in the vertical oscillator.
8. If operation is not restored, inject the drive signal at the output of the vertical amplifier. If normal operation returns, the defect is in the vertical amplifier.
9. If operation didn't resume, inject the drive signal at the output of the vertical output stage. Should normal operation be restored, the defect is in the vertical output stage. If there's no change, the defect is in the deflection yoke. The yoke can then be tested with the ringer on the Analyzer.

Vertical module checks (Magnavox T-809)

The T809 chassis uses discrete transistors and a push-pull vertical output system. There are three power supplies for the vertical circuits, the +24 volt regulated and +14- and –22-volt scan-derived supplies on the vertical module.

Vertical problems have always been some of the more difficult faults to isolate in TV receivers. Unstable sync and no deflection or intermittent roll has given many technicians a rough time. Once again, however, signal injection can come to your rescue. All you need to do is inject the signals into the different stages and test points to isolate the problem area. To locate an unstable sync problem, use the vertical and horizontal composite sync signal from the VA62 Analyzer and perform the following test procedures:

1. Inject the multiburst bar sweep pattern into the IF stages. On some sets, this can be via the UHF input on the VHF tuner.
2. Inject a 1.0-volt peak-to-peak composite sync signal into the junction of R2 and C1, which is part of the vertical integrator on the vertical module. If the set now locks in vertically, proceed to step 3. If the set will not lock in vertically, proceed on to step 4.
3. Inject the composite sync signal into TP8 on the motherboard. If the picture locks in, check the sync separator on the IF module or IF alignment for compressed sync. If the picture will not lock in, check the connector on the vertical module and motherboard.

4. If the picture doesn't lock in with the signal injected at R2 and C1, inject the signal into the junction of R3 and C2. If the picture now locks in, check R2 and C1. (See Fig. 7-5.)

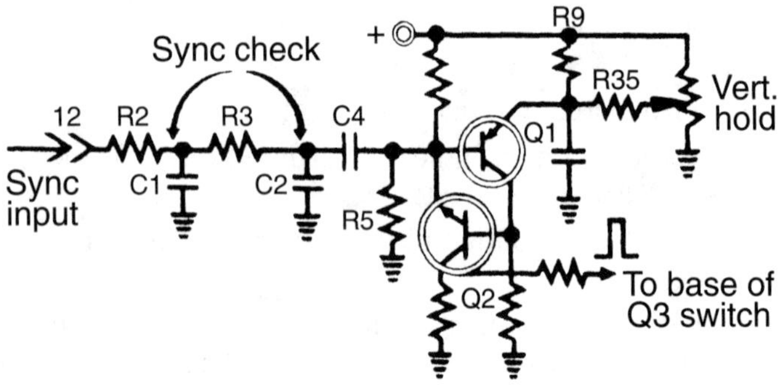

7-5 Integrator, oscillator, and separator vertical sync check points.

5. Should the picture not lock in with the signal injected into the junction of R3 and C2, check the vertical oscillator transistor and associated components.

A no-vertical deflection problem is usually tougher to locate because the circuit is direct-coupled, and the defect can be in any stage. Let's now look at the following steps in order to make vertical sweep deflection troubleshooting easier. Refer to the circuit in Fig. 7-6 for these checks:

1. Inject the multiburst bar sweep pattern into the set's VHF terminals to sync-in the set.
2. Inject a 1.0-volt peak-to-peak positive vertical signal pulse into the base of Q5. If about two-thirds of a raster deflection is obtained, the predriver, driver, and output stages are probably good. If there's no deflection, go on to step 4.
3. Inject the signal into the base of Q4. If deflection is restored, check the vertical oscillator circuit. If there's no deflection, Q4 is at fault.
4. If there's no deflection with the signal injected into the base of Q5, inject a 30-volt positive signal at the collector of Q7. Deflection here indicates that the output stage is operating. No deflection points to the output transistors or scan-derived voltages of +14 and –22 volts.
5. If there's deflection with the signal injected at the collector of Q7, inject a positive 1.0-volt signal at the base of Q7. Deflection at this point indicates that Q7 is good and the fault is with Q5 or Q6. No deflection at this point indicates that Q7 is defective.

Signal injection from the VA62 can help you to quickly isolate troubles in the vertical sweep system.

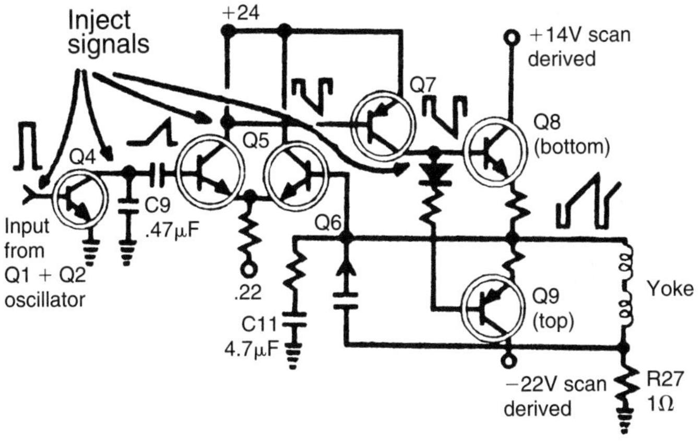

7-6 Signal injection points for checking the vertical deflection stages.

GE vertical circuit operation

Let's now go through the operation of a GE TV set's vertical sweep operation for a solid-state system. An important signal test point in this set is at pins 8 and 12 of IC501. At pin 8 you should find the horizontal sync pulse output and at pin 12 the vertical ramp signal. Both of these signals are then fed to the horizontal and vertical sweep circuits. Refer to Fig. 7-7 for this circuit and waveforms.

Vertical sweep circuits

From pin 12, the vertical ramp signal is fed into the base of Q601. The inverted ramp output is fed to the base of Q603, where it's amplified and corrected for proper linearity. Linearity correction is accomplished by a signal fed back from the vertical output circuitry to the emitter of Q603 at the junction of R617 and 601. The transistor will amplify only the signal difference between the base and emitter, so any difference in waveshape at this point will cause correction to occur, as shown in the Fig. 7-8 waveform drawing.

Vertical "S" correction is also accomplished in the circuit of Q603 by use of feedback to the base of Q603 via R602 and C600. Resistor R635 and capacitor C623 are the actual "S" correction components. Defects in R635 or C623 can cause severe nonlinearity with foldover, or vertical bounce. With large, flat-faced CRTs, a perfectly linear, vertical sawtooth sweep voltage will cause the raster to stretch at the top and bottom. With the edges of the CRT being further away from the point of deflection than the middle, "S" correction is necessary. A capacitor or resistor-capacitor in series or parallel with the yoke is normally used to correct the waveform for linear picture reproduction. R635 and C623 do this in this General Electric EM chassis. Refer to the vertical sweep output circuit that's shown in Fig. 7-9.

Transistor Q603 supplies the necessary linearity and "S" corrected signal to Q605, the output driver. Q605 drives the output pair, Q2607 and Q2609, which in turn supply the vertical yoke coils with sawtooth deflection current through C619. In

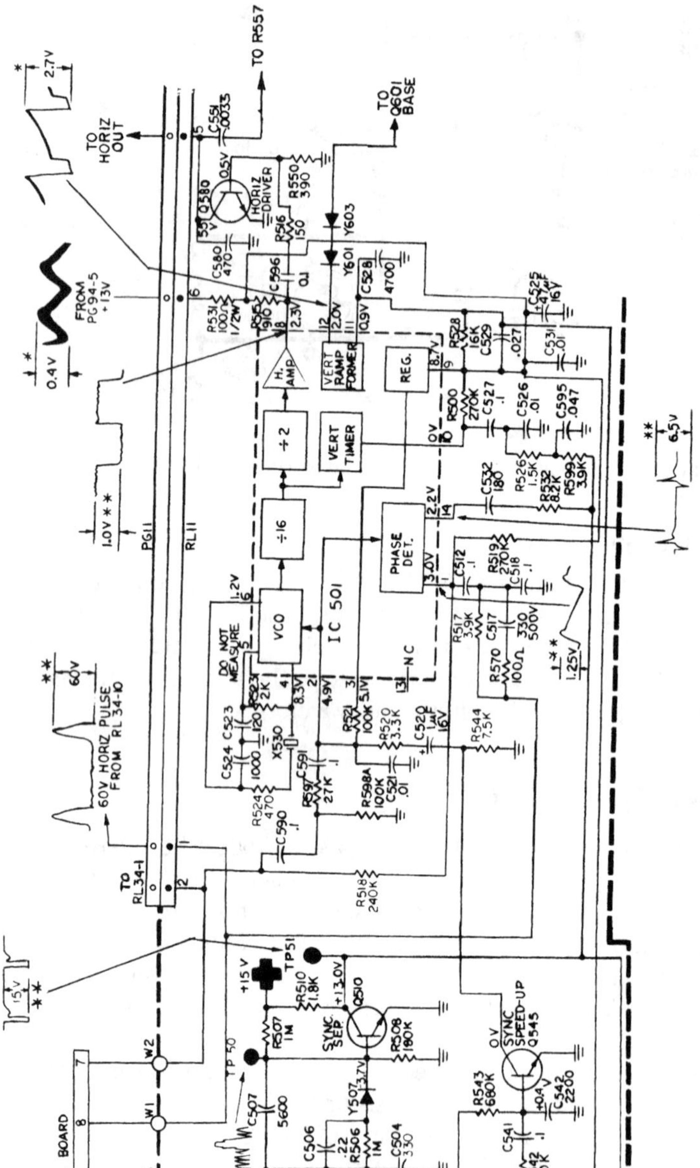

7-7 GE countdown sweep system diagram. GE

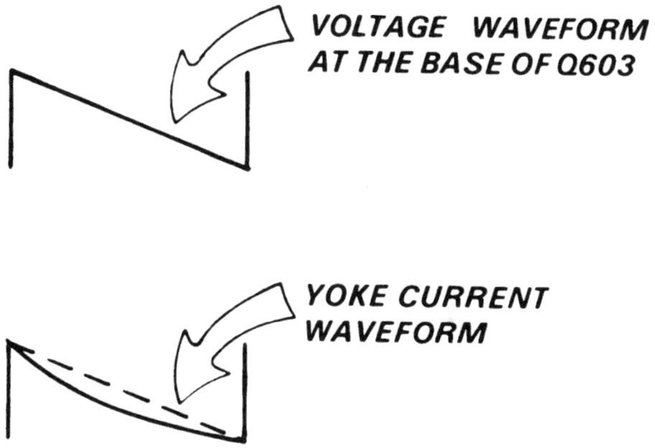

7-8 Vertical waveshape correction via feedback signal.

some circuits using an output of this type, half-screen pictures can be obtained if one of the output transistors is defective. This is not always the case in this EM chassis. However, there's usually some difference in output. If Q2607 is defective, a band of at least two inches of vertical deflection should appear on the screen. However, if Q2609 is faulty, a thin, bright horizontal line might be all that's seen. The vertical output is a Quasi-complementary circuit with each transistor operating one-half of the vertical sweep time; both are off during retrace time.

Though not as straightforward in operation as the complementary output circuit, the Quasi-complementary circuit is used because of the increased reliability gained from using NPN transistors over PNP devices. The vertical yoke pulse waveform is shown in Fig. 7-10. Note that it varies around an approximate 28-volt dc level. During vertical, when both output transistors are off, yoke coupling capacitor C619 charges to approximately 100 volts through R645, the pincushion transformer, deflection yoke, C619, Y613, the base-collector junction of Q2607, R639, and the 125-volt source. Source of current at this time is the collapsing magnetic field about the yoke, which causes the anode of Y613 and the b-c junction of Q2607 to be more positive than the 125-volt source, thus causing the seemingly incorrect direction of current flow.

Now, as C619 is charged and retrace is completed, Q605 begins to provide forward bias to Q2609. Transistor Q2609 is biased on slightly, causing the emitter voltage on Q2607 to go negative enough for forward bias to occur. Q2607 saturates, and the yoke capacitor begins to discharge through Q2607, Y607, the yoke, and pincushion transformer. Observe in the waveform drawing (Fig. 7-10) that the yoke pulse voltage drops to approximately 28 volts at screen center.

At screen center, transistor Q2609 is receiving enough bias from Q605 to saturate, and the yoke capacitor continues to discharge. Capacitor C619 now discharges through Q2609, Y611, and Y613 until the voltage, as shown in Fig. 7-10, is approximately 4 volts. When both transistors are turned off by the drop in bias signal from Q2605, it in turn removes forward bias from Q2607 and flyback (retrace) occurs.

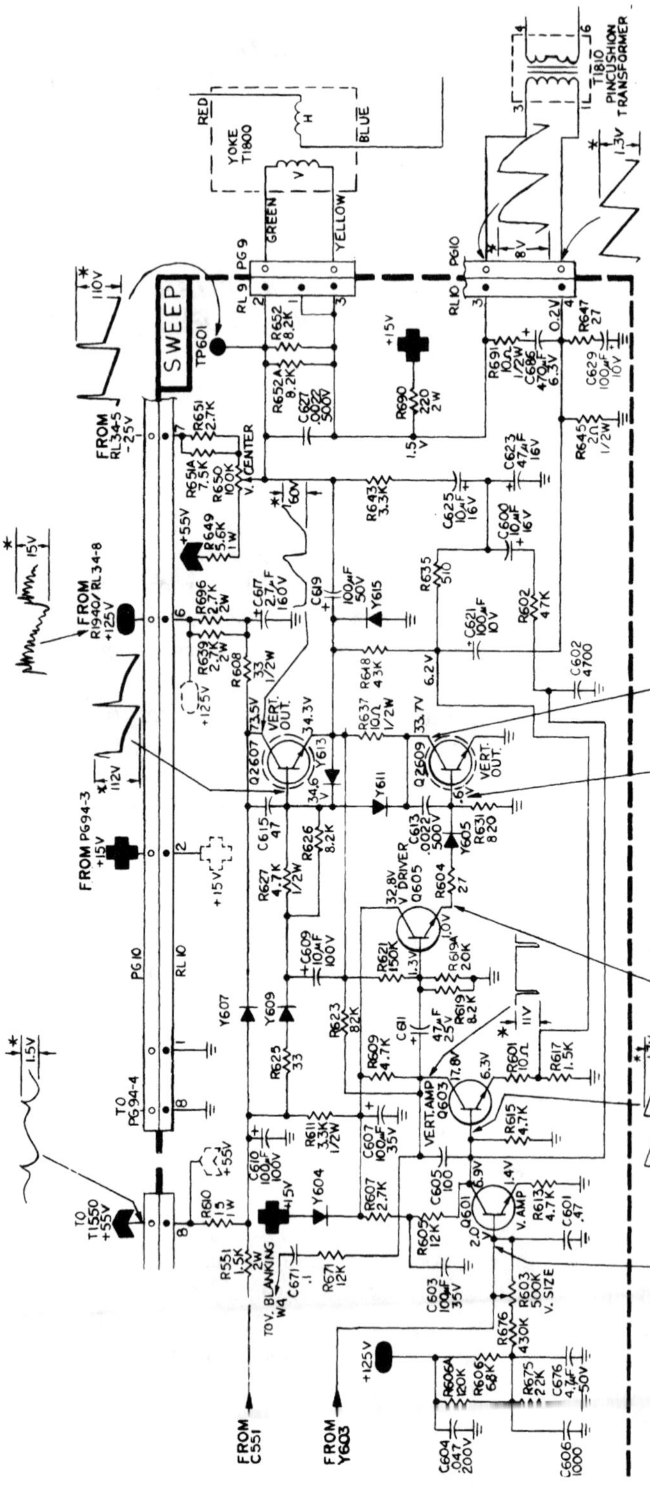

7-9 GE vertical sweep output circuit.

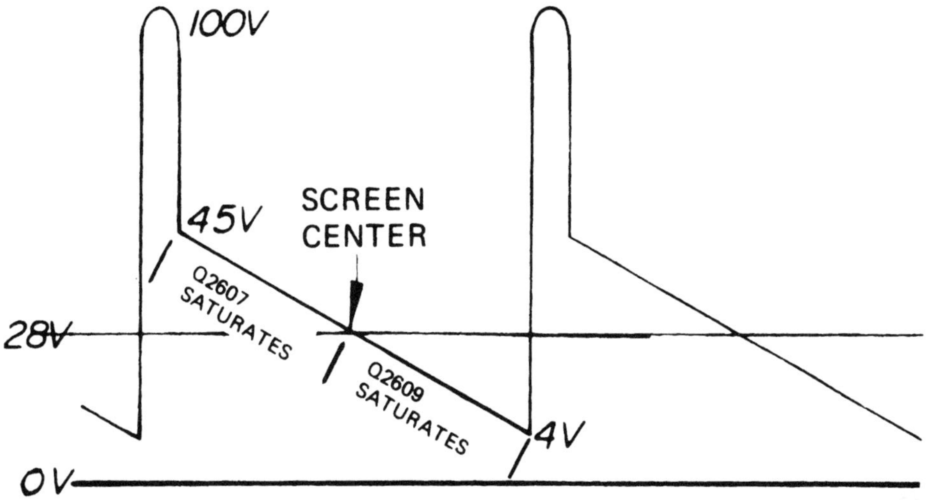

7-10 Vertical yoke pulse waveform.

Sandcastle circuit troubleshooting (RCA)

In this section, you'll see how sandcastle circuits operate, and you'll look at ways to isolate problems in these circuits.

Starting with the RCA CTC108 chassis, you'll find it contains an integrated circuit that processes the luminance and chroma signals. The IC requires a unique signal called the *sandcastle*, which is a mixture of horizontal-keying, vertical-blanking, and horizontal-blanking pulses. The three signals mix outside the IC, feed into a single IC pin, and then reseparate inside the IC to perform a number of functions.

The set will show one of two symptoms if some or all of the sandcastle signal is missing. The most common symptom is a total loss of raster. The second symptom is vertical retrace lines at the top of the screen. You can use the VA62 to confirm that the IC is not the problem by substituting the composite sandcastle signal or any of its individual parts. This dynamic test saves the unnecessary cost of replacing the expensive IC when the problem might well be outside the chip.

Checking out "no picture" problems

Some technicians report servicing TV receivers that have no picture because of sandcastle circuit problems. Of course, other faults (such as IF or detector problems) can cause the same symptoms as a defective luminance/chroma IC.

Make sure the luminance and chroma signals are present at the IC before suspecting the sandcastle circuits. Feed the VA62A color bar pattern to the set's antenna terminals. There will be a picture if only one of the two signals is missing; of course it will not have either color or luminance detail. In the case of a blank screen, confirm the signals are at the IC with an oscilloscope. Don't begin the sandcastle checks listed in the following if both signals are missing, as you must have the two video inputs before the sandcastle affects the IC. Use conventional signal injection or signal tracing procedures to confirm if the problem is in the tuner, IF stages, detector, comb filter, etc. If, on the other hand, the signals are at the correct IC pins, proceed with the following steps.

The VA62A "SCR gate" drive signals provide a suitable substitute for the composite sandcastle signal. If the problem causing the no-raster symptom is outside the IC, the raster should return when injecting this signal at the sandcastle IC pin. Use the positive drive polarity. Check the service notes to determine the correct peak-to-peak value. The CTC108 chassis, for example, uses a 10-volt peak-to-peak signal, while the CTC111 chassis uses a 7-volt peak-to-peak signal. As for all signal injection tests, feed the VA62 RF signal into the set's antenna input (and set the VA62 channel program and TV tuner to the same channel) to hold all circuits in proper synchronization while injecting the test signal. Connect the VA62's ground lead to the set's chassis ground.

Don't be concerned about retrace lines that appear on the CRT when subbing in the sandcastle signal. The Analyzer "SCR gate" signal doesn't include the vertical blanking signal normally supplied by the sandcastle generator. The first step is isolating the no-picture symptom.

If the raster doesn't return with the "SCR gate" signal injected, the problem could be the IC, the output stages, or the CRT. Make one more injection step before condemning the IC chip.

Move to any of the three IC video outputs and inject the "video pattern" signal from the VA62. Again, use the same peak-to-peak voltage as shown on the schematic. If a picture returns, you know the problem is related to the IC. (Note: the picture will only have one color because you're feeding only one output amplifier.) Check the dc voltage that feeds the IC. You've confirmed that the IC is defective if the voltage is normal. The problem is after the IC if the picture didn't return when feeding the output stage. Check the output circuits, the CRT, or the CRT bias voltages.

Sandcastle generator signals

Any of several problems in these sets can cause a no-picture symptom. It could be that the sandcastle signal might not be reaching the IC because of a poor solder connection or defective coupling component, or either of the horizontal signals needed to form the sandcastle might be missing because of problems in one of the two coupling circuits. The VA62 provides signals to isolate the specific problem within the sandcastle generator. Refer to the typical schematic in Fig. 7-11 as you briefly look at the way the generator creates the sandcastle waveform.

The top circuit path forms horizontal keying pulses needed to separate the color burst from the composite video signal. Most TV receivers use a pulse supplied by the sweep transformer to separate burst. RCA uses a different method.

The horizontal sync from the sync separators feeds a delay circuit. The delayed sync pulses line up with the color burst on the back porch of the horizontal blanking interval. The delayed signal then mixes with the two other signals.

The IC requires a pulse from the horizontal output transformer, but this pulse is used for horizontal blanking rather than keying functions. The transformer pulse comes through the circuit in the lower, right-hand portion of the schematic. The

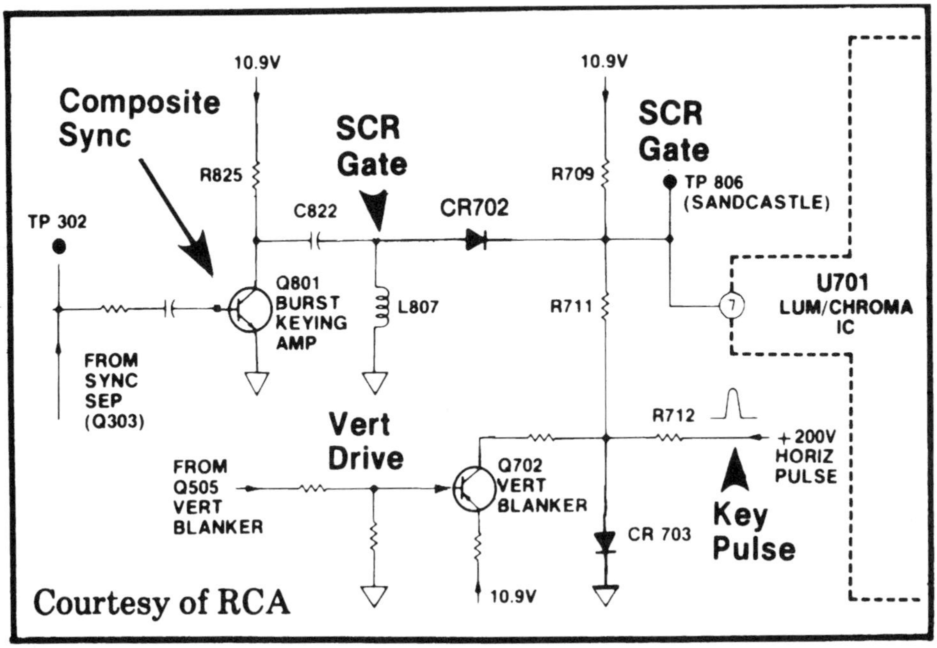

7-11 RCA sandcastle generator circuit. RCA

screen remains blank if either the keying pulse or the sweep transformer blanking pulse is missing or low in amplitude.

The third path provides vertical blanking. A signal from the vertical oscillator IC feeds through the blanker transistor and mixes with the first two signals. Problems in this circuit cause retrace lines, rather than a blank raster, and these will be discussed later.

Isolating picture problems

The VA62A will substitute for each signal used to form the composite sandcastle. Remember, you only use the following procedures if the picture returned when injecting directly at the IC pins. If the picture didn't return, you should analyze the IC or the output stages.

You must first determine if the problem is in the keying pulse circuit or the horizontal blanking circuit. Use the following procedure to isolate the correct circuit. (Refer to the functional analyzing troubleshooting guide blocks shown in Fig. 7-12.)

1. Inject the VA62A "SCR gate" signal, positive polarity, in place of the keying pulse (after the delay circuit following the sync separator). If the picture returns, this is the defective path.
2. If the picture didn't return, inject the VA62 "horiz key pulse," positive polarity, at the point coming from the horizontal flyback transformer. If the picture returns, this is the defective path.

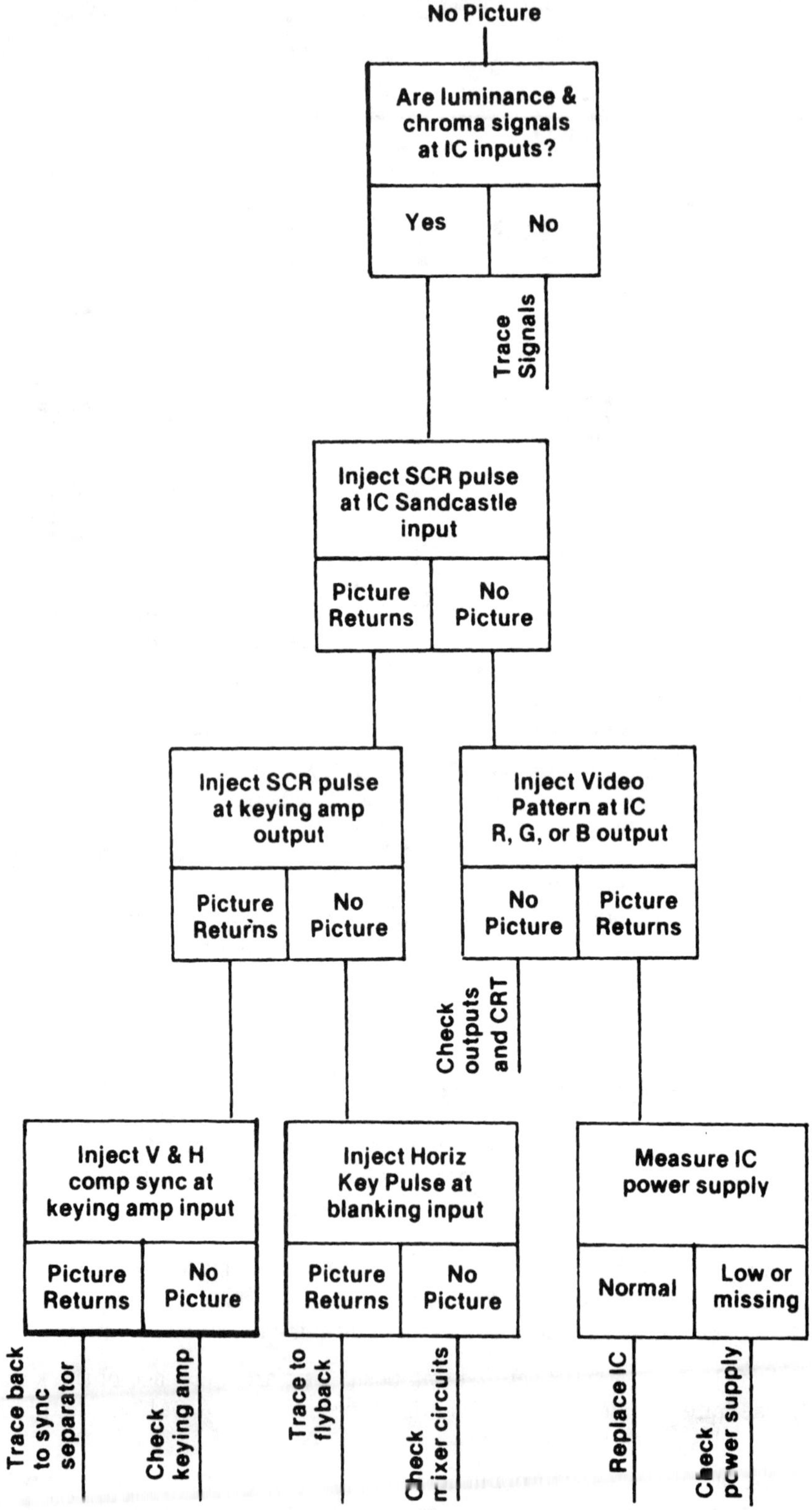

7-12 Functional analyzing troubleshooting guide for isolating no picture symptoms in an RCA CTC108 chassis.

Keying pulse path

If the picture returned in step 1 above, you must then narrow the problem to one stage. Move back towards the sync separator one stage at a time. Inject composite sync, positive polarity, at the "burst keying amp" input. If the picture is blank, the keying amp is defective. If the picture returns again, continue moving back towards the sync separator until you isolate the defective stage.

Flyback pulse path

The flyback feeds directly to the mixing resistor. If the picture returned in step 2 above, move back towards the flyback. If the picture returns as you inject at each test point, you know the circuit's good from there to the IC chip. Should you get all the way back to the flyback and the picture returns when the signal from the VA62 is substituted for the circuit signal, you then have a defective sweep transformer winding or a poor connection between the sweep transformer and the sandcastle circuit.

How to locate retrace line problems

The second symptom related to the sandcastle is retrace lines at the top of the picture. These might be caused by a defective chroma/luminance IC or missing vertical blanking pulses in the sandcastle signal. There will still be a picture if these pulses aren't present.

Inject the VA62A Analyzer's "vertical output" drive signal, negative polarity, at the input of the vertical blanker transistor. If the retrace lines disappear, the problem is in the circuit feeding the transistor. If the lines remain, move to the output of the blanker transistor. If the lines disappear, the problem is in the blanker amplifier. If they remain, the problem is in the luminance/chroma IC.

Detailed sandcastle troubleshooting

Many of the newer TV chassis contain an integrated circuit that processes the luminance and chroma signals. As previously discussed, this IC requires a unique input signal called the sandcastle. The sandcastle is a special signal used to inject three signals into a single IC pin. The three signals mix outside the IC and form the waveform shown in Fig. 7-13. The IC separates these signals and uses them for various internal functions.

The three signals that make up the sandcastle are:

1. Flyback pulse.
2. Delayed horizontal sync pulse.
3. Vertical blanking pulse.

The three are mixed to from the sandcastle signal, as shown in the typical circuit of Fig. 7-14. The top circuit path produces horizontal keying pulses needed to sepa-

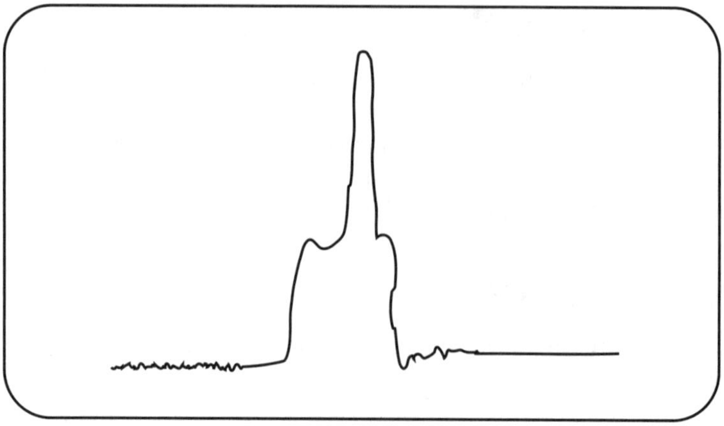

7-13 A drawing of the sandcastle composite waveform.

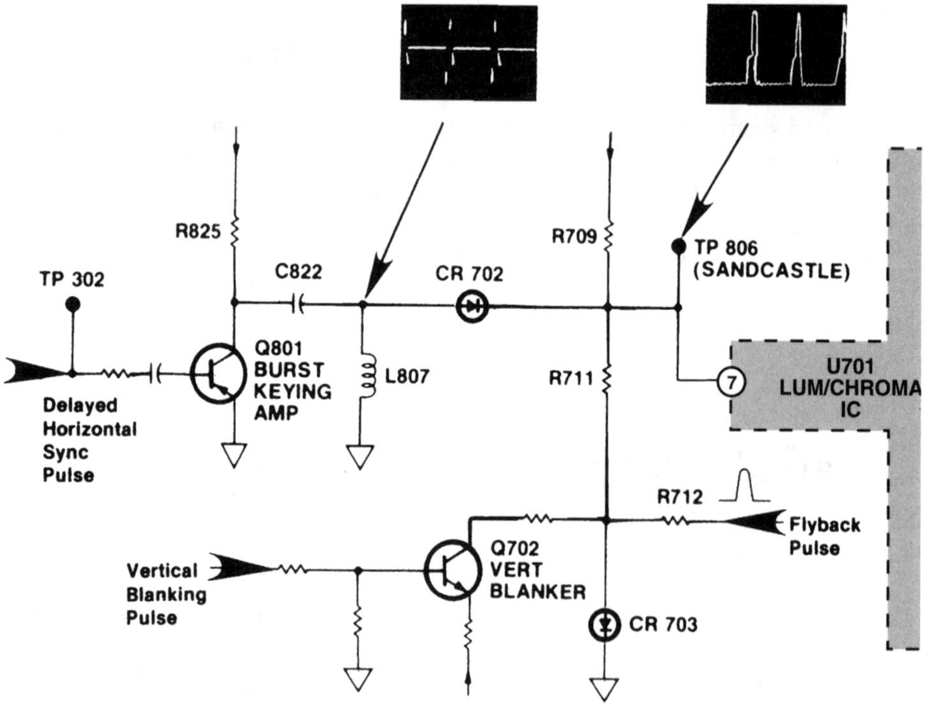

7-14 The sandcastle, named for its shape, is formed by flyback, vertical blanking, and horizontal sync pulses.

rate the color burst from the composite video signal. Some TV receivers use a pulse supplied by the flyback transformer to separate burst. TV receivers with the sandcastle signal use a different method.

The horizontal sync from the sync separators feeds a delay circuit. The delayed sync pulses line up with the color burst on the back porch of the horizontal blanking interval of the composite video signal. The delayed signal then mixes with the two other signals.

The IC also requires a pulse from the flyback transformer, but this pulse is used for horizontal blanking rather than keying functions. The transformer pulse comes through the circuit in the lower, right-hand portion of the circuit in Fig. 7-14.

The third path provides vertical blanking. A signal from the vertical oscillator feeds through the blanker transistor and mixes with the first two signals.

Isolating sandcastle problems

The luminance/chroma IC produces no output signal if the sandcastle signal is missing altogether, which results in a blank CRT. This same symptom occurs if the flyback pulse ("blanking" pulse) or the delayed sync pulse ("keying" pulse) is missing, even though the other two sandcastle signals are normal. This one symptom applies to nearly any problem related to the luminance/chroma chip. Of course, other symptoms such as IF or detector problems cause the same symptoms as a defective luminance/chroma chip.

Confirm that the luminance and chroma signals are present at the IC before suspecting the sandcastle circuit. Feed the VA62A color-bars pattern to the antenna terminals and note the results on the CRT. There will be a picture if only one of the two signals is missing, although it will lack either color or luminance detail. In the case of a blank screen, use an oscilloscope or the VA62A signal tracing meter to confirm that the signals are at the IC. Don't troubleshoot the sandcastle circuits if both the luminance and chroma signals are missing, as you must have the two video inputs before the sandcastle affects the IC. Use conventional signal injection or signal tracing procedures to confirm whether the problem is in the tuner, IF, detector, comb filter, etc. If, on the other hand, the signals are at the correct IC pins, proceed with the following tips for sandcastle troubleshooting.

Isolating "no picture" problems

Whenever the VA62A's drive signals are used for signal substitution troubleshooting, the VA62A's RF signal must be injected at the antenna terminals (and have the VA62A and TV receiver tuned to the same channel). The antenna signal causes all of the good stages to operate normally while the substitute signal is injected into one suspected stage at a time. The sequence shown in Fig. 7-15 gives the best troubleshooting efficiency in sandcastle circuits.

One of the main symptoms associated with sandcastle signal is "no picture." Start troubleshooting by substituting for the sandcastle signal. You should use this point first because the sandcastle comes from three diverse circuits. Starting at the sandcastle signal eliminates more questionable stages than starting at the luminance input.

If there's a problem in the sandcastle generator circuits, injecting the VA62A's signal in place of the sandcastle signal will unblank the luminance/chroma IC, putting

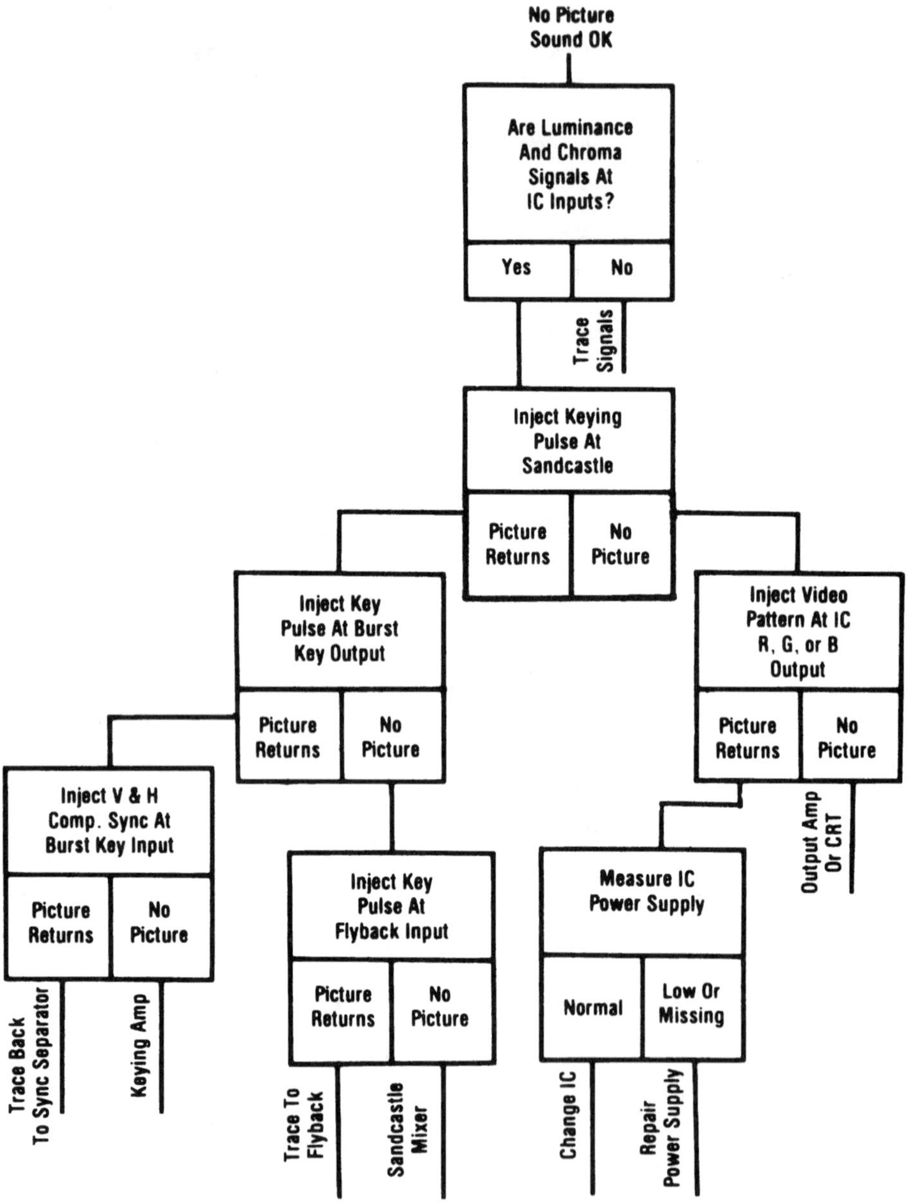

7-15 Trouble tree for troubleshooting sandcastle-related problems.

a raster on the CRT. Simply inject the VA62A's "horiz key pulse" at the sandcastle input of the IC, using positive polarity and the same peak-to-peak amplitude shown on the schematic.

Note: When substituting the sandcastle signal, don't be concerned about retrace lines that appear on the CRT. The VA62A's "horiz keying pulse" signal doesn't include the vertical blanking signal normally supplied by the sandcastle generator. Your first step is isolating the no-picture symptom.

If the picture returns, then the luminance/chroma IC works properly and the problem is in one of the sandcastle generator circuits. Isolate the bad sandcastle circuit by moving back one stage, substituting for one sandcastle input and then the other (if necessary) to determine which path has the problem.

First, move back through the path that supplies the keying (delayed sync) pulse to see if it's the path with the problem. If the picture again returns, you know you've found the correct path because the substitute signal takes the place of the missing signal.

If the picture returned when injecting directly at the IC input, but yielded no change when substituting for the keying pulse, move over to the second sandcastle generator input. Simply inject the VA62A keying pulse down the path that supplies the flyback pulse. Continue to inject signals at earlier and earlier stages until you identify the source of the problem.

8
Troubleshooting horizontal sweep systems

IN THIS CHAPTER I SHOW YOU THE VARIOUS WAYS TO TROUBLESHOOT THE horizontal oscillator, horizontal driver circuits, and the horizontal sweep systems when using the VA62A Analyzer. This includes the conventional horizontal sweep system and the RCA-type SCR horizontal deflection system. I also touch on the Zenith System 3 electronic sweep and HV regulation system, including tips on checking out the new type of master scan-and-count down/divider driver circuits.

Most technicians agree that the horizontal sweep section of TV receivers is the toughest section to troubleshoot because it seems to have more problems than any other section, and faults can be rough to pin down. Thus, the VA62A was designed with you in mind to help pin down troubles in these complex sweep circuits and eliminate the suspected stages or components one at a time.

The horizontal oscillator

The first section you want to check out for correct operation is the horizontal oscillator stage itself. The horizontal oscillator has just one purpose: to drive the horizontal sweep output stage. A good way to check the oscillator is to feed the horizontal oscillator drive signal from the VA62A to the input component of the horizontal sweep stage. You can then continue substituting on back to the oscillator circuits. These tests can weed out suspected coupling capacitors, the driver transformer, and other components.

To check all types of sweep systems, the Analyzer provides three different kinds of horizontal output drive signals because there are three different kinds of horizontal output stages now being used: a tube type, a transistor type, and an SCR system. You'll have to determine which type of horizontal sweep stage you're troubleshooting. The only problem is identifying the SCR sweep system from the transistor type. This will have to be carefully determined because you usually can't tell the difference if you don't know the TV or you don't look at the circuit schematic.

For the first check, the horizontal output substitute signal is fed into the input of the horizontal output stage, and the drive signal amplitude from the Analyzer is increased until the raster returns to normal. The first time you make this check, you'll be very pleased to see the picture pop into sync and you'll appreciate the VA62A's phase-locked signals.

You'll want to monitor the peak-to-peak drive signal with the drive signal digital meter, but you must be sure that the drive signal waveform is of the correct polarity.

No tube-type sweep output stage should be operated very long without a drive signal because the lack of a bias will cause excessive output current, and the horizontal output tube can be damaged. It's best to check the service data for polarity and drive signal amplitude and set the drive signal on the VA62A to about that level. Most conventional solid-state horizontal sweep output stages will just shut down with the loss of a drive signal and will not be damaged. However, to be on the safe side, don't operate any sweep circuit for any length of time without a drive signal.

In most sets, almost no circuit element, even the horizontal output tube itself, is designed to burn up instantaneously, even when the drive signal is not present. Also, all tube-type sweep systems have a positive drive signal, and most transistor and SCR sets have a positive drive signal as well.

Underdriving can do some damage, too, but the chance of circuit damage by overdriving is nil. Thus, from a practical standpoint, when you don't have a schematic (or don't take the time to look it up), start with the drive level turned up to the most positive setting and notice whether or not the raster comes on. If it doesn't, turn the control back through zero, all the way to the most negative setting to see if the picture sweep now comes on.

Caution: Don't leave the Analyzer connected to the horizontal output with the wrong polarity, as you might do some damage to the TV sweep system (but not to the VA62A). Very often, the problem is a short at the point of the test signal feed-in, and at this point you don't want to go any further. This is quickly identified by the Analyzer's digital drive signal monitor meter because the short reduces the drive signal output, making the meter show a drop in its reading just as soon as the test point is connected. This, then, is a very good troubleshooting check, and it works in other circuits as well, where there are shorts or high load conditions caused by faulty circuit components.

Horizontal drive checks

The Sencore Analyzer provides the following horizontal drive signals for troubleshooting the horizontal sweep stages:

1. A transistor drive signal as shown in the Fig. 8-1 waveform photo.

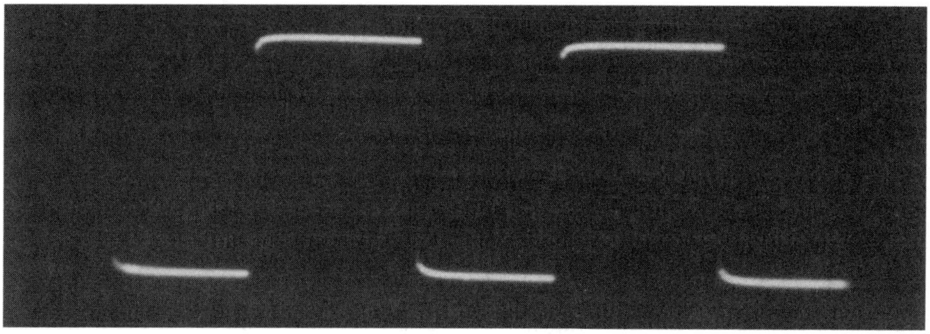

8-1 Correct horizontal drive signal pulse.

2. A drive signal (Fig. 8-2) as found in a tube-type set.
3. An SCR horizontal drive signal as shown in Fig. 8-3.

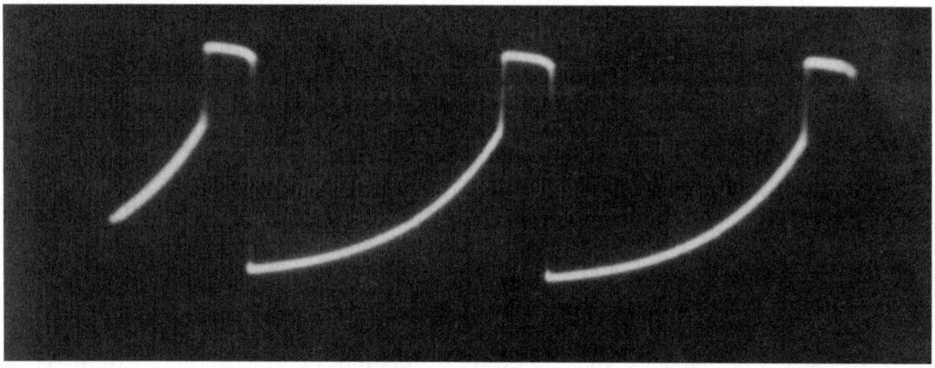

8-2 Horizontal drive signal for a tube-type set.

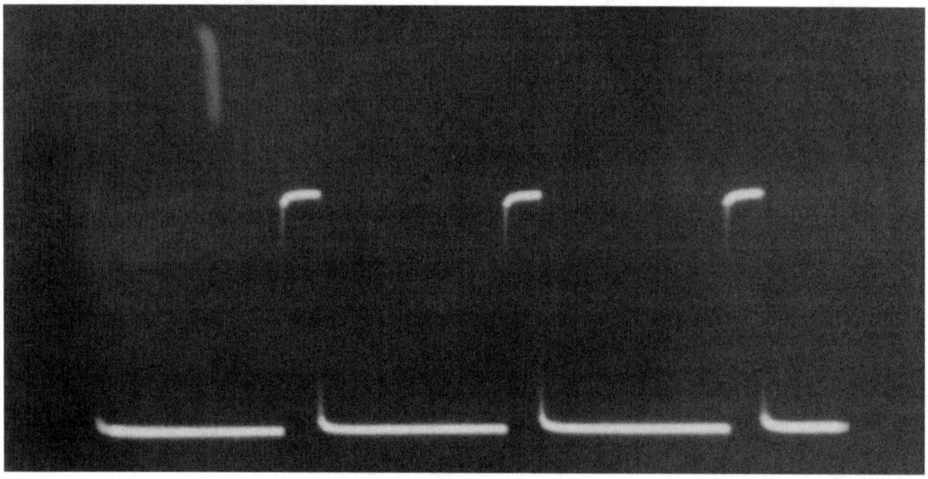

8-3 The SCR horizontal drive signal pulse.

Each of these signals is specially shaped for the type of circuit each is designed to be used with. The horizontal drive signals are designed to be fed directly into the input element (transistor base, tube grid, or SCR gate) of the horizontal output device.

If the horizontal defect is in the horizontal oscillator, a substitute for these signals will reestablish both the horizontal sweep and high voltage normally generated in the horizontal sweep output (flyback) transformer.

Caution: The horizontal output device usually has B+ boost present at its output. To prevent any electrical shock, use extreme caution when injecting a test signal into the output amplifiers.

Refer to Fig. 8-4 for the Analyzer control positions and injection points for the horizontal drive signal substitution. Use the following procedures for horizontal

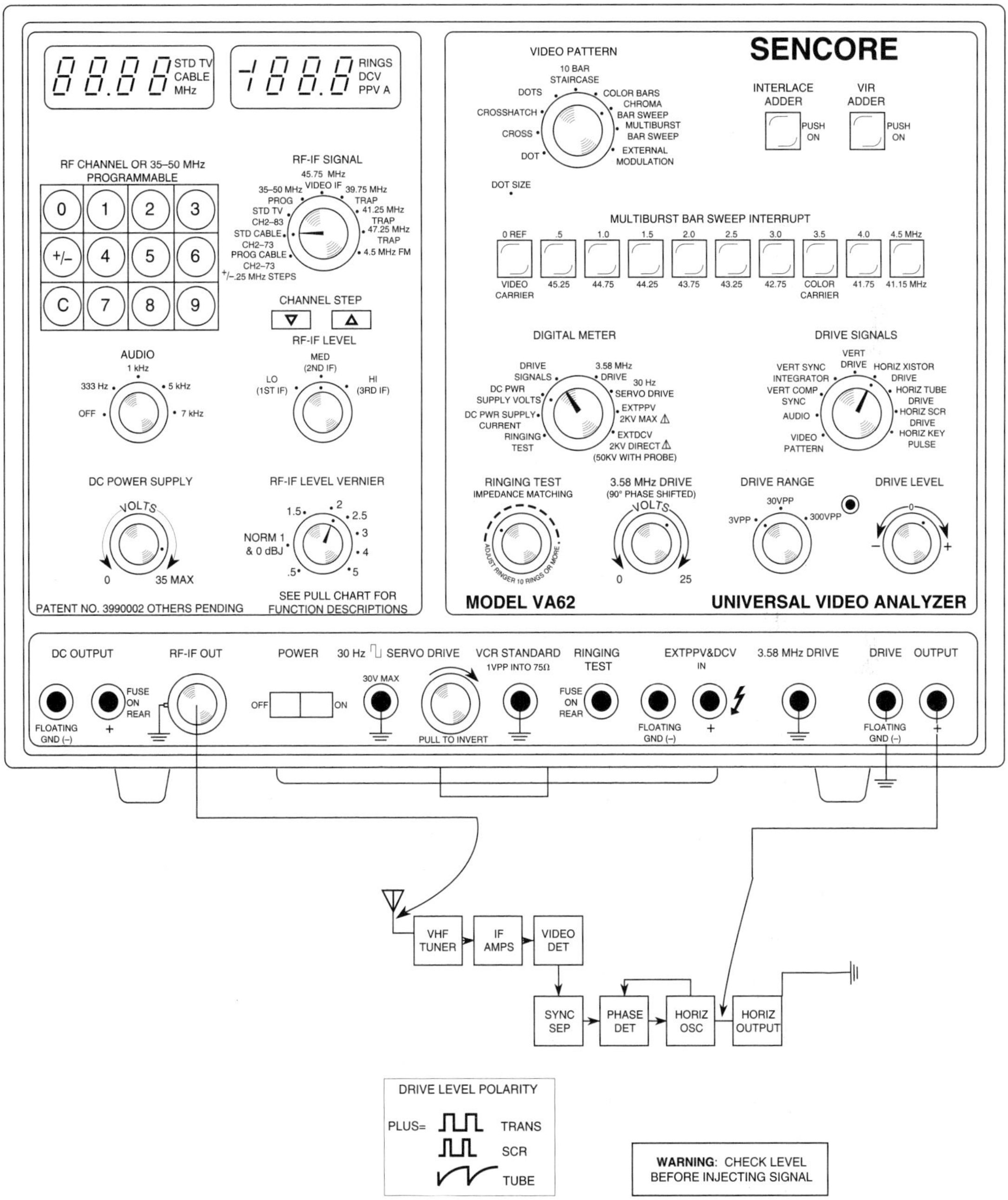

8-4 Analyzer control positions and horizontal drive signal injection points.

drive signal injection:

1. Remove the power to the TV receiver under test. Don't apply power until the test leads have been connected.
2. Connect the RF output of the Analyzer to the set's VHF antenna input terminals.
3. Touch-tune the same channel on the Analyzer and the TV receiver under test.
4. Determine the proper signal level and polarity for the substitute signal. Caution: Observe proper polarity and level of the drive signal before injecting to prevent damage to solid-state devices.
5. Select the drive signal's monitor position on the digital meter switch.
6. Select the proper horizontal output on the drive signal's switch: Choose the transistor base, SCR gate, or tube grid.
7. Connect the signal to the injection point.
8. Turn the receiver on and wait for the filament of the CRT (and other tubes in a tube-type receiver) to heat up.
9. If the set starts operating properly, the defect is in the horizontal sweep oscillator stage.
10. Before removing the test leads, remove power from the TV receiver under test.
11. If horizontal deflection doesn't return, the problem is in the sweep transformer, deflection yoke, or the HV tripler device. The yoke and sweep transformer can be tested with the ringer checker to determine if they're the cause of the sweep circuit failure. If these components check out well, then the fault is in the horizontal output HV tripler or HV rectifier section.

Horizontal output transistor testing

The horizontal output transistor requires a squarewave driving signal. The signal should go positive at the same instant that the horizontal sync pulse is present in the video circuit. It then should go negative to turn the transistor off and prevent it from running too hot. Thus, the correct duty cycle is very important for proper sweep transistor operation. The waveform depicted in Fig. 8-5 shows the correct turn-on and turn-off pulse drive signal that's produced by the VA62A Analyzer.

This drive signal is fed in between the base and the emitter leads of the horizontal output transistor. The emitter is usually grounded, so the ground lead from the VA62 can be connected directly to the chassis of the TV receiver under test. Some of the model TV sets with the electronic power systems use a split-sweep transformer system, as shown in Fig. 8-6. The emitter is connected to a portion of the sweep transformer primary before being connected to ground, which places the emitter about 400 volts dc above ground. Because the VA62A has a floating ground, you can directly feed the drive signal into the emitter, and no damage will occur to the set or to the Analyzer. You won't need a driver signal isolation transformer with

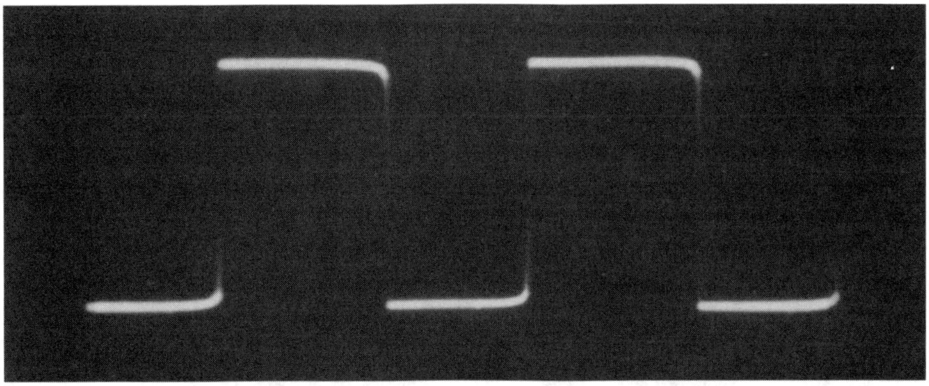

8-5 Correct horizontal pulse produced by the VA62A.

8-6 The split-flyback system places the output transistor in the electrical center of the sweep transformer with the base at nearly 400 volts peak-to-peak above ground. Thus, the VA62 with floating ground can feed in the drive pulses directly without using an isolation transformer.

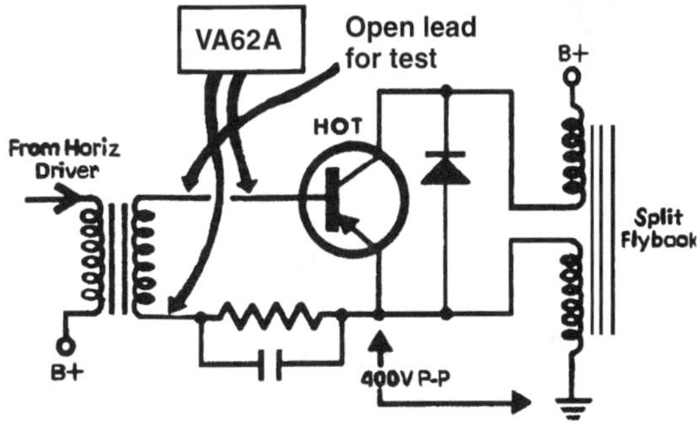

the VA62A as you do with the VA48. However, you should use an ac line isolation transformer to prevent shock hazards and possible damage to the TV chassis being serviced.

Horizontal output SCR gate

An SCR requires a narrower pulse than that required for a conventional transistor sweep output stage. Of course, the Analyzer does produce this special SCR pulse. The second trace of the SCR must be fired by the timing circuits of the output stage. I'll take a brief look at these special SCR horizontal sweep circuits a little later in this chapter. The correct SCR gating pulse is shown in Fig. 8-7.

This SCR signal can also be used to substitute for the horizontal keying pulses fed from the sweep transformer to low-impedance solid-state circuits such as keyed AGC, sync separators, or the color burst amplifiers.

Horizontal output tube grid

The Analyzer tube grid signal is identical to the one fed from the horizontal oscillator found in a tube set. The waveform found at the grid of the horizontal output tube

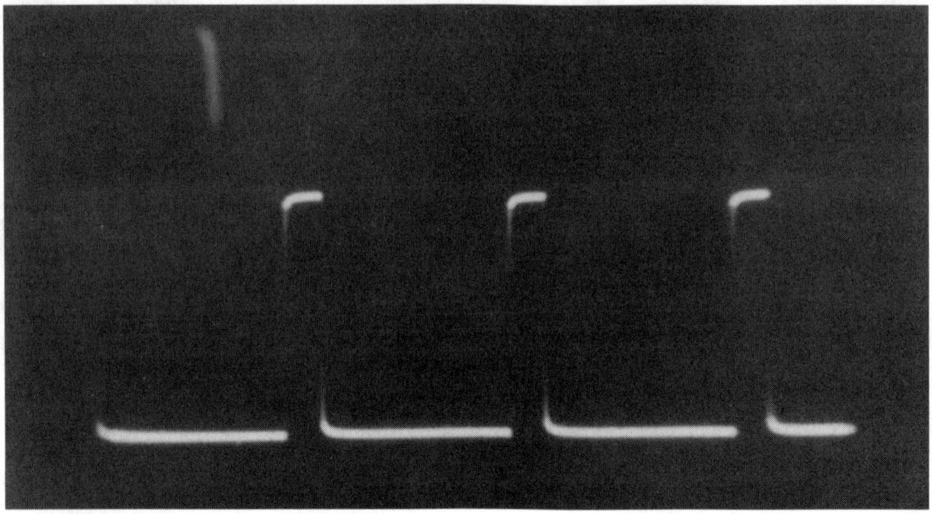

8-7 SCR drive gate signal pulse.

in Fig. 8-8 compares directly with the signal from the VA62A Analyzer output tube test signal shown in Fig. 8-9. The amplitude of this signal from the VA62 is fully adjustable to 300 volts to drive any output tube in black-and-white or color TV sets.

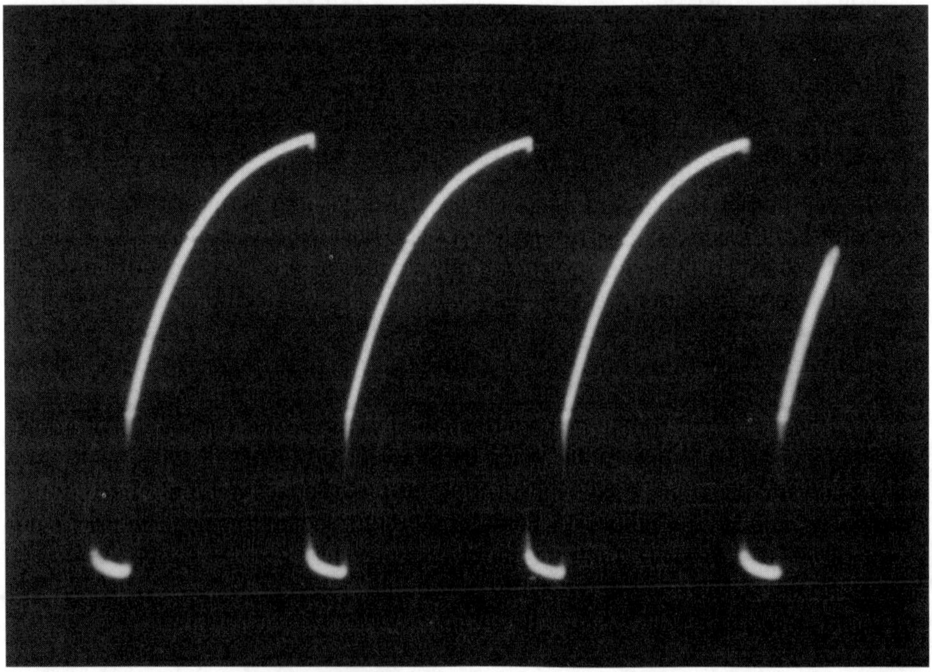

8-8 Waveform found at the grid of the horizontal output tube.

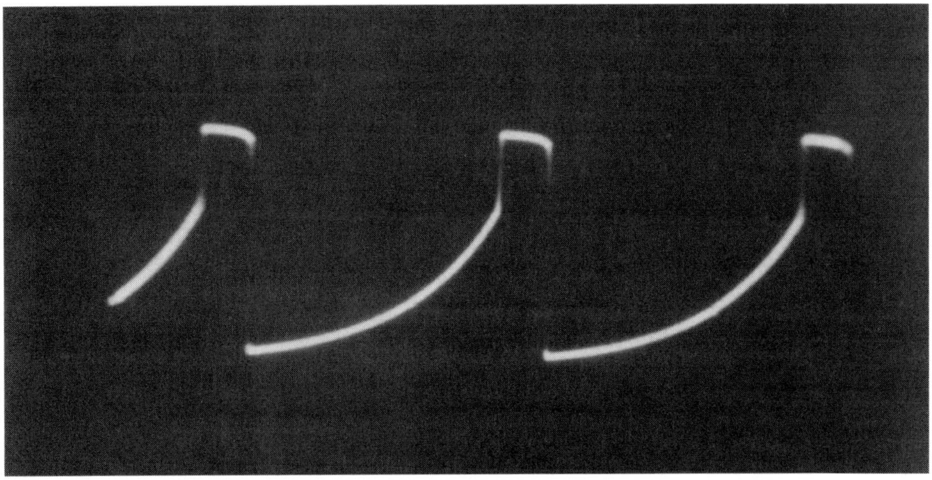

8-9 The tube-type horizontal drive test signal from the VA62A Analyzer.

Split and "hot" chassis flyback testing

The three distinguishing features of the split-flyback system, as shown in Fig. 8-10B, are as follows:

1. The emitter of the horizontal output transistor is connected to a winding on the flyback transformer rather than ground.
2. The series yoke cap returns directly to the output transistor emitter and not to ground, as in a conventional sweep circuit.
3. The output transistor base waveform is generally shown as a large negative retrace spike of 300 to 400 volts peak-to-peak when measured between the base and chassis ground.

The split-flyback base waveform has probably confused many TV technicians and led them to believe that this is the driving signal for this type of circuit when, in fact, it's a spike that's generated during retrace time and has nothing to do with the driving waveform.

Actually, the 300- to 400-volt negative pulse is not the driving signal but a flyback pulse from a portion of the sweep transformer that's connected between the horizontal output transistor emitter and chassis ground. The secondary of the "horizontal driver" transformer is in parallel with the deflection yoke, through the base-to-collector capacity, which is in series with the sweep transformer windings between B+ and ground. If you place a scope probe at a point between the base and chassis ground, you'll measure up to 400 volts peak-to-peak because you're actually measuring across the secondary of the "horizontal driver" transformer in series with the sweep transformer winding to ground, and not the actual driving pulse that exists between the transistor base and emitter. The actual driving signal from the transformer is a square wave that's always less than 30 volts and so small in compar-

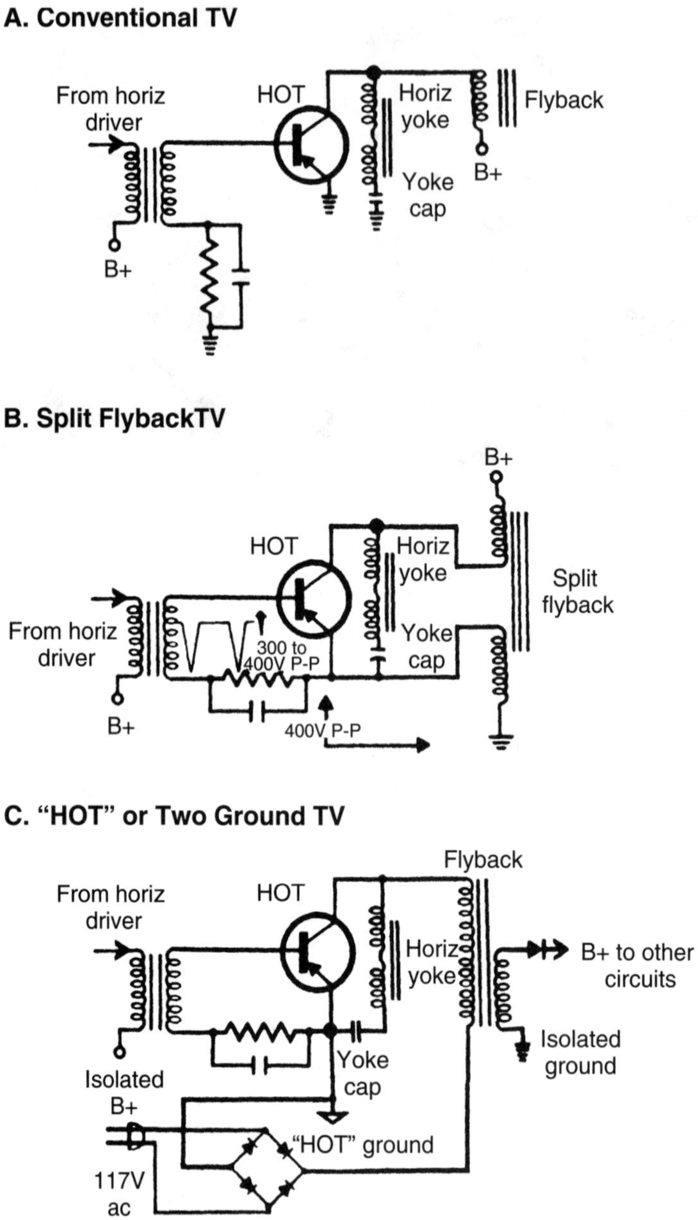

8-10 The three basic sweep system circuits: A. Conventional sweep circuit. B. The split-flyback sweep circuit. C. The "HOT" or two-ground flyback circuit.

ison to the large flyback pulse that it's lost in the overall scope waveform if the scope probe is connected between the output transistor base and ground.

It's necessary to drive the input to the horizontal driver transistor to isolate problems in these horizontal sweep systems. The floating ground of the VA62 enables you to drive the output transistor directly, isolating the problem to the driver transistor or transformer. The floating ground of the VA62A isolates the drive test signal from the horizontal deflection ground that it's driving. If the set doesn't have a power transformer, an isolation ac line transformer must be used for your safety as well as to prevent damage to the equipment. The floating ground of the Analyzer doesn't eliminate the need for an ac line isolation transformer.

Horizontal sweep and B+ regulator circuit checks

The horizontal output transistor and SCR B+ regulator on the RCA CTC-97 chassis is mounted on a separate PC board because the ground for the horizontal output stage is tied back to the ac power line via the bridge rectifier. This makes the ground "hot" or 65 volts RMS above earth ground at all times. The main chassis ground is isolated from the ac power line by the sweep transformer. If the two grounds are connected together, the main chassis also becomes 65 volts above earth ground. This presents a danger to the technician who might touch the chassis and a ground such as the conduit used in the shop's electrical system. This is a good reason why all of your ac outlets on the service bench should be supplied via an isolation ac line transformer.

The horizontal output in this chassis will operate if there's a problem in the oscillator, driver, or start-up circuit. You can use the VA62A Analyzer to isolate these problems with the horizontal drive signal. You need only inject the drive signal at the base of the horizontal output transistor in order to locate the problem.

To drive the horizontal output transistor, just connect the two leads from the Analyzer's drive output jacks. Set the drive level control to a positive 30 volts peak-to-peak and drive switch to the horizontal output transistor base position. The signal level is set to 30 volts to prevent underdriving the horizontal output stage. If too little signal is used, the output transistor can go into a class A operation, dissipate too much heat, and destroy itself in a very short period of time. You can't overdrive the transistor stage. Connect the (+) red lead to the base of the horizontal output transistor. Now connect the (–) black floating ground lead to the emitter.

After the drive signal level is set to 30 volts, turn the TV receiver on. If the high voltage returns, as noted by the sound coming on or a raster on the screen, the output transistor, SCR regulator, and sweep transformer are good. The problem that has caused a loss of high voltage is either the horizontal oscillator, driver, or start-up circuit.

You can quickly isolate the problem to the oscillator and driver stages or to the start-up circuit by removing the (+) red lead from the Analyzer's drive output jack. If the TV set remains on, the problem is in the oscillator or driver stages. The start-up circuit's required only to start the horizontal sweep circuit operation; once it's working, the startup circuit remains idle until the set is again turned on. The oscillator, on the other hand, must start and continue to operate for the set to stay on.

If the set goes off, indicating that the oscillator or driver stages aren't operating, turn the set off and disconnect the driver signal leads. Now connect the driver output from the Analyzer directly to the base of the horizontal driver and the ground lead to the main chassis ground. A 4-volt peak-to-peak drive signal will be needed at the base of the driver. Turn the receiver on and observe the high voltage or sound. If they return, that indicates that the driver stage is operating properly and is allowing the VA62A drive signal to drive the output transistor. This, then, indicates that the oscillator signal is missing and the oscillator is faulty.

Now, should the high voltage not return when the signal is applied to the base of the driver, the problem is in the driver stage. At this point, you should check out the driver transistor, transformer, and associated components. The screen on the receiver might not show the correct brightness when you're subbing in the VA62A test drive signal. These conditions are normal and are caused by the X-ray protection and brightness limiter circuits within the TV chassis.

If the application of a horizontal drive signal to the base of the horizontal output transistor doesn't restore the high voltage, the sweep transformer might be defective. In this case, you can use the ringing checker built into the VA62 for yoke coil and transformer checks. With this feature you can check the sweep transformer and yoke, in or out of circuit, and determine if they're good or bad. Note the ringer test setup in Fig. 8-11. To check, connect the ringing test leads to the primary of the sweep transformer, rotate the impedance match switch through its six positions, and watch the digital meter for 10 or more counts, which indicates a good coil. On the CTC97 chassis, leads going to pins 3, 8, and 12, and the CRT socket must be removed. The circuits connected to these pins and CRT socket will load the transformer and make it appear to be defective.

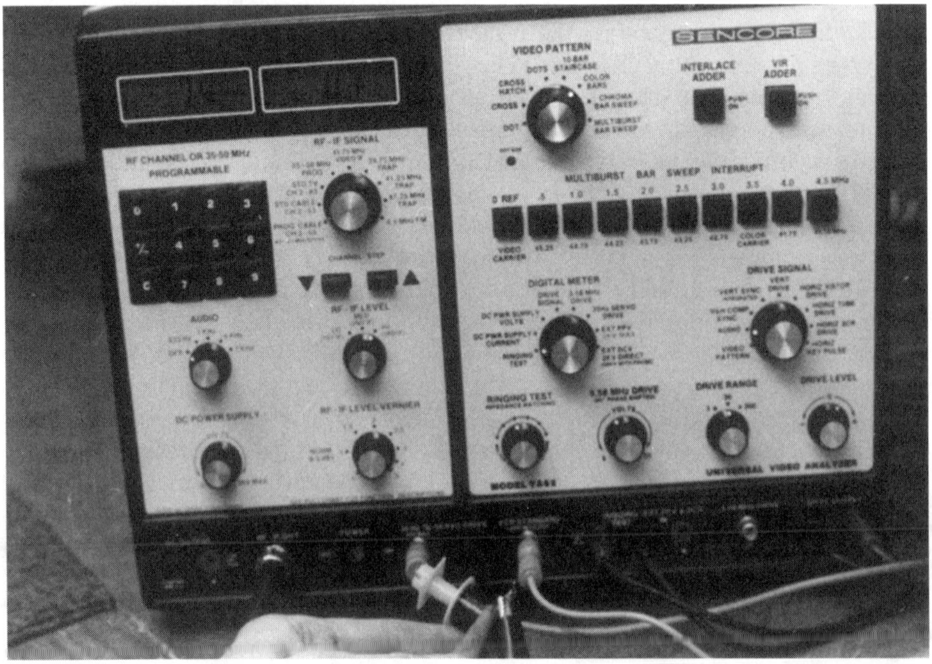

8-11 The VA62 being used for a coil ringer test.

The sweep transformer in the CTC97 chassis is one of the newer types now found in many TV sets. It's called an integrated flyback and has built-in tripler diodes. This design eliminates one extra part and reduces the overall chassis costs. The diodes have a high turn-on threshold and can't be tested with an ohmmeter. Let's now see how the VA62 can be used to test both the integrated flyback and the diodes.

To test the diodes in the integrated flyback, discharge the second anode lead and remove it from the CRT. Connect a high-impedance meter (DVM) to the second anode lead and the common lead to "hot" ground. Using a meter with a lower impedance than the 150 megohm found in some meters will load the diodes and give you inaccurate readings. Connect the drive signal from the VA62A to the primary of the sweep transformer. Do not turn the TV set on for this test! Set the drive signal switch to the "horiz. key" pulse position and adjust the drive level control for 200 volts peak-to-peak on the digital meter.

Note the test setup drawing shown in Fig. 8-12. The voltmeter will read about 340 volts dc if the sweep transformer is good. If the dc voltage is 20 percent or more lower or the drive output control can't be adjusted to 200 volts peak-to-peak, then there's a defective diode and the entire flyback transformer unit must be replaced.

SCR drive circuit operation

Before you start to troubleshoot SCR drive circuits, you need to see how they operate. You'll find that they operate differently than the conventional transistor or tube-type sweep output stages.

First, you need to keep in mind that the SCR drive circuits must do the same things as a tube or transistor—provide a drive signal to move the electron beam across the CRT screen. In order to understand how the SCR circuit works, you need to know two things.

1. How an SCR operates.
2. What function each SCR serves in that particular operation of the sweep circuit.

SCR operation

An SCR is really just a diode that has a third control element called a *gate*. You can think of an SCR as a solid-state relay (Fig. 8-13) that will latch, once a positive current is applied to the gate, and will continue to conduct as long as the anode/cathode junction is forward biased. This diode junction will continue to conduct until one of the following two things happens:

1. The voltage across the diode drops to zero.
2. The diode is reverse-biased.

If either of these conditions occurs, the diode junction becomes an effective open circuit until the gate lead is again fed a positive bias current. The operation is similar to a tube that's biased into cutoff (class C bias), with one exception. The tube will turn itself off when the grid signal goes negative, while the SCR will remain in the

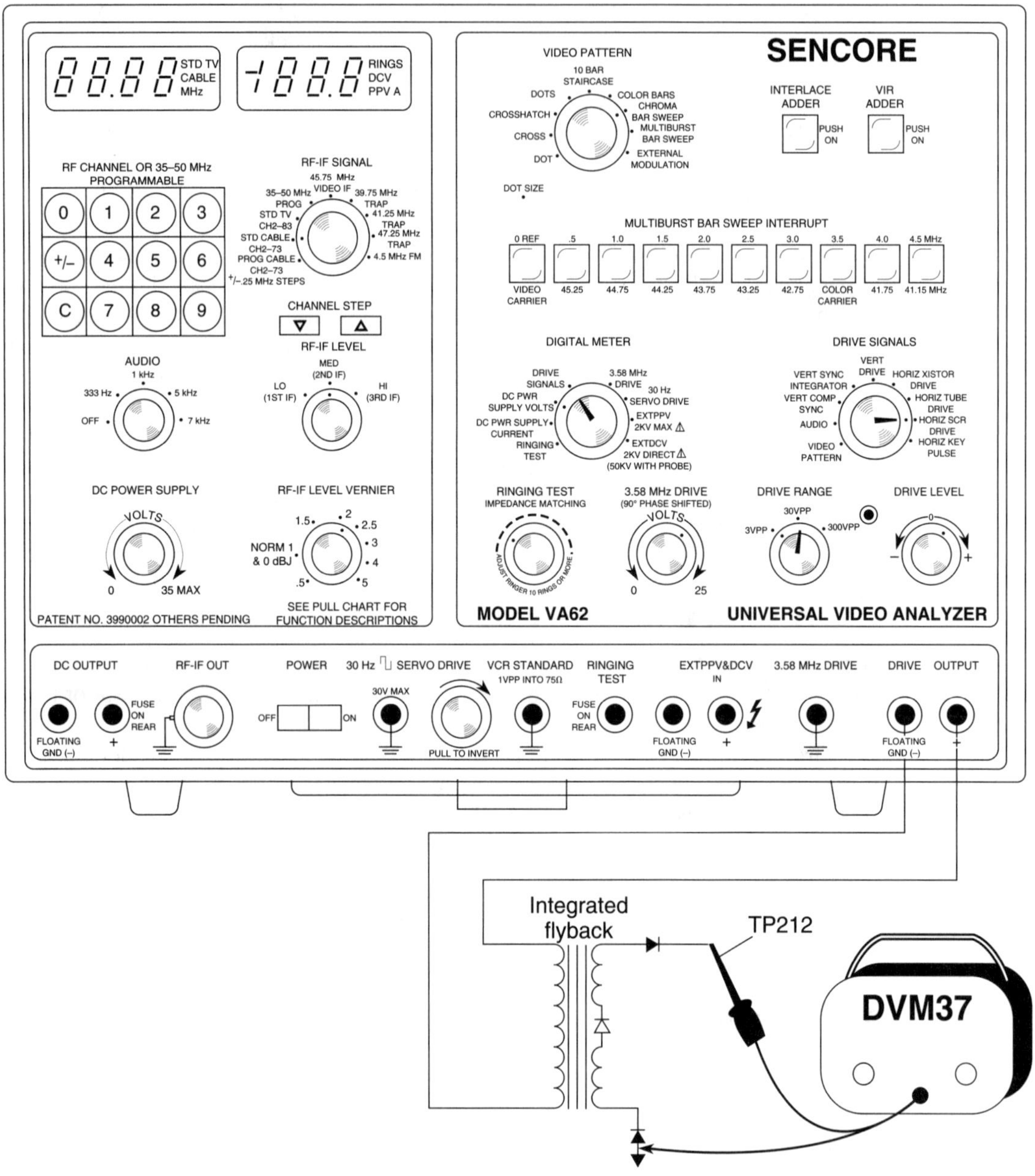

8-12 Test setup for checking the integrated flyback system.

on state until the diode junction is reverse-biased. This is why two SCRs are required in the horizontal drive circuit. The first SCR is turned on by the horizontal oscillator. This SCR, fed through a timing network, turns the second SCR on. The second SCR produces a signal that will turn the first SCR off, and finally the timing network causes the second SCR to turn off.

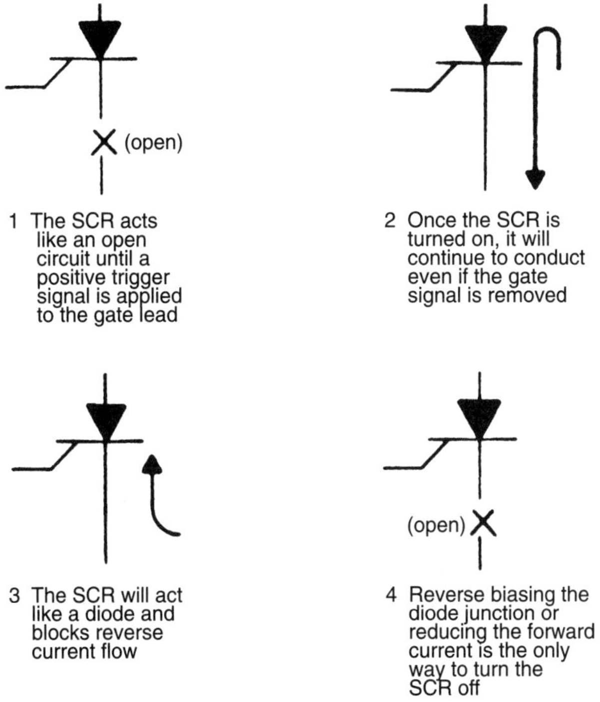

8-13 How an SCR acts like a solid-state relay.

There are two important things you need to know about the gate of the SCR to keep from burning up the SCR. First, you need enough current to fire the SCR. However, once the SCR is fired, you need to remove the gate signal quickly to prevent excessive gate current and possible gate circuit burn-up. The VA62 Analyzer provides a specially shaped SCR drive signal that prevents this from happening. This signal provides sufficient driving current but is of a short duration to prevent damage to the gate circuit.

Function of each SCR during sweep operation

Let's now move on to an explanation of the SCR circuit for a better understanding of how the SCR is used in this circuit. Refer to the simplified SCR sweep circuit in Fig. 8-14. Notice that two SCRs are used in these horizontal output stages; one is called a *trace SCR* (SCRt) and the other a *retrace SCR* (SCRr). Both SCRs have diodes in parallel with them that are connected in the opposite polarity. These diodes, in conjunction with the energy stored in the commutator coil (LC), are crucial to the operation of the sweep circuit and work in conjunction with the SCRs to give proper timing to the horizontal sweep.

The horizontal SCR sweep circuit consists of two tuned (resonant) circuits that operate at approximately 10 and 40 kHz. (This is no different than for tube or tran-

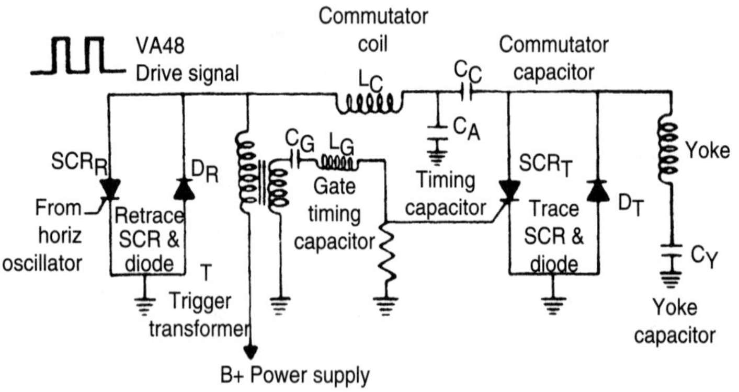

8-14 Simplified SCR horizontal sweep output diagram.

sistor sweep circuits.) The 10-kHz tuned circuit is made up of the trace SCR (SCRt), trace diode (DT), yoke, flyback, and the yoke capacitor (CY). This 10-kHz circuit is the one that gives you trace as it causes the CRT electron beam to scan across the CRT (from left to right) at the proper rate. The 40-kHz tuned circuit is made up of the retrace (SCRr), diode (DR), and some special inductors and capacitors (LC, CC, and CA). The main function of the 40-kHz circuit is to move the electron beam from the right side of the screen all the way back to the left for retrace.

Let's begin our explanation with the CRT beam at rest in the center of the screen when you first turn on the set. Refer to the drawing in Fig. 8-15. CY will charge from B+ through LC and CC. This capacitor will be charged from the rest of the circuits for all following lines. Current will flow through the transformer primary (T) and the commutator coil (LC) as the capacitor (CA) charges from B+. The current in the transformer primary induces a positive polarity spike in the transformer secondary, which is passed to the gate of the trace SCR (via CG and LG), causing it to turn on. The trace SCR now acts like a closed switch, and current flows through the SCR, the yoke, and its 10-kHz tuned circuit. The current flow through the yoke creates a magnetic field that begins to move the electron beam to the right of the screen.

As the trace nears the right side of the screen, resonance is reached in the 10-kHz trace circuit. You might recall that when a coil reaches resonance, the current change is minimum. The voltage across the coil is determined by the amount of change in current through the coil (if we ignore the small amount of resistance) and drops to zero. With zero voltage across the anode/cathode junction of the SCR, the SCRt turns off.

As soon as current through the yoke reaches this maximum current condition, the field around it begins to collapse, which induces a voltage of the opposite polarity. This reverse voltage provides a forward bias for the trace diode (DT across the trace SCR), the trace diode begins to conduct, providing a current path for the 10-kHz trace circuit, and the electron beam continues to move to the right. This is illustrated in Fig. 8-16.

At approximately the same time that the trace SCR stops conducting, a positive pulse from the horizontal oscillator is applied to the gate of the retrace SCR (SCRr). This positive pulse turns the retrace SCR on and it begins to conduct. You now have two paths for current flow: a forward current through the trace diode (DT), yoke,

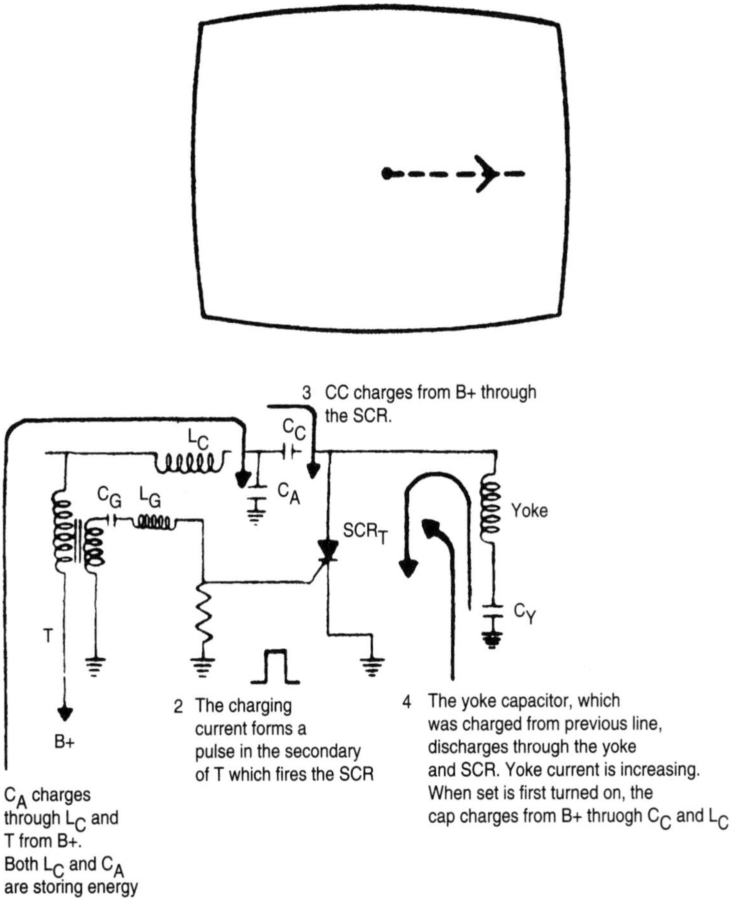

8-15 Illustration of the scan cycle beginning, as the CRT electron beam moves from the center of the screen to the right side.

and its capacitance (CY) forming the 10-kHz trace circuit, and a reverse-current path through the retrace SCR (SCRr), trace diode (DT), and the added capacitance and inductance in the 40-kHz retrace SCR circuit.

The reverse current rapidly builds a voltage of the opposite polarity across the trace diode. When the reverse voltage reaches the same level as that developed in the forward polarity across the yoke, the trace diode will no longer be forward-biased, and it will stop conducting. With the trace diode and the trace SCR turned off, the circuit now has only one path left for current flow: through the retrace SCR and the added capacitance and inductance in its associated circuit, which forms the 40-kHz retrace circuit. Refer to the diagram in Fig. 8-17 for this portion of the sweep operation.

The energy built up in the yoke is now discharging into the 40-kHz resonant circuit, which contains the commutator coil (LC), capacitor (CC), SCR5, and the yoke. This quickly returns the electron beam to the center of the screen, where the 40-kHz

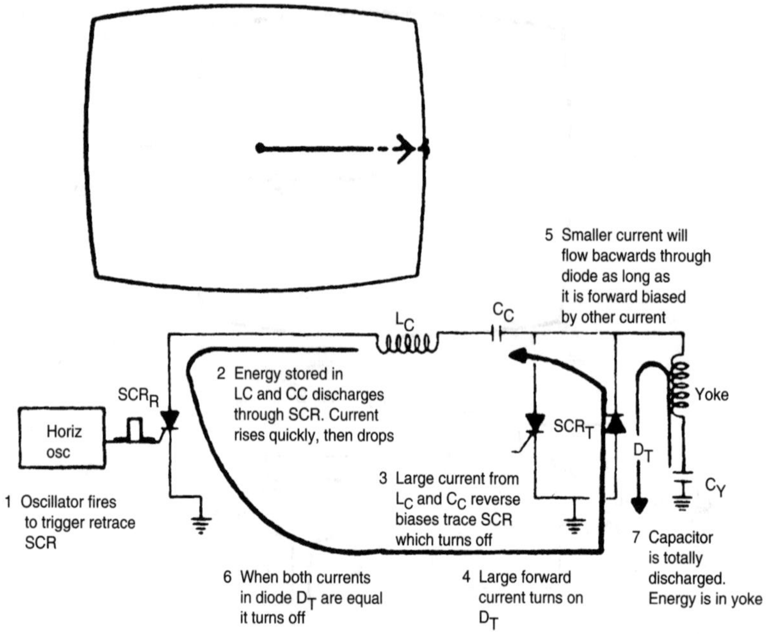

8-16 The end of the trace and initiation of the retrace portion of the cycle.

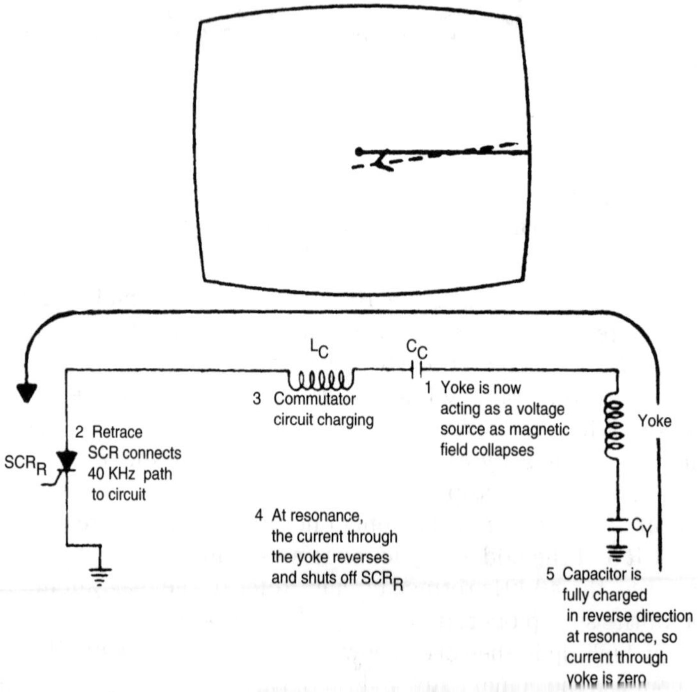

8-17 Current path and flow for the first half of retrace.

circuit reaches resonance and the voltage reverses. At this point, the dc potential (on the anode of the retrace SCR) is low enough to stop conduction, and the retrace diode (DR) begins to conduct until the beam reaches the maximum opposite potential and the beam is all the way to the left side of the screen. This builds up a negative potential across the deflection yoke. Refer to the operational drawing in Fig. 8-18.

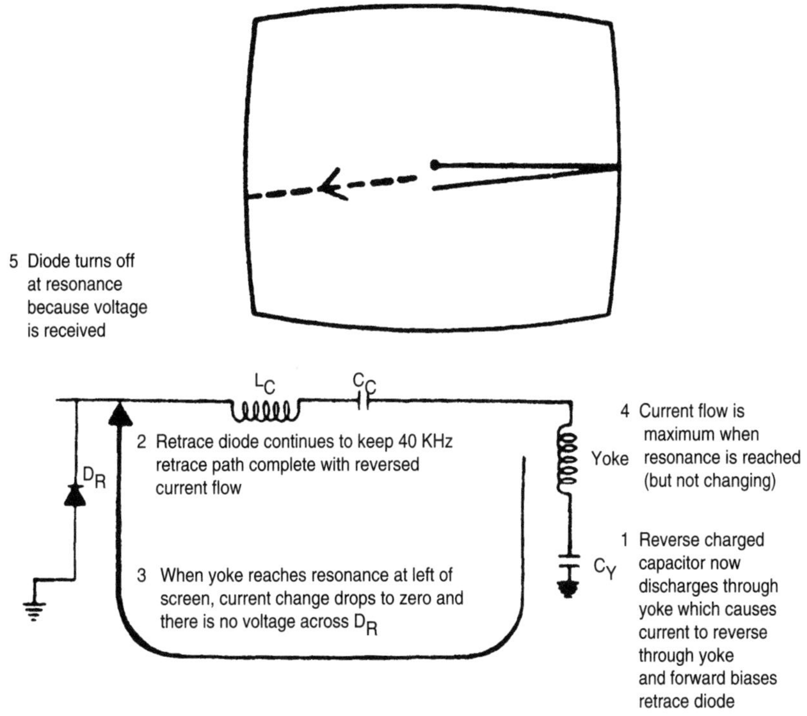

8-18 Current path flow for the second half of retrace.

The high negative potential on the yoke forward-biases the trace diode (DT) and it begins to conduct, placing the 10-kHz trace circuit back into operation. The trace diode is now functioning exactly like a conventional damper diode. The magnetic field begins to collapse, and the CRT beam moves from the left to the center of the screen at the same rate of travel that it moved from the center to the right because both rates are controlled by the 10-kHz circuit time constants. You now have one complete scan line across the face of the picture tube. Note Fig. 8-19.

Recall that the beam started from the center of the screen when the TV set was turned on, which charged the capacitor (CA) and turned on the trace SCR. You must turn on the trace SCR to have it start the second cycle of the scan. This is accomplished by charging the capacitor (CA) from B+ again because it had discharged through the retrace SCR during retrace. This fires the trace SCR to repeat the process for the next line. This repeats itself at the rate of 15.750 kHz. The complete SCR horizontal sweep circuit for an RCA CTC-40 chassis is shown in Fig. 8-20.

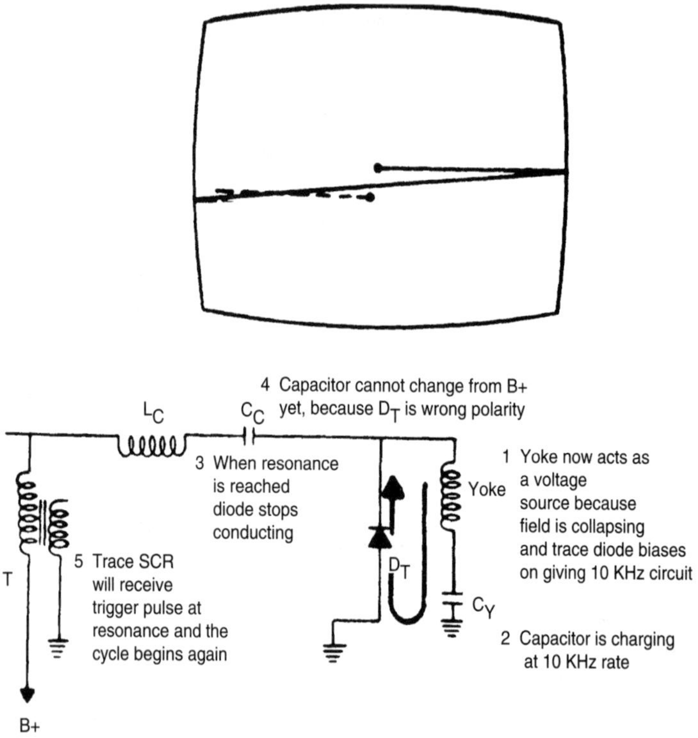

8-19 Current flow and electron beam movement from the left of the screen to the center.

SCR sweep-circuit troubleshooting

You've probably noticed, as you analyzed the circuits in the SCR system, that there really isn't much difference between the operation of the SCR system and the conventional tube or transistor drive circuit. The key difference is that the signal that controls the trace SCR comes from the other circuits and not from the horizontal oscillator, as it does with tube and transistor horizontal output circuits. For this reason, you shouldn't attempt to drive the trace SCR directly with a substitute signal. Any attempt to substitute this gate signal will upset the resonance of this circuit and cause too much current to be drawn from the power supply. This will usually cause the circuit breaker in the set to trip out.

As noted earlier, the SCR drive signal provided by the VA62A has the proper current capabilities and the correct pulse duration to prevent the possibility of damaging the SCR gate. This signal is designed to substitute for the horizontal oscillator output and is connected directly to the gate of the retrace SCR. Let's see how to use the SCR drive signal to locate problems in the sweep output stage.

Loss of horizontal sweep could be caused by no output from the horizontal oscillator, or a defect in the trace or retrace circuits. You need to be able to isolate the actual cause of this defect. The first step is to substitute for the horizontal oscillator output. First, select the horizontal output SCR gate position of the drive signals switch

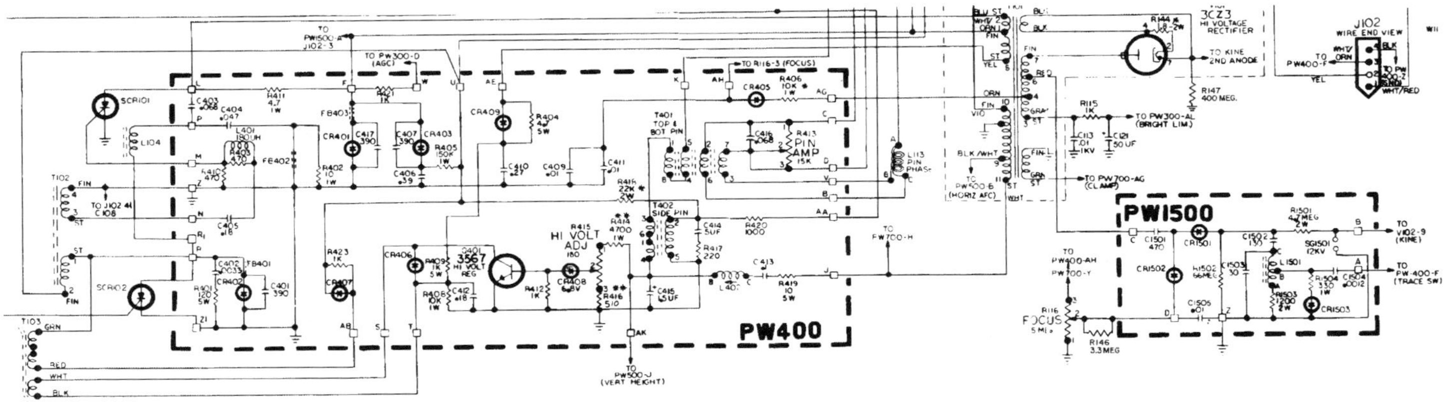

CTC-40 Schematic Diagram—Preliminary

8-20 Schematic of the complete SCR sweep system used in the RCA CTC-40 color TV chassis.

on the Analyzer. An SCR always requires a positive drive polarity, so the only other thing we need to know is the signal amplitude. Check the service schematic data for this information. If the correct waveform is not shown on the diagram, keep in mind that underdriving the retrace SCR can be worse than overdriving, so you should start with the drive level control in the 30-volt maximum positive position. Before turning on the receiver, connect the RF output of the Analyzer to the set's antenna input as a synchronizing signal. This signal will give you vertical/horizontal sync and a video pattern to observe to make sure that all signal components are present when feeding in the substitute signal.

Don't make connections to the retrace SCR until after the set has been powered up. This will prevent the possibility of damaging the SCR with a gate signal with no other bias on the SCR. Now, inject the test signal pulse and see if the horizontal sweep and high voltage return. If the set operates properly, you know that the defect is in the oscillator circuit because a known-good drive signal has been fed in from the VA62A Analyzer.

The digital drive signal monitor meter, built into the Analyzer, enables you to determine if the retrace SCR gate is shorted. If the SCR gate is shorted, the signal level of the meter will drop appreciably when the test lead is connected to the retrace SCR gate because the shorted SCR will load down the drive signal output. If the set operation doesn't return to normal with the substitute signal applied, and the meter doesn't indicate a loading problem, then you don't have a defective horizontal oscillator or a shorted retrace SCR gate.

Sometimes you can't check the horizontal oscillator drive circuit because the fast-acting horizontal output circuit breaker continues to trip. This is not all bad, however, because the circuit breaker of the receiver under test can give you a good clue as to where the defect is located. When the horizontal drive signal is removed from the gate of the retrace SCR, you should be pulling no current at all through either SCR. Thus, the first step (if the circuit breaker continues to trip) is to disable the signal coming from the horizontal oscillator, if you've determined that the horizontal oscillator is good. If in doubt, use the SCR test signal that you know is good for signal injection.

The horizontal oscillator is easily disabled with plug-in module TV sets just by pulling the horizontal oscillator module from the chassis. If the set is not modular, just open the gate lead to the retrace SCR. If the circuit breaker continues to trip with no signal applied to the retrace SCR, suspect a defect in the retrace circuits. The retrace SCR or diode are most likely shorted. This can be quickly determined with an ohmmeter check across the components that you suspect are bad.

If moving the drive signal from the retrace SCR has stopped the circuit breaker from tripping out, you know that the retrace SCR is good because it acts as an open circuit when there's no signal applied to its gate, but it draws current when a signal is applied.

Let's go a step further and see how to determine if the trace SCR is good. At this point, you should still have the drive signal disconnected from the retrace SCR gate lead. The circuit breaker, for this example, doesn't trip out when power is applied to the set.

To check the trace SCR, you must now reconnect the horizontal oscillator or apply the SCR drive signal from the Analyzer to the gate of the retrace SCR. Note in Fig. 8-21 the drawing test instrument setup and test point connections for driving the SCR. You must now once again observe the operation of the circuit breaker. If the breaker trips, you know that there's an open circuit in the 10-kHz trace circuit.

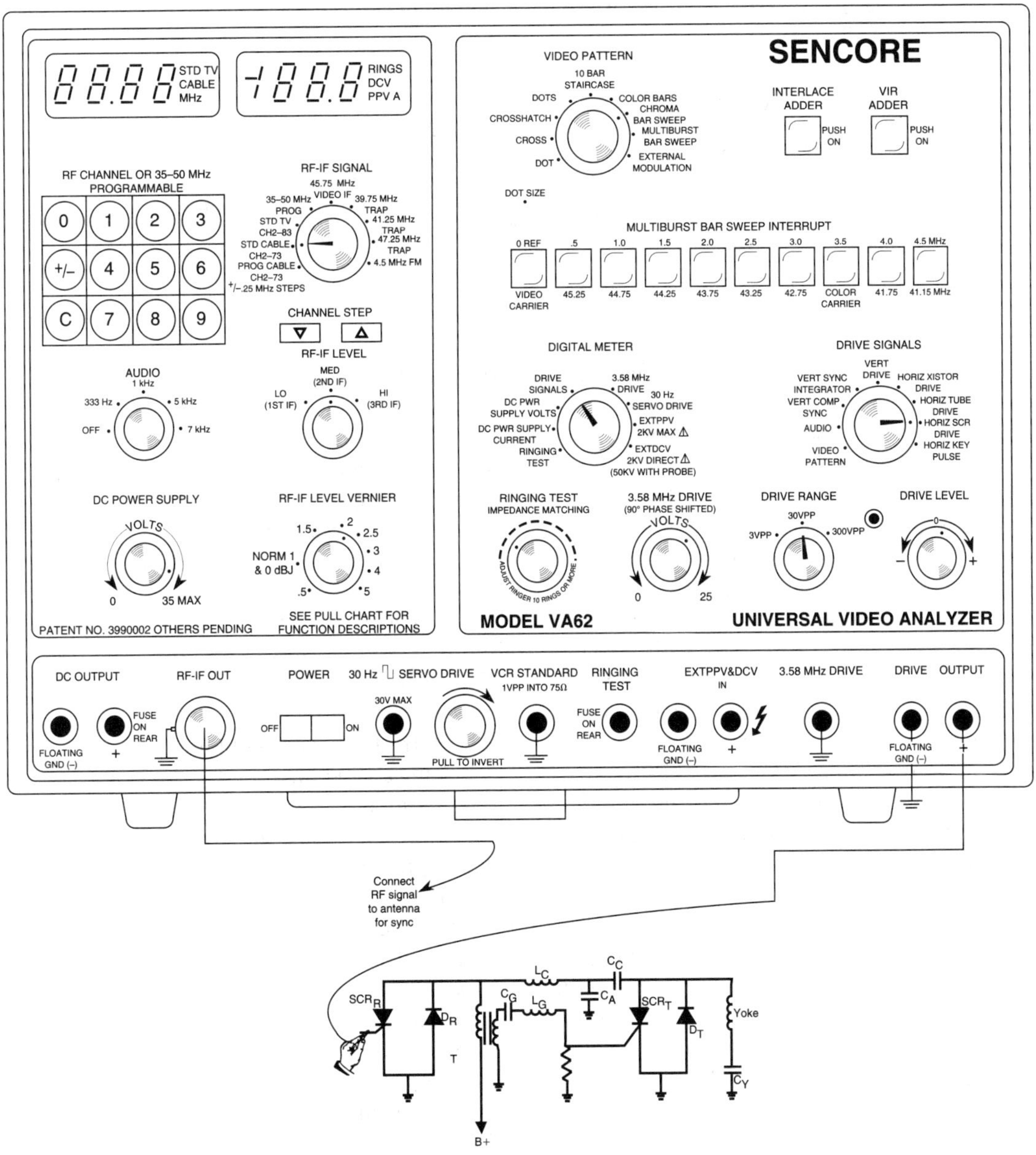

8-21 Analyzer setup for test signal injection into the retrace SCR.

The reason you know the trace circuit is open is because the trace SCR and its associated components are necessary to turn the retrace SCR off at the proper time. If the trace circuits are open, there's no turn-off pulse, so the retrace SCR will continue to conduct (as long as B+ is applied) after a trigger signal has been applied to the gate lead. This condition will load the B+ line, which will cause the circuit breaker to trip open. The open circuit might be the SCR itself, but could be one of the components in the associated circuit. Refer to the detailed circuit in Fig. 8-22.

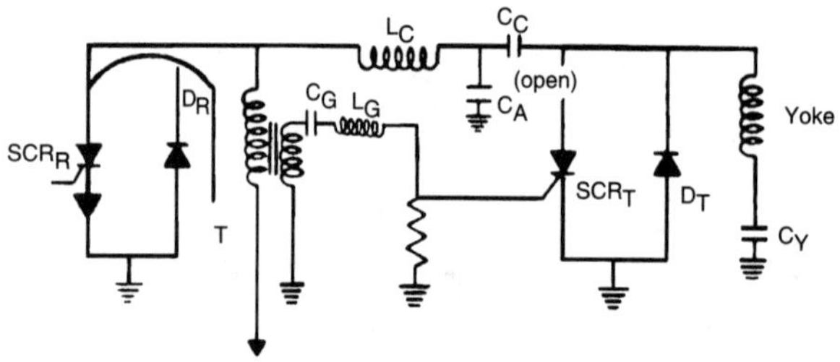

8-22 An open trace SCR will cause the circuit breaker to trip because there is no signal to turn the retrace SCR off, and it will load down the power supply.

You'll find that if the trace SCR were shorted, driving the retrace SCR with the Analyzer would not reestablish operation of the horizontal sweep circuit, and the circuit breaker would not trip out either. The reason is that even though the trace SCR is shorted, you still have enough reverse EMF from the commutator circuit made up of LC and CC to turn the retrace SCR off for each cycle. As long as the retrace SCR is turned off during each cycle, you won't draw enough extra current to trip the circuit breaker. There's no sweep, of course, because the shorted trace SCR is shorting out the yoke.

Let's now look at a way to quickly check out the SCRs. The dc power supply built into the Analyzer provides you with a voltage source that can be used to test the sweep SCRs. You need only to remember that an SCR will act like an open circuit until a positive signal is applied to the gate lead. Once the SCR fires, it should continue to conduct until the B+ power is removed to the SCR. With this background information, you can reliably test the questionable SCR by placing a 100-ohm (½-watt) resistor in series with the SCR and the Analyzer's dc power supply.

Start the check by connecting the negative lead of the power supply to the cathode of the SCR. The cathode is usually the circuit that's grounded to chassis. Next, set the voltage of the Analyzer power supply from 20 to 30 volts to be sure of enough voltage to fire the SCR. Now place the 100-ohm resistor in series with the positive test lead of the power supply and connect the other end of the 100-ohm resistor to the anode of the SCR. Next, connect a voltmeter across the 100-ohm resistor. The resistor now limits the current to prevent the possibility of damaging the SCR and also acts as a current shunt so that the meter effectively reads the current through the SCR itself. At this point, the meter should show a very low reading, indicating that there's little or no current passing through the resistor and SCR because the gate is not connected to anything, and the SCR is wide open.

Now, momentarily short the anode to the gate of the SCR. The meter should now show a large voltage reading across the 100-ohm resistor, indicating that the SCR is turned on. If you don't obtain an increased reading, you know the SCR is open. Next, remove the short between the anode and the gate to see that the meter continues to read the higher voltage. If it does, this indicates that the SCR has latched properly. One final test is to momentarily open the positive lead of the power supply and then reconnect it. The meter should drop to almost zero to indicate that the SCR has turned itself off. The SCR has now been tested for opens, shorts, triggering capability, and latching action. The setup for this test procedure is shown in Fig. 8-23.

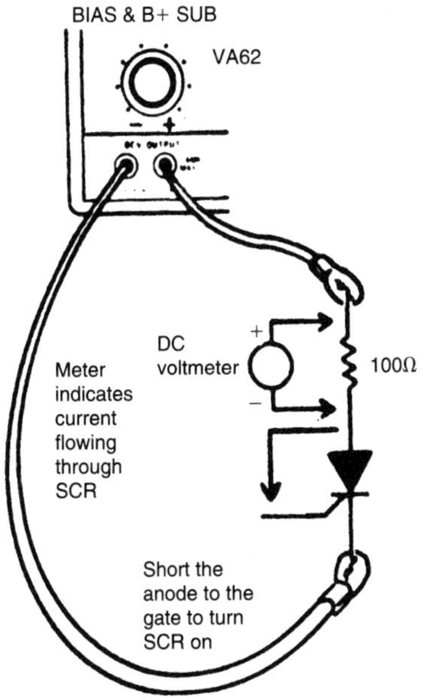

8-23 A test setup for checking suspected defective SCRs with a VA62A power supply.

If the problem has not been located up to this point, the sweep transformer and deflection yoke would be the prime suspects. The ringing tests, built into the VA62, will check out these SCR drive system coils with the same, reliable good/bad test that's used with a tube or transistor TV receiver. All that's required is to: connect the ringing test leads across the coil to be tested (after removing power to the set), switch the meter control to the ringing test position, and rotate the impedance matching switch for the highest digital meter readout. If the coil is good, it will always show a readout of 10 cycle counts or more. If it's defective, it will indicate less than 10 ringing counts or no counts at all.

Zenith electronic sweep system

The following information is courtesy of Zenith Radio Corporation. A block diagram of the Zenith Electronic Power Sentry system is shown in Fig. 8-24. You'll be looking at the operation for some of these blocks.

A line-operated 150-volt dc power supply provides the energy to a horizontal pulse-locked chopper. The chopper regulates the voltage and drives the sweep and

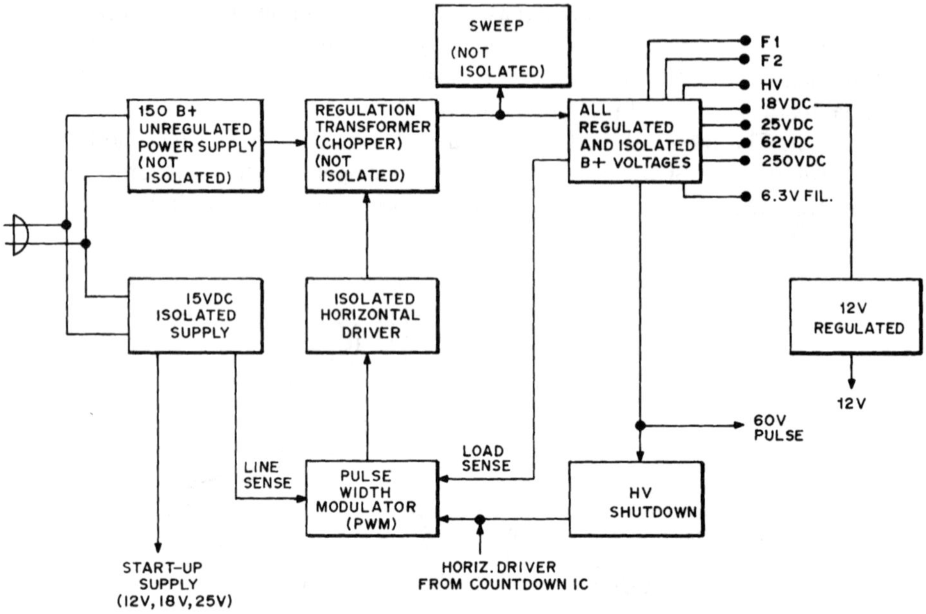

8-24 Zenith electronic power supply block diagram. Zenith Radio Corp.

high-voltage transformer with a constant pulse. The HV transformer also supplies all the regulated voltages for set operation.

The "key" to this regulation system is the pulse width modulator (PWM). The line and load voltages are sensed by the PWM which controls the chopper rate by varying the "on time" of the chopper, and thus, the supply voltage.

The 150-volt dc B+ is regulated by the pulse width modulator and horizontal output transistor. As the B+ increases, the PWM decreases the "on time" of the horizontal driver, decreasing the stored energy and maintaining the sweep voltages constant. As B+ decreases, the PWM increases the driver "on time," increasing the stored energy and maintaining a constant voltage at the sweep outputs.

Horizontal sweep

Let's review the conventional sweep circuit that's shown in Fig. 8-25. These waveforms show the horizontal output transistor starting the system. The "S" capacitance is discharged through the yoke, and the B+ supply through the high-voltage transformer. The yoke current scans the picture tube from the center to the right-hand side and stores the energy to scan the left-hand side. The transformer current stores energy used to produce the high voltage. The output transistor current is limited by the yoke and transformer induction and the B+ level. The base current to the output transistor is reversed, turning off the transistor, and then beginning retrace.

As shown in the Fig. 8-26 waveform, the damper diode is off. The output transistor is turning off, and the collapsing magnetic field of the yoke and flyback transformer is trying to maintain the current flow. The continuing inductive current

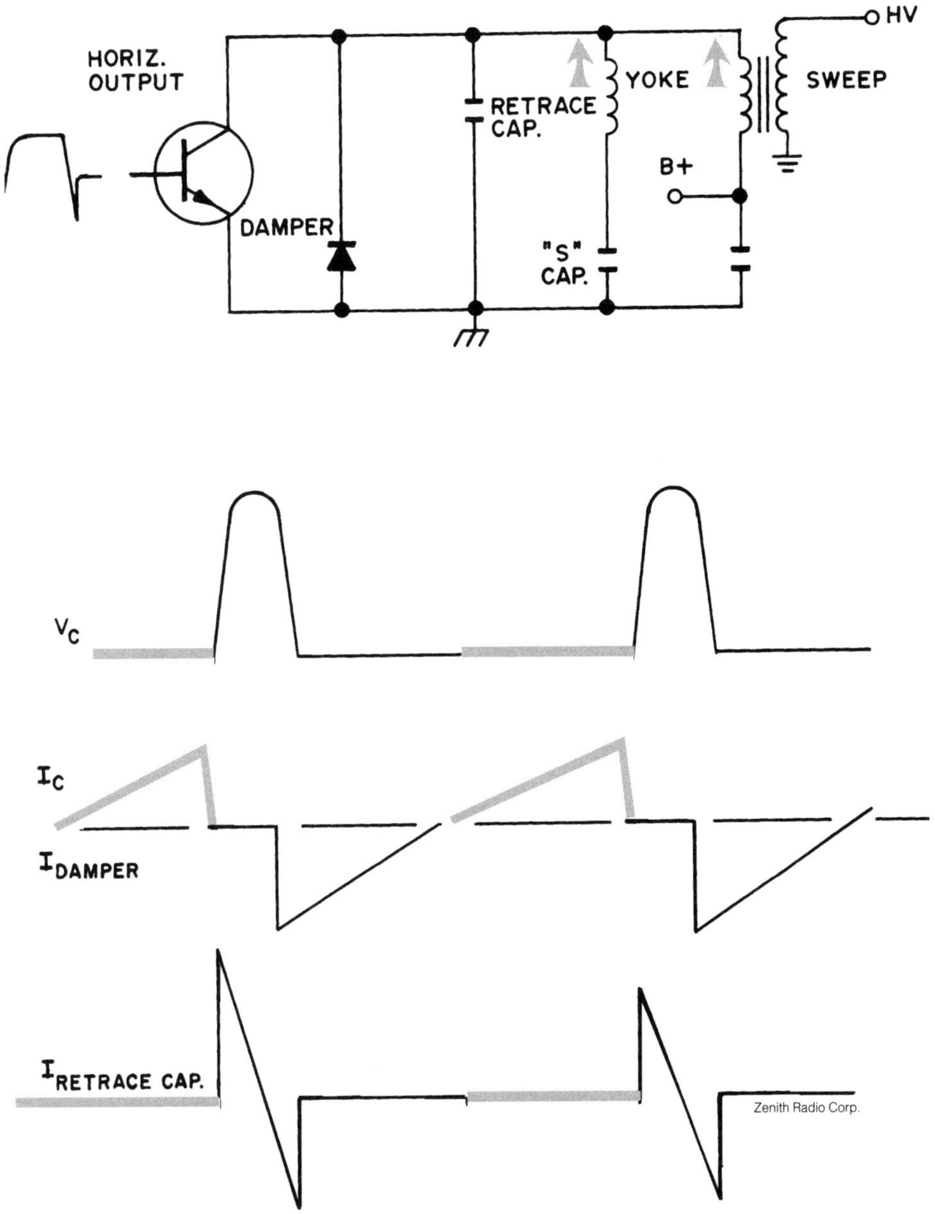

8-25 Zenith electronic sweep system. Right side of raster.

charges the retrace capacitance. The transfer of the inductive energy to the retrace capacitance forms the first half of the retrace pulse.

The waveform in Fig. 8-27 shows, starting at the middle of the retrace pulse and the peak yoke voltage, that the current in the circuit is at zero and starts to increase again in the negative direction. The voltage across the "LC" network decreases as the

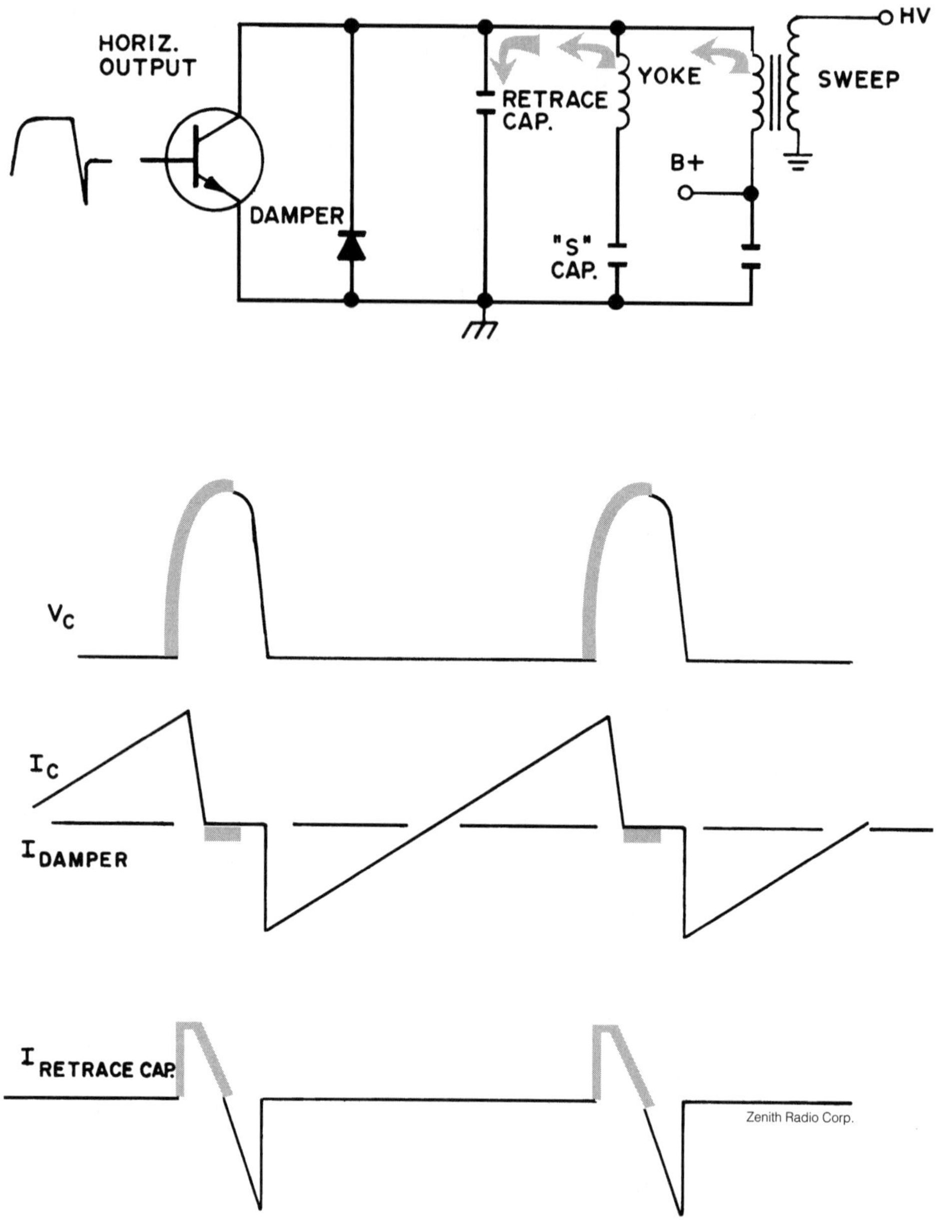

8-26 Sweep system waveform-start of retrace.

retrace capacitance discharges through the inductors, recharging the magnetic fields. When the retrace capacitance current reaches zero, retrace time is ended.

Referring to the Fig. 8-28 waveform, the output transistor and retrace capacitance are inactive as the circuit starts damper time. The inductors again try to maintain the negative current by charging the retrace capacitance in the opposite

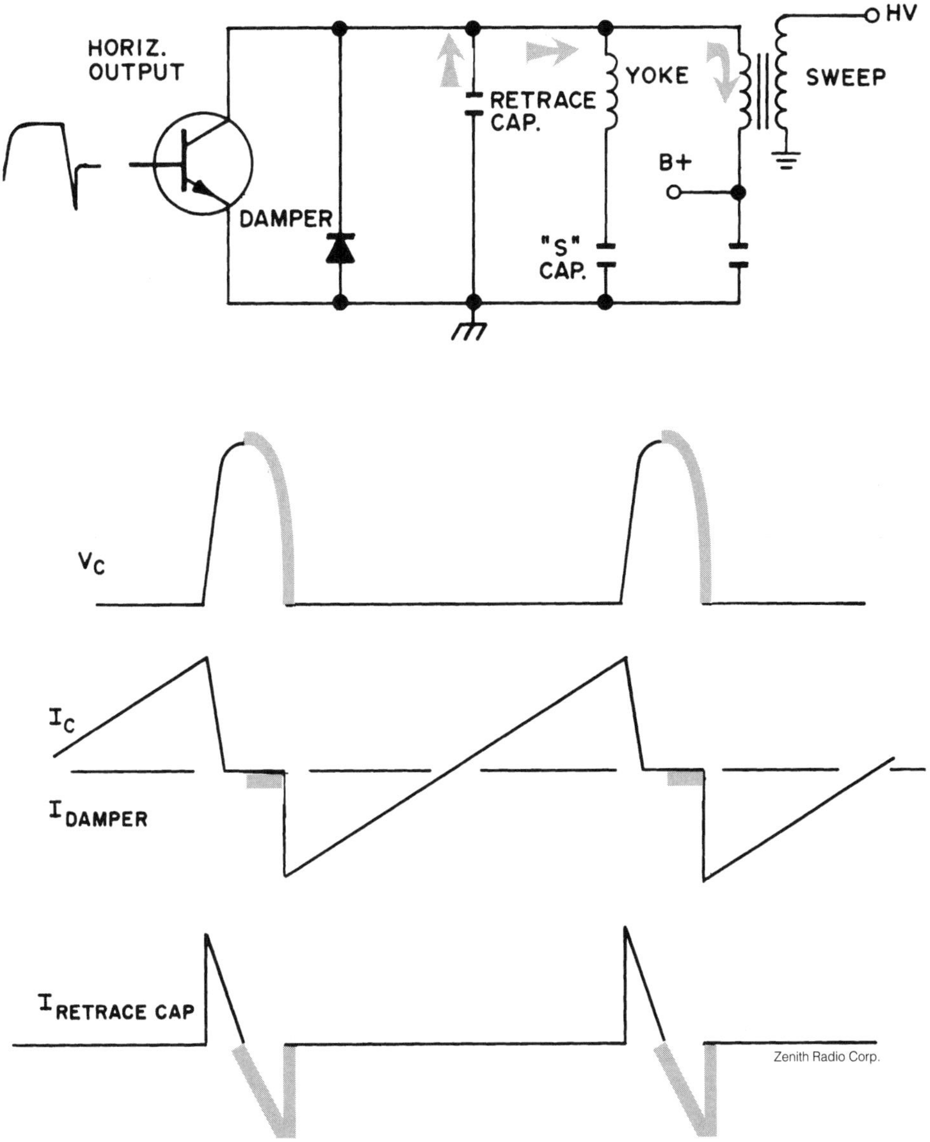

8-27 Sweep system-end of retrace.

direction. Essentially, the top of the retrace capacitance starts to go negative on the second half of the "LC" oscillation. A negative voltage at the retrace capacitance forward biases the damper diode. This starts the scan from the left-hand side. The damper diode rectifies the stored energy of the inductors, recharging the "S" capacitance and the B+ supply. As the scan reaches the center, the yoke current will go to zero. At this time, the output transistor is turned on, which starts to draw positive yoke current again, thus starting the cycle all over again.

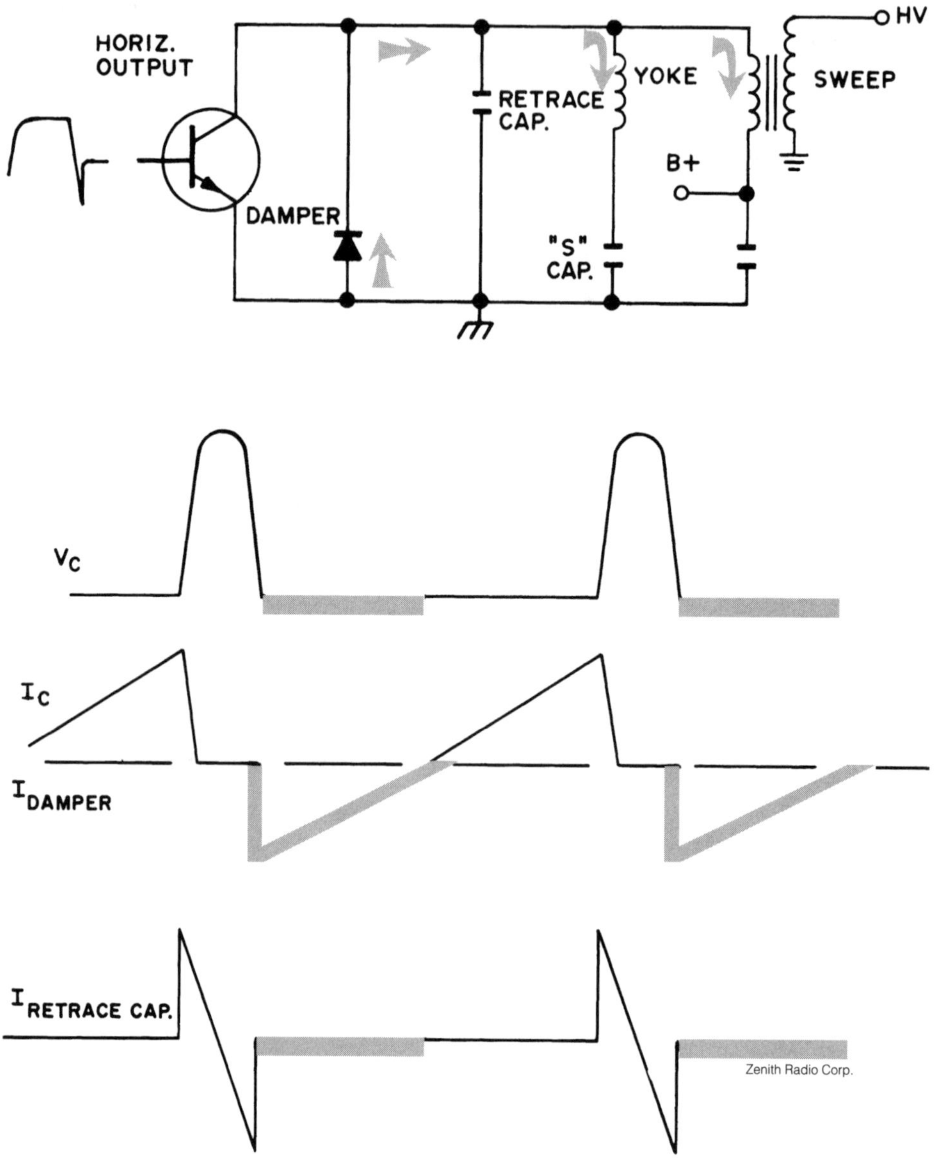

8-28 Left side of raster sweep.

Chopper/circuit operation

Refer to the circuit and waveforms in Fig. 8-29 as I discuss the chopper circuit used in the Electronic Power Sentry system. The simplified diagram describes the basic chopper circuit. This type is a pulse-width modulated energy storage setup convertor/regulator. The ac line-connected bridge rectifier and filter provide a voltage to the primary of a transformer. The horizontal output transistor, QX3326, which is in

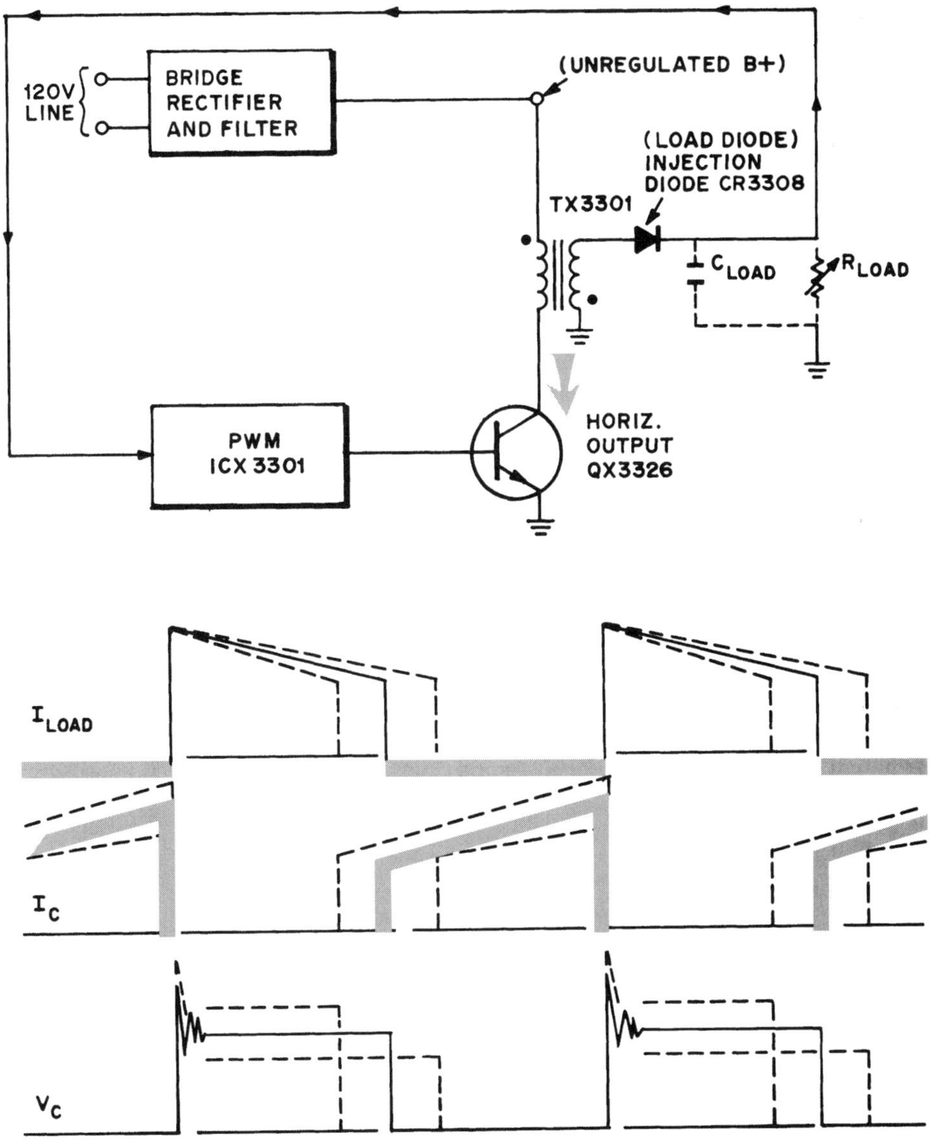

8-29 Chopper system and waveform. Zenith Radio Corp.

series with the primary of TX3301, controls the stored energy by varying the "on time" inversely to changes in B+. If the B+ decreases, the "on time" must decrease to maintain a constant current level. For a constant energy, current must be constant. The secondary winding of the transformer provides the step-up and phase inversion. During conduction of the transistor, the polarity across the secondary is minus to plus. The negative voltage at the anode of the load diode keeps it off.

After the transistor turns off, the voltage on the primary reverses. Note the waveforms in Fig. 8-30. Therefore, the secondary voltage also reverses, forward bi-

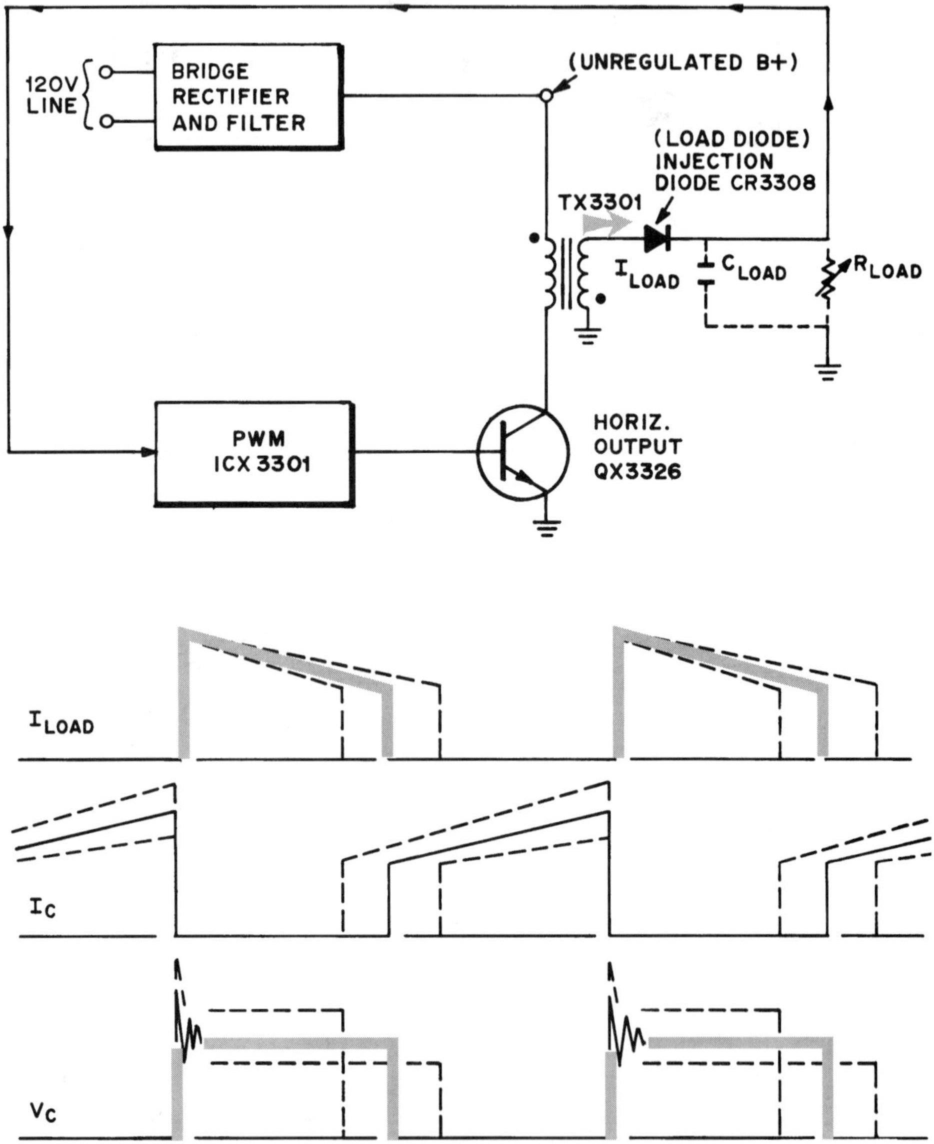

8-30 Chopper system and waveform. Zenith Radio Corp.

asing the load diode and charging the load capacitance. The voltage at the load capacitance is sensed and fed back to the pulse width modulator (PWM). The PWM varies the base drive inversely proportional to the load voltage variations. This provides a regulated load voltage. The peak "off" voltage of the transistor is also inversely proportional to the "on time."

The in-phase driver

The conventional sweep circuit is an out-of-phase system. For example, the driver is on while the output transformer is off, and vice versa. The electronic power sentry is an energy storage system, storing energy during conduction, and supplying output base drive during the off time of transistor QX3326. Refer to the circuit diagram and waveforms in Fig. 8-31.

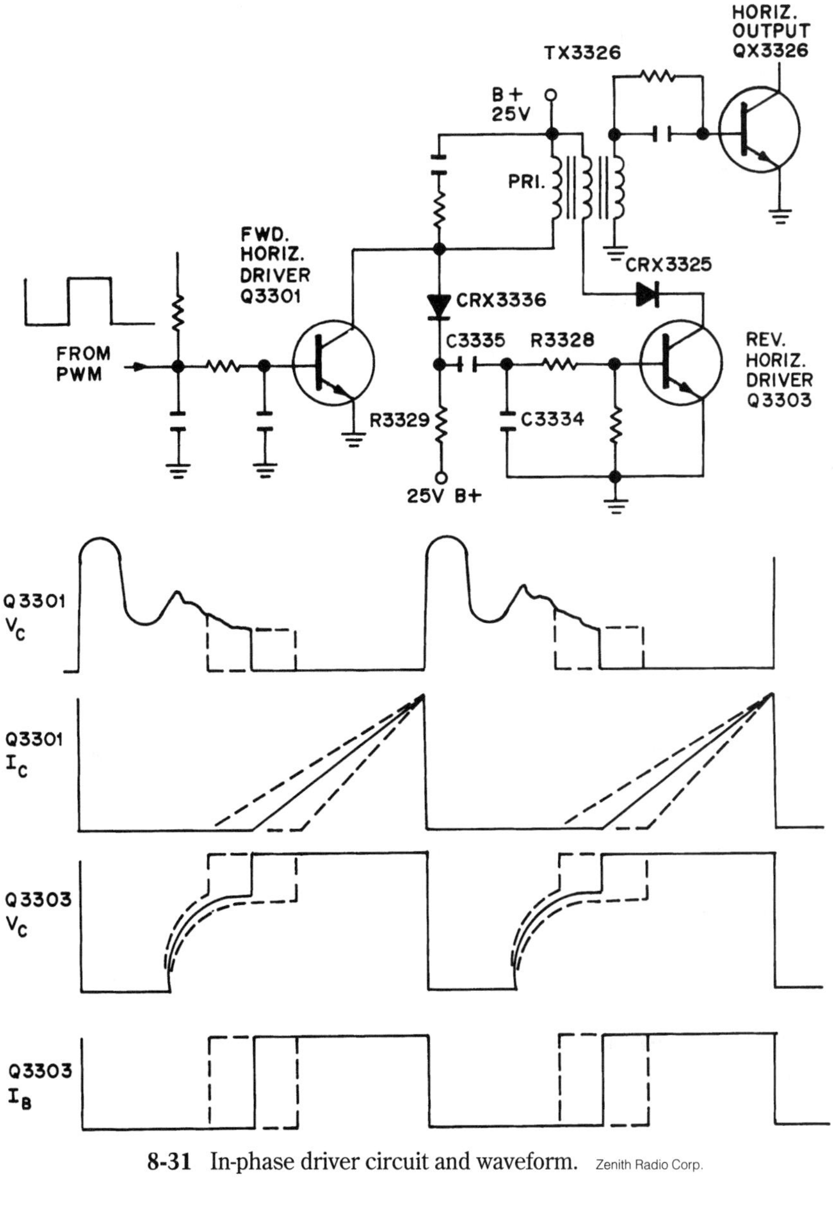

8-31 In-phase driver circuit and waveform. Zenith Radio Corp.

In the electronic power sentry, the driver is part of an in-phase system. When the forward driver Q3301 turns on, the output transistor QX3326 turns on. The driver circuit is powered by the 25-volt supply, which allows for a low-resistance primary winding in the driver transformer TX3326. The closely coupled secondary winding is directly driven by a current ramp. The inductive impedance and low resistance provides for a very efficient driver circuit. The direct drive produces a constant drive level that's independent of the duty cycle. The conventional out-of-phase drive system requires a minimum "on time" to turn the output transistor "on" at low-line voltage and produces excessive drive current at high-line voltage.

A disadvantage is the negative drive current, which is not acceptable for a fast fall time in the output transistor. In order to obtain a fast fall time in the output transistor QX3326, a reverse driver transistor, Q3303, is added to the circuit. The forward driver transistor Q3301 is driven by the PWM ICX3301 chip, and the reverse driver transistor Q3303 is driven by the collector voltage of Q3301.

A diode and resistor form a clamp circuit that clips (or clamps) the peak of the collector voltage of Q3301. The waveform is decoupled by capacitor C3335. A capacitor and resistor form an integrating network that shapes the signal that drives the base of Q3303. Transistor Q3303 turns "on" after Q3301 turns "off." The current through Q3303, flowing through the primary of TX3326, drives the base of the output transistor, QX3326, negatively providing a "clean" output pulse to switch QX3326 in about .5 microsecond.

The pulse width modulator (PWM)

At the bottom of Fig. 8-24, there's a block marked PWM. This is the pulse width modulator, the main control element for line and load voltage regulation. It's a specially designed 8-pin IC chip.

Figure 8-32 shows the circuit schematic of the PWM. Pin 3 is connected to a reference voltage, while pin 4 is driven with a triangle waveform. When the triangle wave reaches a larger amplitude than the reference voltage, Q9 and Q10 will steal current from Q11 and Q12 from the current source Q13. The action of Q10 conducting will cause Q6 and Q7 (a positive feedback switch) to snap on. Transistor Q5 is an in-phase current mirror of the Q7 collector current. Therefore, Q5 will also turn on, supplying a large current into the base of Q3 which, in turn, will also overdrive Q1, the output device. The result will be a square wave output, starting and ending where the triangle waveform crosses the reference voltage that's fed to pin 3. A simple transfer characteristic of the device is shown in Fig. 8-33.

Let's now go into more detail on how this system is used in the electronic power sentry. Referring to the circuit in Fig. 8-32, pin 7 of the PWM receives the "H" drive pulses from the "countdown" IC chip located in another module. If you suspect that the drive pulse is missing or is not correct, you can use the transistor drive test signal from the VA62 Analyzer. These pulses are coupled to the base of Q14, an overdriven amplifier that amplifies and clamps the "H" pulses to ground and B+, producing a 12-volt peak-to-peak square wave. This square wave exits the PWM at pin 6 and is then shaped through R3317, R3312, and R3310 to a triangle and coupled to pin 4, along with a dc reference voltage from R3317, R3312, and R3310 to the

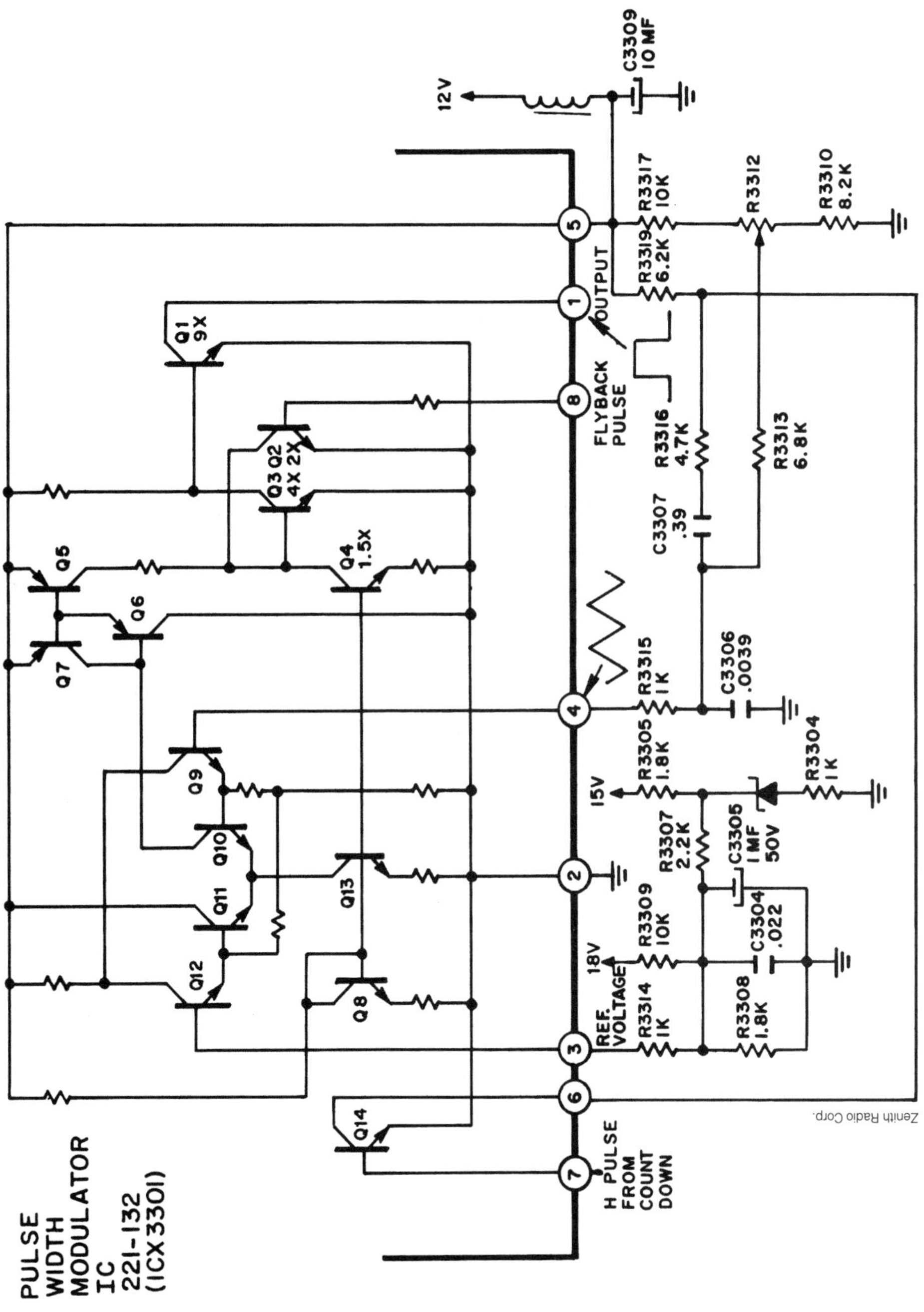

8-32 Pulse width modulator chip, part #221-132.

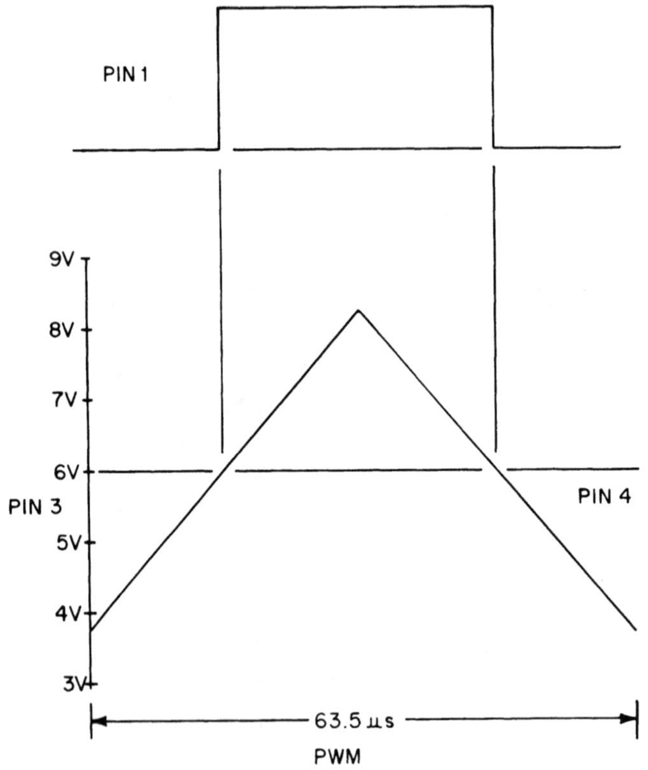

8-33 Transfer characteristic of the pulse width modulator.
Zenith Radio Corp.

same point. (Do not attempt to adjust this voltage, as it is factory set.) On the other half of the differential comparator (pin 3), a sample of the line voltage from the 15-volt, 60-Hz power supply, and a sample of the 18-volt sweep-derived power supply are mixed. If either supply varies, the resultant square wave output will vary in duty cycle, providing pulse modulation. Actual transfer graphs for typical voltage changes are shown in Fig. 8-34.

Now, refer again to Fig. 8-32. Pin 8 of the chip is marked flyback pulse. This input can stop the square wave from appearing at the output of the PWM at pin 1. It's used to prevent the PWM from turning on the driver and, hence, the horizontal output transistor from turning on during sweep flyback when the collector voltage is high. This input is also used to shut down the system if the high voltage gets too high.

Service caution

In these sets, there are two grounds that use electronic power systems due to the fact that certain portions of the circuitry are ac line-connected while others aren't. Thus, when making checks, be sure that the proper circuit ground is selected, or readings will be incorrect. Also, for your protection, a line-isolating transformer should be used between the 120-volt supply and the set's power plug.

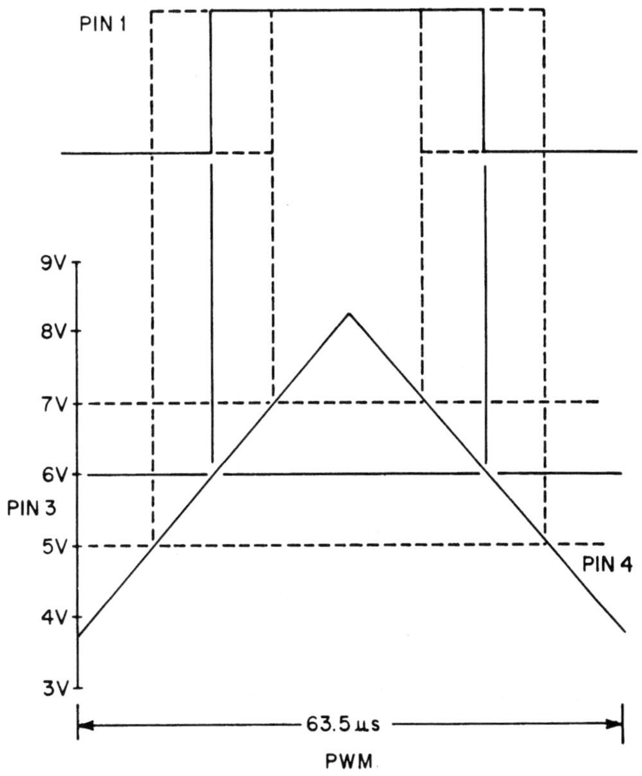

8-34 Variations of transfer characteristics of the pulse width modulator. Zenith Radio Corp.

Safety tips

Please use the following safety tips when you're repairing TV receivers using the electronic power sentry.

1. Don't create solder splashes on the module surface or around plug connections.
2. Maintain physical spacings and barriers between ac line-connected and isolated components.
3. Don't leave any objects such as screws, nuts, or washers loose on the module.
4. Reinstall all barriers and separators.
5. Make certain that all leads are correctly dressed.
6. After repairs, make certain that the back is reinstalled on the set, with all screws in place.
7. Don't try to measure the sweep transformer's ac voltages.
8. Don't attach test instruments to the nonisolated ground of the TV receiver.

Zenith electronic power system (HV shutdown operation)

Before checking out the shutdown system, let's take a look at the block diagram (Fig. 8-35) operation of the Zenith Electronic Power Sentry. In this power system, a line-operated 150-volt dc power supply provides the energy to a horizontal pulse-locked chopper. The chopper regulates the voltage and drives the sweep and high-voltage transformer with a constant pulse. The high-voltage transformer also supplies all the regulated voltages for the TV receiver operation.

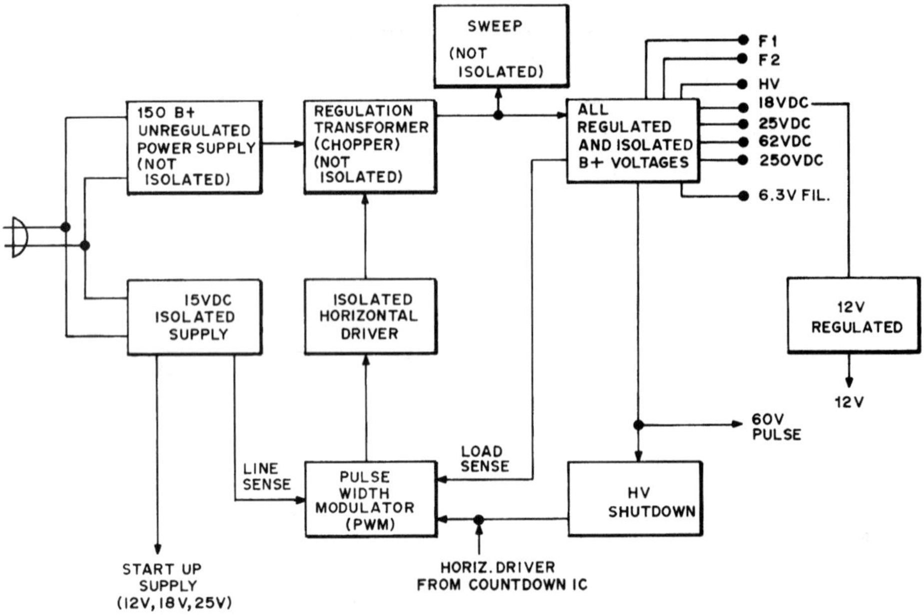

8-35 Block diagram of the electronic power sentry.

The "key" to this regulation system is the pulse width modulator (PWM). The line and load voltages are sensed by the PWM, which controls the chopper rate by varying the "on time" of the chopper, and therefore, the supply voltage.

HV shutdown operation

The HV shutdown block (see Fig. 8-35) is used for X-ray protection and operates only if the high-voltage increases by 4 to 6 kilovolts, or if an excessive load (current) exists. Its effect is to shut down the entire receiver by causing the horizontal stages to be inactive.

The high-voltage shutdown circuit, as shown in Fig. 8-36, receives an input from the sweep transformer, the 60-volt pulse that's used as a high-voltage reference.

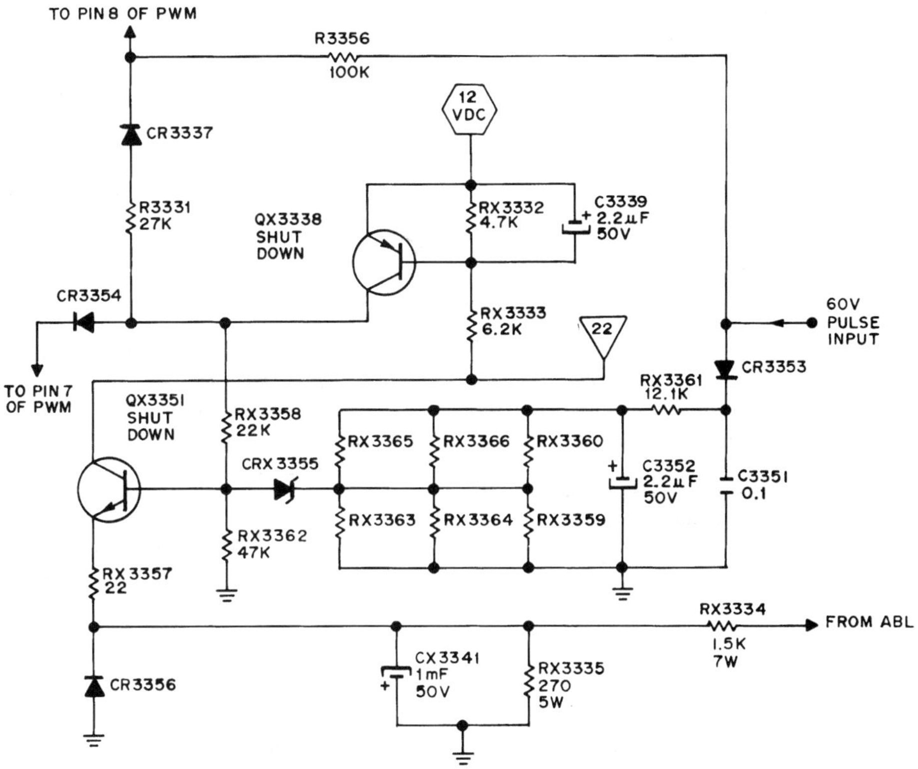

8-36 The high-voltage shutdown circuit.

There are two outputs, either of which can trigger the TV receiver into its shutdown mode.

Again referring to Fig. 8-36, the 60-volt pulse is rectified by diode CR3353 and charges C3351, which discharges through a multiresistor divider (factory-selected values). If the voltage across the zener diode, CRX3355, reaches the zener breakdown level, the diode will conduct. When this occurs, the QX3351 transistor will turn on, also turning on QX3338. The two transistors form a positive feedback switch, which will saturate and stay on once it's triggered. Should this happen, pin 8 of the PWM is deactivated via CR3337, together with pin 7 of the PWM and the "H" output of the countdown IC through diode CR3354. While that action will shut down the entire power sentry system, the 15-volt isolated power supply will maintain the shutdown potentials.

The Zenith "Z" chassis

The 9-186 module in the "Z" chassis contains the horizontal driver circuits, which are the same as found in the 9-160 series power sweep modules. Also, the high-voltage shutdown circuit is basically the same in operation. (Refer to the block diagram shown in Fig. 8-37.)

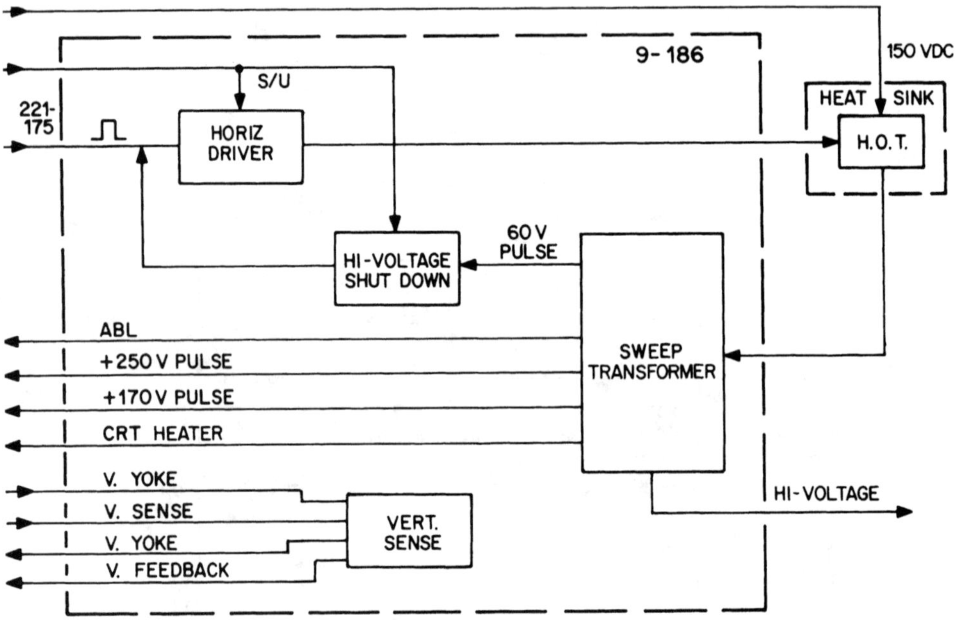

8-37 The horizontal assembly.

The shutdown circuit doesn't clock the PWM circuit. The shutdown circuit consists of three transistors. The standard two are used for the latch, and the third is added to clamp the low-impedance base drive to the horizontal driver transistor 121-1037.

The "Z" chassis has a new sweep transformer that's much smaller in size. This was accomplished with a new winding technique in the high-voltage winding. The coils are wound in opposite directions to reduce stray capacitance, but they're connected in a series-aiding configuration. This method allows the use of only two high-voltage diodes in the sweep transformer.

High-voltage shutdown checks

In order to test out the high-voltage shutdown for the "Z" chassis, go to the 9-181 main circuit board and look for test point #33. The TP #33 is connected to ground through a 100K potentiometer (95-volt adjust pot). Test point #33 is actually pin 6 of the 221-175 vertical/horizontal countdown chip. As the resistance of the pot is decreased, the duty cycle will increase until the high-voltage shutdown circuit is activated. Should the set operate for an hour or more and shut down for no reason, measure the 95-volt TP and reset if not correct. This simple check will eliminate many intermittent problems on the "Z" chassis.

The high-voltage shutdown circuit is shown in Fig. 8-38. If you suspect trouble in the shutdown circuit, look for leakage in the transistors and diodes or resistors that have changed values.

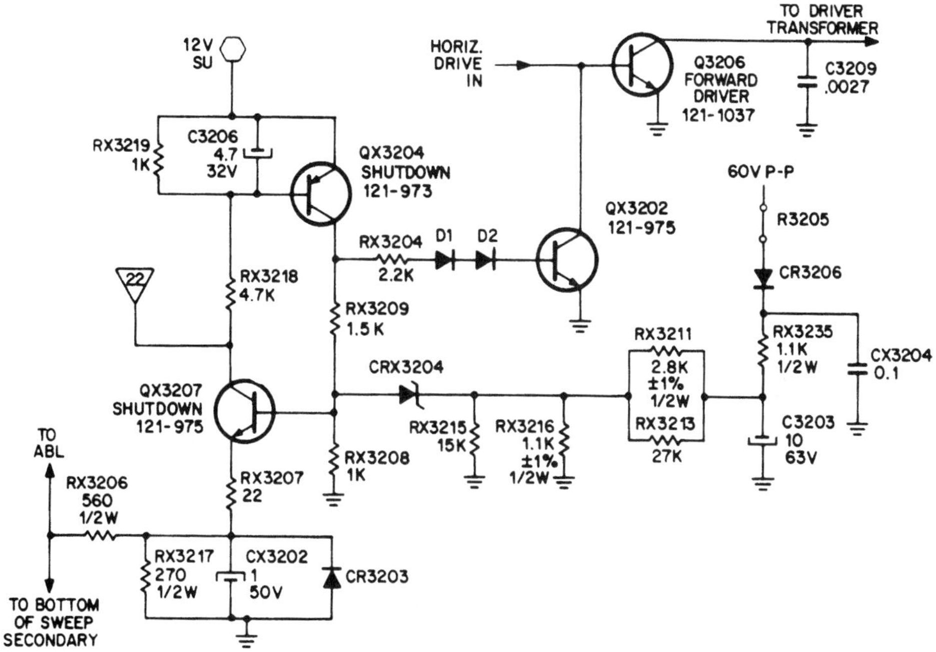

8-38 High-voltage shutdown circuit.

RCA CTC-108 shutdown circuit checks

Should one of these chassis not operate, but the oscillator works on the start-up dc voltage and no other faults can be found with the high-voltage sweep system, suspect a shutdown circuit problem. Keep in mind that nothing can unlatch the shutdown mode except turning off the power to the set. Two different sets of conditions can activate this shutdown mode latch: One is too much high-voltage current; the other is too much peak pulse amplitude from the high side of the HV sweep transformer. Note that diode CR409 obtains horizontal sweep pulses from winding pin 10 of the flyback transformer and rectifies these pulses.

One technique for finding out which component is the troublemaker is to measure the voltage drop across R433, a 1000-ohm resistor. Refer to Fig. 8-39 for the shutdown circuit. When the circuit is latched in the shutdown mode, you should measure between 1 and 2 volts dc across R433. For normal set operation, the voltage drop should read zero.

Now, should these checks indicate that shutdown has occurred, you now have to determine if the problem is caused by overcurrent, high voltage, or a shutdown circuit defect. Should excessive CRT current be the problem, turn off the set for a few seconds, which will unlatch Q413 and Q414. Now ground the Q415 collector and turn the set back on. The set should now operate if the CRT current or a fault in the Q415 circuit is the problem.

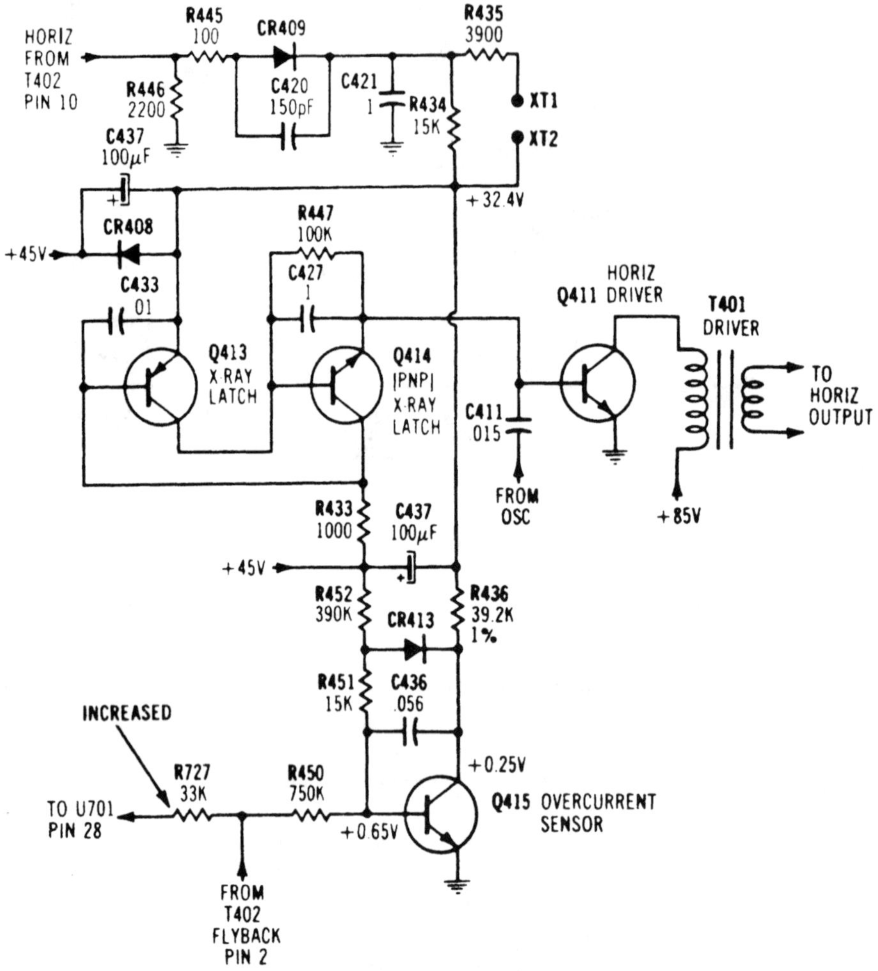

8-39 RCA CTC-108 high-voltage shutdown circuit.

Should the set be in shutdown because of excessive high voltage (high sweep transformer pulses), lower the ac line voltage with a Variac and then short together the base and emitter of Q413. If the set now operates, this indicates that excessively high voltage is the probable cause of the shutdown condition.

Zener diode CR407 has caused erratic shutdown by reducing the Q413 base voltage and making the shutdown action too sensitive. Resistor R436, should it increase in value, might cause the set to not start up at all. Increased resistance of R727 will affect the operation just as though the high-voltage current is excessive. It's a good habit to check the value of all of the resistors in the shutdown circuit.

Make sure you check out the shutdown circuit for proper operation regardless of what service was performed on repair of the chassis. Check these receivers by shorting together the XT1 and XT2 test points, which will activate the normal shutdown if all systems are correct. However, should the shutdown not occur, repair the circuit so it will go into the shutdown mode.

9

The NT64 NTSC color pattern generator and VC63 VCR test accessory

THE NT64 PROVIDES TWO SPECIAL VIDEO PATTERNS THAT OPERATE EXACTLY THE same as the patterns generated inside the VA62A (except VIR can't be added with the VA62A "VIR adder"). These patterns are phase-locked and sync-locked to all of the other VA62A signals. After connecting the NT64, simply move the VA62A video pattern switch to the "external modulation" position any time you want to use the NT64 patterns. The switch on the NT64 selects between the full-field and split-field patterns.

The NT64 receives its power from the VA62A. The NT64 will be turned off and on by the power switch on the VA62A. An LED indicator light on the NT64 shows when it's receiving power. Check the cable connectors if the LED fails to light when the VA62A is turned on.

The NT64 produces the two standard EIA color bar patterns: the full-field color bars and the split-field color bars. Simply move the NT64 function switch to the position that corresponds to the pattern needed for a particular test.

The NT64 patterns

The two patterns supplied by the NT64 were designed by the Electronic Industries Association (EIA) for use in broadcast applications involving color cameras, tape decks, etc. The patterns consist of seven bars. Three bars are the primary colors (red, blue, and green). Three bars are the secondary colors (yellow, cyan, and magenta). The seventh bar is grey (75 percent of peak white).

A secondary color is made of equal amounts of two of the primary colors; yellow is red and green, cyan is green and blue, and magenta is red and blue. White (grey) is made of equal amounts of all three colors. The "full-field" pattern consists only of these bars, extending from the top to the bottom of the screen. I'll discuss the details of the "full-field" and the "split-field" patterns later in this chapter.

Use with a vectorscope

The NT64 color patterns are generally used with an NTSC vectorscope. Note: A standard oscilloscope with a "vector" input is not an NTSC vectorscope. This is a special instrument that separates the chroma from the composite video signal and then graphically represents the color saturation (amplitude modulation) and the color tint (phase modulation) of the chroma signal. The vectorscope screen has markings (small boxes) for each part of the split-field color bar pattern. The boxes show the correct position of the vectors for the color burst, all six of the primary and secondary color bars, and the –I and Q bars. The color saturation and tint are correct if the vectors fall inside the correct boxes on the vectorscope graticule. Note this pattern shown in Fig. 9-1.

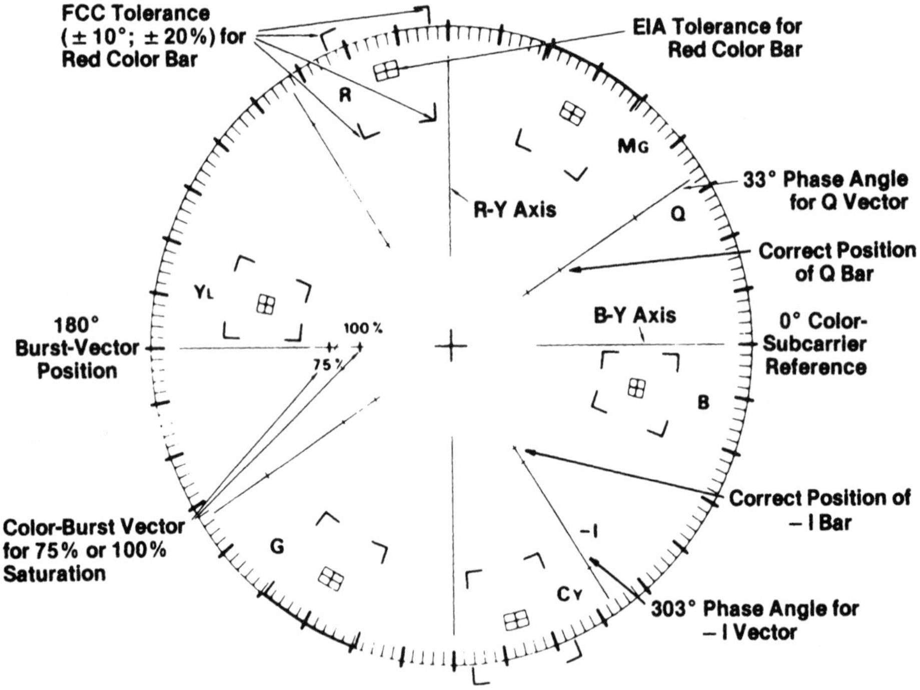

9-1 The graticule of an NTSC vectorscope has a box for each part of the split-field pattern. Each vector is within FCC tolerance if it falls in the large outside box, and within EIA tolerance (RS189) if it falls into the smaller box. The cross in the center of the small box represents "zero" error.

Most applications requiring a vectorscope involve video devices (such as cameras and tape recorders) that have controls that affect color phase and amplitude. Most broadcast equipment, for example, has controls that affect one or both parts of the color signal. These controls are adjusted until the dots representing the color signals on the vectorscope fall inside the corresponding boxes.

Other vectorscope applications involve recording a signal and then playing it back to see if the color signals are distorted in the recording/playback process. The applications of a color generator, in this case, are only in the recording circuits, as any playback adjustments should be made using the color bar pattern from a reference tape recorded on a perfectly adjusted recorder. After the playback circuits are properly adjusted, it will be necessary to record a segment of tape, and then play it back. The color phase or amplitude control in the recording circuit is then turned a small amount, and a new test is recorded and played back. The results on the vectorscope indicate whether the adjustment made the signal better or worse. This is repeated until the control is within the correct tolerance.

Color frequency response

The EIA color bar patterns produced by the NT64 (or any other "NTSC" color bar generator) test the color circuits over a narrow band of frequencies. Some technicians and engineers believe that the color circuits are tested over their full frequency range because the pattern includes an I signal, and the I signal has an ideal frequency response of 1.3 MHz. While it's true that the I signal has a wider allowable bandwidth than the Q signal, the color circuits of a monitor don't require 1.3 MHz to pass the I signal of the EIA color pattern. (See Fig. 9-2.)

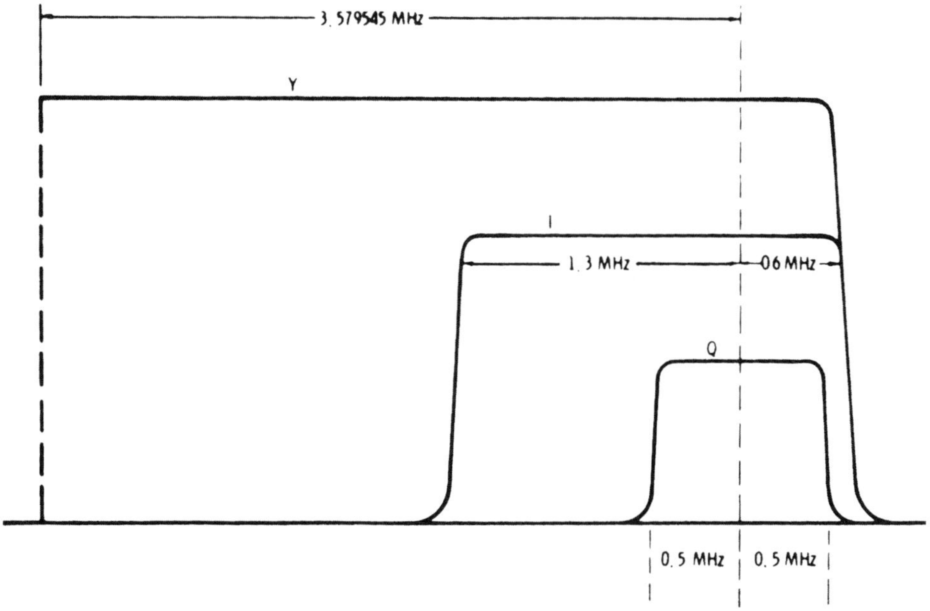

9-2 A broadband monitor allows up to 1.3 MHz of chroma response for the I channel, but only 0.5-MHz response for the Q channel. This allows smaller objects of colors (represented by +I or –I signals) than a standard receiver, which has only 0.5-MHz response.

The reason for this is that the bandwidth of the color circuits in a receiver or monitor determine how small an object can be in the picture and still contain color information. Smaller objects require higher bandwidth. The I signal has a maximum frequency response of 1.3 MHz, compared to only 0.5 MHz for the Q signal. Colors represented by the I signal are reddish-orange (+I) and greenish-blue (–I). Colors represented by the Q signal are reddish-blue (+Q) and yellowish-green (–Q). Others colors are made of a mixture of the I and Q signal, meaning the maximum frequency response is somewhere between 0.5 and 1.3 MHz for other colors. Thus, a monitor with sufficient chroma frequency response can show color on objects represented by a +I or –I signal that are between one-half and one-third the size of objects represented by a +Q or –Q signal.

If, however, the I and Q signals represent fairly large objects in the picture, they both will show color, whether or not the monitor has a full 1.3 MHz I frequency response. In the case of the EIA color bars, both the I and the Q signals (as well as all the other color bar signals that contain both I and Q components) represent objects the size of one-sixth to one-seventh the width of the picture. The frequency response needed to carry objects this large can be determined by multiplying 9,531 (the horizontal sweep frequency of 15,734 Hz times 1.2, to compensate for retrace time, divided by 2, because the bars only show an "on" time, and no "off" time) by the reciprocal of the fraction of the screen covered by the bar (7 for the top part of the split-field pattern, or 6 for the lower part of the split-field pattern). The results are that only a 66.7-kHz bandwidth is required to pass the NTSC full-field color bars correctly, and only a 57.2-kHz bandwidth is needed for the I and Q portions of the NTSC split-field pattern.

The VA62A chroma bar sweep pattern, on the other hand, requires a full 500-kHz bandwidth to pass the 3.0- and 4.0-MHz bars at full amplitude. Use this pattern any time chroma frequency response is in question. Even though the chroma bar sweep doesn't test the full 1.3-MHz bandwidth needed for a wideband color monitor, it tests the color circuits over a wider frequency range than any other test pattern.

NT64 "NTSC" color patterns

The optional "NTSC" color patterns might be helpful to you should you have trouble relating scope waveforms to those shown on VCR service information. The NT64 "NTSC" Video Pattern Generator supplies a standard EIA split-field or full-field color bar pattern to the VA62A's external modulation jack. This pattern (see Fig. 9-3) is phase-locked and sync-locked to all of the other VA62A systems signals, so it can be used like any of the internally generated patterns.

Of course, you can feed your own "NTSC" generator into the VA62A instead of using the NT64. Any standard 1-volt peak-to-peak, 75-ohm signal source can be fed into the VA62A's external modulation jack. Thus, any "NTSC" generator will modulate the VA62A RF and IF signals, the signals supplied from the VA62A's VCR output jack, the video pattern position of the drive signal section, and the companion VC63 tester. Any other generator, however, will not be phase locked to the other VA62A drive signals or to the 30-Hz servo drive output because the video pattern must be

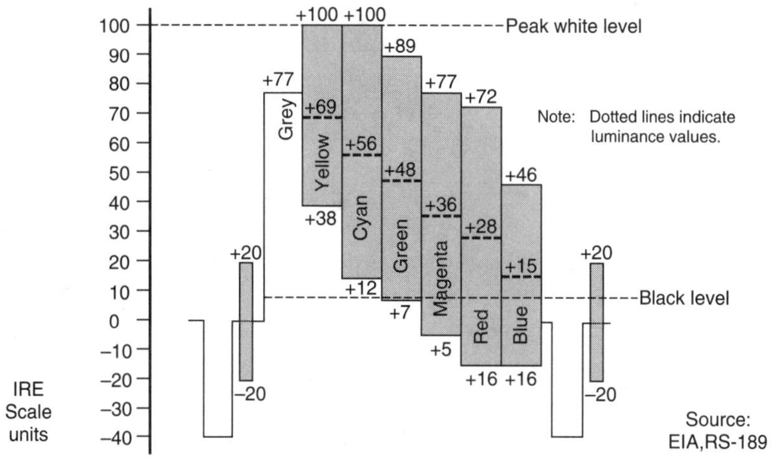

9-3 Each color bar has a specific luminance level plus a specific chroma level. The peaks must not extend higher than a peak white bar.

generated inside the VA62A to hold all the other signals in sync. The NT64 offers the advantage of providing all phase-locked and sync-locked signals, just like any of the internal VA62A video patterns.

"NTSC" color pattern comments

TV-station engineers use a number of test patterns for checking and aligning their station equipment. This will include everything from the camera to the video tape units, special pattern generators, etc. One test pattern, the "NTSC" color pattern (Fig. 9-3) is used for a majority of this broadcast work. This same pattern is also the fundamental pattern used on most manufacturer's VCR schematics and test/alignment tapes.

The pattern shown in Fig. 9-4 is the scope waveform of the "NTSC" pattern drawing in Fig. 9-3. This pattern consists of seven bars; six color bars and one white bar. Each bar has a different luminance level, ranging from black to white. In addition to the luminance levels, the six color bars have a 3.58-MHz color carrier. The carrier phase is different on each bar and therefore produces a different color on each bar, including the three primary colors (red, blue, and green) and the three secondary colors (yellow, cyan, and magenta). The carrier phase remains the same across the entire bar width, so the color remains constant across each bar. The broadcast engineer almost always uses this pattern with a special type of oscilloscope called the "NTSC" vectorscope.

Luminance circuit checks and alignment can also be performed with this pattern. The luminance "stair-step" portion of the "NTSC" color bar pattern produces seven luminance steps, spaced at equal intervals between pure black and white levels. Any changes in the spacing of these steps indicates a luminance problem, such as compression or clipping.

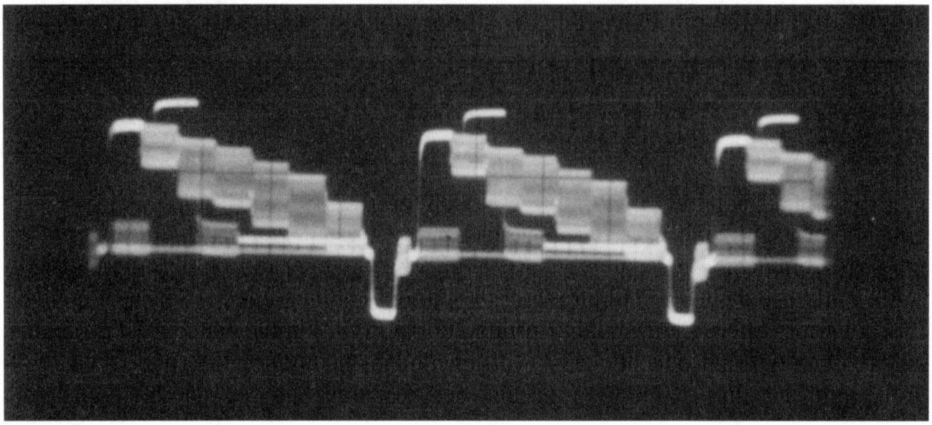

9-4 This is the scope waveform of the "NTSC" pattern drawing shown in Fig. 9-3.

The NT64 patterns

The two patterns from the NT64 generator consist of seven bars, as shown in Fig. 9-3. Three bars are the primary colors (red, blue, and green). The other three bars are the secondary colors (yellow, cyan, and magenta). The seventh bar is grey (75 percent of peak white).

A secondary color is made of equal amounts of two of the primary colors; yellow is red and green, cyan is green and blue, and magenta is red and blue. White (grey) is made of equal amounts of all three colors. The "full-field" pattern consists only of these bars, extending from the top to the bottom of the screen.

Full-field pattern

Each color bar is to be 75 percent saturated. This means that if all three colors were mixed together to make white, the resulting white level would be only 75 percent of the amplitude of a true (100 percent) white signal. This is, in practice, the largest amplitude combination of color signals that can be applied to a TV transmitter without overmodulating it.

Split-field pattern

The split-field pattern is identical to the full-field pattern for the top 75 percent of the screen. The bottom quarter of the screen has a different combination of signals for special tests.

The last half of the adder split-field information is divided into three areas; –I, white, and Q. The right half of this area is pure black. The white bar is 100 percent white, as opposed to the 75 percent white bar on the left-hand side of the color bars. The black area is slightly lower in amplitude than the smallest step of the staircase pattern formed by the luminance content of the color bars.

The two remaining signals are samples of the I and Q subcarriers used to develop the different colors. The I signal is 180 degrees phase shifted from true "I" (thus the term "minus I," which gives it the same amount of blue content as the Q bar for special applications in setting up color studio monitors).

The 75 percent color bar waveform signal that's recorded on VCR alignment tapes is shown in the top drawing of Fig. 9-5, while the color bar pattern produced from the VCR alignment tape is shown on the bottom drawing of Fig. 9-5.

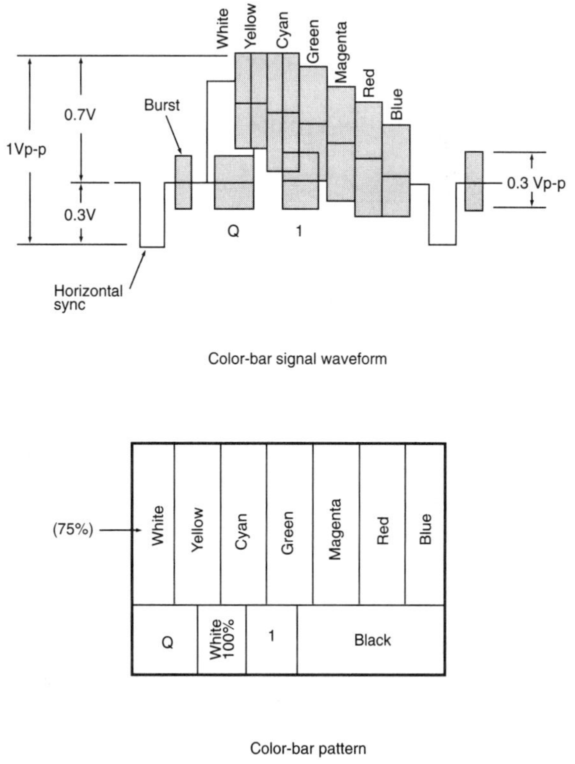

9-5 The 75 percent color bar signal recorded on the alignment tape.

VC63 VCR test accessory

The VA62A brings the speed and effectiveness of signal substitution to VCR servicing. VCRs have two special signals that are quite different from signals found in TV receivers. For this reason, the VC63 VCR accessory to the VA62A is the main signal source for VCR servicing. Let's now look at why VCRs have special signals, what these signals are, where to use the special VCR signals provided by the VC63, and how to use the VC63 signals effectively to isolate VCR problems.

Special VCR signals

Special VCR signals were developed to overcome the limitations of the magnetic tape recording process. A basic problem with using magnetic tape to store information lies in transferring information onto and off of the tape. When signals are read back off of the magnetic tape, the amplitude at the playback head increases as the frequency of the recorded signal increases, as Fig. 9-6 illustrates. This occurs because the amount of flux produced across the playback head gap varies with the wavelength of the recorded signal. For each octave or doubling in frequency of the recorded signal, the output voltage increases by 6 dB (quadruples). At some point, determined by the size of the playback head gap, the output sharply drops to zero.

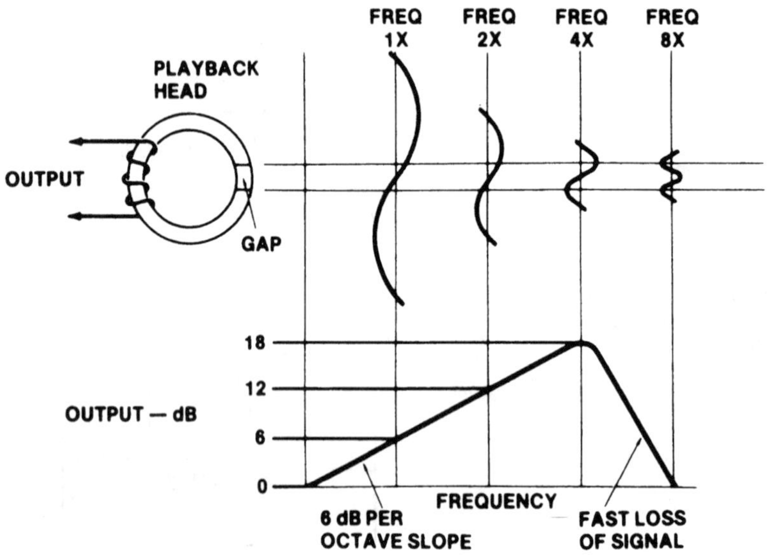

9-6 The voltage induced in the playback head increases at a 6 dB/octave rate as the frequency of the recorded signal increases. When the wavelength equals the head gap, the induced voltage is zero. Sencore

If only a relatively narrow range of frequencies is to be recorded, this limitation can be tolerated. Audio engineers have worked around this problem for many years. Recorded audio frequencies range from about 40 Hz to 20 kHz, encompassing 10 octaves. The 6-dB/octave playback response produces a 60-dB change in output. Although recorders use elaborate equalization circuits to minimize the effects of tape playback response, 10 octaves is the limit for tape recording.

FM luminance signal

The answer to this dilemma is to reduce the octave range of the recorded signal. This is accomplished by using the composite video signal to frequency modulate a carrier. Recording the luminance video information using FM is the first special VCR signal. In this FM conversion, the sync tips are the "at-rest frequency" of the carrier and the peak white video signals are at maximum deviation. How fast the carrier deviates corresponds to the frequency of the video signal. The range of video frequencies modulating the carrier is limited to about 2.5 MHz. This FM signal is recorded directly onto magnetic tape.

Both VHS and beta formats use FM to reduce the number of recorded octaves. With the FM modulation sidebands extending from about 1 to 6 MHz, less than 3 octaves are recorded. Though slightly different FM frequencies are used in VHS and beta, frequencies around 4 MHz were selected as the best compromise between head gap size, writing speed, and bandwidth.

The FM signal offers a further advantage. An FM system is not affected by moderate signal amplitude changes. Thus, small amplitude variations in the FM signal picked up off the tape (caused by erratic tape-to-head contact) do not affect picture quality. During playback, the FM signal is converted back to a standard video signal.

Down-converted color

Using FM to record the luminance solves bandwidth problems, but it creates problems for recording color. The reason is conventional NTSC video signals use phase-reference color information. The phase must be accurately maintained, or portions of the video picture will be the wrong color or continually change in color. All phase information, however, is totally lost in the FM process. The color information in a VCR, therefore, can't be recorded as FM along with the luminance signal. Recording the color information directly as it appears in the 3.50-MHz NTSC signal would cause very noticeable and undesirable interference beats between the 4-MHz luminance and the 3.58-MHz color signals and sidebands.

Because neither FM nor direct recording of color will work, the alternative is to convert the frequency of the recorded color, for example, placing it either higher or lower than the FM luminance frequency. Placing the color frequency above the FM luminance causes problems with the head gap, so the color information is lowered in frequency. This becomes the second special VCR signal—down-converted color.

The color information in a VCR is converted down to about 600 kHz, placing it below the luminance frequency (thus the term "color under"), as illustrated in Fig. 9-7. This signal is still amplitude- and phase-modulated for color saturation and tint. The down-converted color signal is recorded directly onto the magnetic tape using the same heads that record the FM luminance signal. As with FM luminance, the color signal is converted back to the standard video signal during playback.

Down-converted color answers a second problem of recording color information on magnetic tape. As the tape moves through a VCR, variations in tape speed occur that cause severe problems for the crucially phased color signal. During the playback process of converting the 600-kHz signal back to 3.58 MHz, these errors are easily corrected.

Color recording requires one other change to the conventional color signal besides frequency conversion. To reduce color cross-talk from adjacent tape tracks, the phase of the recorded color information is shifted. During playback, color signals in the adjacent tracks have the "wrong" phase and are removed by a comb filter. Thus, only the correct color for each picture element is recovered off the tape.

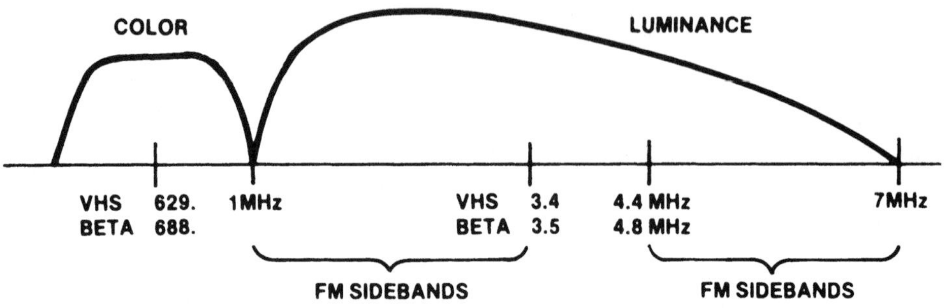

9-7 Color information is converted down to 600 kHz, placing it below the FM luminance frequency. Sencore

Special signals cause problems

The FM luminance signal presents a major problem for VCR technicians who rely on signal tracing. To begin with, the FM signal picked off the tape by the spinning video heads is a minute 500 microvolts—well below the measuring capabilities of an oscilloscope. Yet, unless this signal is good, the VCR will not play back a good picture. Also, you can't determine very much by looking at it, but you can determine a lot with an FM signal with a scope.

Most of the symptoms caused by a problem in the circuits before the FM detector are the same—a raster filled with snow or a raster that's completely blank. How do you troubleshoot the luminance stages between the video heads and the FM detector? After measuring a few voltages and using a little head cleaner, you're faced with the possibility that the heads could be defective. So, to put an end to your doubt, you swap heads. If the VCR is now repaired, the heads were defective. However, what if it still won't work? At least you know that the heads probably were not the problem. Head swapping is a 50/50 proposition, at best.

Using the recorded signal of the tape as the reference when troubleshooting VCRs leaves too many variables. You don't know at what point the signal goes bad, or if the signal is even being properly picked up off the tape. Is the problem electrical, or is the tape path alignment severely off, preventing the heads from picking up any signal? Perhaps the tracking is completely out of whack or the heads aren't switching properly. The answer to these problems is to inject a known-good reference signal into the circuits rather than trying to follow impossible-to-trace signals. This is the same efficient method of signal substitution that the VA62A makes possible for television servicing.

With signal substitution, you inject a good signal into the circuits. If you inject the good signal after the defect, the VCR operates properly, telling you that the circuits from that point forward are working. Then simply trace backward, step by step, into preceding circuits until the VCR stops functioning properly. At that point, you're injecting into the defective stage.

Where to use the VC63

As you've learned, VCRs have special FM luminance and color-under signals. The VC63 VCR Accessory works with the VA62A Video Analyzer to provide these special signals for VCR troubleshooting. Not all the circuits in a VCR require the use of the VC63. The signals after the FM detector are baseband video and are substituted for by using the VA62A. Most of the chroma circuits involve important timing relationships and require the use of an oscilloscope for waveform checks.

For review summation, the FM luminance and down-converted color signals begin at the spinning video heads during playback. The signal picked up by the heads is coupled through a rotary transformer to a head preamplifier. The output from the head amps is switched by the A/B head switcher so that only the signal from the head that's in contact with the tape continues. This keeps the noise picked up by the other head out of the picture. After the switcher, the 4-MHz FM luminance signal is separated from the 600-kHz color-under signal by a combination of high-pass and low-pass filters. Both special VCR signals continue down separate paths until they're converted back to standard NTSC format signals and combined by the Y/C mixer. All of these circuits, from the video heads through to the FM detector and color-under frequency converter, require you to use the special signals provided by the VC63.

You troubleshoot a VCR by first connecting it to a television monitor and playing back a test tape. Playing the tape can reveal one or more important symptoms on the monitor, depending on the defective stage in the VCR. Carefully analyzing the symptom tells you if the problem is in the servo, luminance, or chroma circuits.

Servo symptoms look like picture tearing or rolling. Sound that's "too fast" or "too slow" also indicates a servo problem. Symptoms other than these are called video playback and are caused by any of the nonservo circuits. Video playback symptoms include no video, poor video, no color, and poor color.

To better understand how to use the VC63, let's use a Panasonic PV1225 VCR and the common VCR playback symptom of a blank raster as an example. A portion of the actual luminance block diagram furnished by the manufacturer for this deck is shown in Fig. 9-8. Always troubleshoot from a block diagram and/or trouble tree to help keep you on the right track.

The first step is to bypass the video mute circuit. This circuit is not used on all VCRs, but it mutes the video and audio if the servos aren't locked in. Bypassing the muting allows you to determine if the problem is servo related or video circuit related. In this example, the monitor remains blank. Does this mean the problem is in the direction of the heads, or the video stages?

To determine which way to go, inject a video signal at the output of the FM detector, point (48). The signal here is a baseband video, rather than one of the special VCR signals, so use the crosshatch video pattern on the VA62A, because this pattern is different from any recorded on the test tape. Next, connect the VA62A drive output to the FM detector output. Increasing the drive level just above 0.3 volts brings a picture onto the monitor. This is a very important clue. It means that the remaining video circuits function correctly and the problem lies in the direction of the heads.

The circuits in the path toward the heads require you to use the special VCR signals from the VC63. Because this is a VHS VCR, set the special signals switch on the VC63 to VHS lum, as shown in Fig. 9-8. This setting provides an FM luminance signal without the down-converted color signal (the presence of the color-under signal could cause misleading luminance symptoms). The lum & chroma position provides the spe-

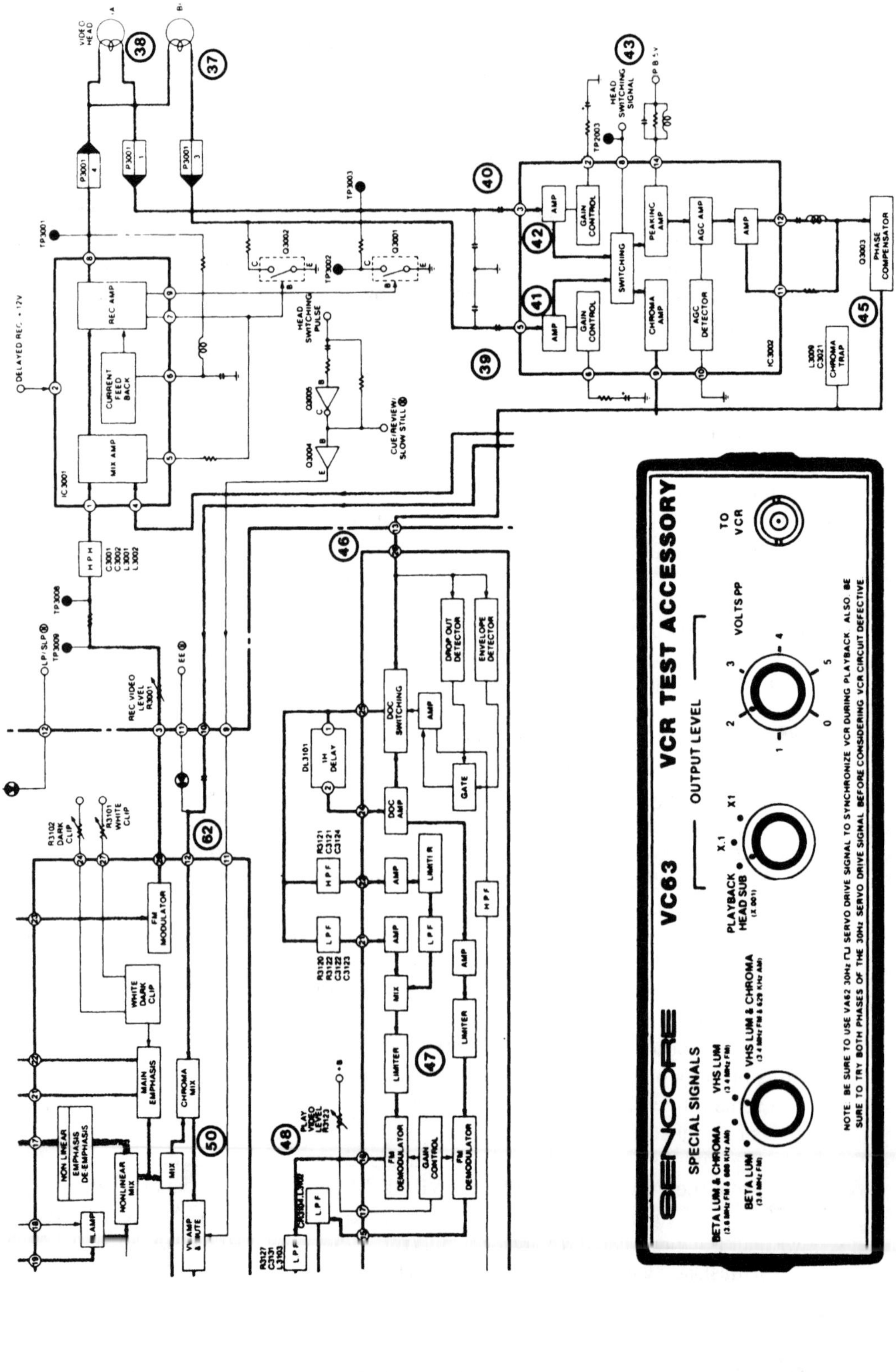

9-8 VC63 VCR test accessory. Luminance playback section of a Panasonic VCR.

cial color-under signal for troubleshooting color circuits. The setting of the VA62A video pattern switch determines the modulation of the VC63 signal that's set to provide a contrasting signal with those recorded on the work tape. This is so you can quickly tell if the picture on the monitor is the tape signal or the VC63 test signal.

Now you want to see which of the circuits between the detector and heads are working, so inject an FM luminance signal into the output of the A/B head switcher, point (45). Injecting here cuts the remaining stages in half, allowing you to quickly narrow down the problem. Point (45) is the combined output of the video heads after the preamplifiers. The substitute signal is supplied by the VC63 with an output level control setting of X–1. Increasing the output level vernier between 1 and 2 in this example brings a pattern onto the monitor. Now you know that all the stages between test points (45) and (48) are also working.

The next logical step is to inject a signal into points (39) and (40). These points are the outputs of the video heads. The signal level here is unamplified and very small. The playback head sub (X.001) setting of the VC63 output level switch provides this low-level signal. With the X.001 setting, some setting of the output level should produce a picture. However, in this example, increasing the vernier all the way to 5 mV doesn't produce a picture on the TV monitor. Can you see that this means the problem lies between points (39), (40), and (45)?

You could further isolate the problem to either the preamps or the A/B head switch by injecting a signal into test points (41) and (42), but as the block diagram for this particular VCR shows, the preamps and the head switch in this VCR are part of IC3002. Before you replace the chip, double-check that it's receiving B+ and the 30-Hz head-switching signal.

What if injecting at points (39) and (41) had returned the picture? This leaves the heads, rotary transformer, or associated connections in question. You can use the VC63 to isolate problems here as well. By using the VC63 and signal substitution to prove what stages work, VCR troubleshooting is reduced to a few, quick signal injections.

10
Servicing video tape recorders (VCRs)

THIS CHAPTER BEGINS WITH A BRIEF LOOK AT WHAT TEST INSTRUMENTS, SPECIAL equipment, and test tapes are required to do a meticulous and efficient VCR service job. Next is a review of videotape recording and operating principles. You'll see how to make your own VCR test tape, and you'll find a thorough explanation of which test patterns from the VA62A Video Analyzer to use in certain troubleshooting procedures. You'll also find ways to use signal injection to isolate RF head-end (tuner), video IF-stage, and coax-switching problems. I also cover record and playback VCR circuits, comb filters, troubleshooting various servo problems, and tips for using the Analyzer and scope for alignment.

The chapter concludes with an in-depth report on using the VC63 VCR test accessory plug-in unit to troubleshoot playback and record video head amplifiers and confirm a faulty video head by injecting a test signal directly into the video heads. I cover both VHS and Beta VCR circuits.

Let's begin with a brief review of the test instruments and tools required for quick and efficient VCR servicing. Of course, the VA62 Video Analyzer and VC63 test accessory top the list for VCR test equipment requirements. Other instruments include a wideband dual-trace scope, digital voltmeter (dvm), frequency counter, transistor tester, and a complete kit of VCR tools for VHS and/or Beta tape machine systems.

Dual-trace scope

Modern service shops will have a dual-trace triggered sweep scope for troubleshooting the advanced electronic circuits of today and tomorrow. One big advantage of a dual-trace scope, compared to a single-trace service scope, is that it allows comparison of two waveforms at the same time. This is very useful for tracing signals from input through to outputs of various stages, but it's also essential for troubleshooting such crucially timed circuits as servo stages, where two signals must be properly timed.

Make sure the scope has a least a 20-MHz bandwidth and a bright CRT trace that will give you good visibility when the trace is expanded to X10. The scope should also have a built-in TV sync separator with triggering circuits for both vertical and horizontal modes. The TV sync mode position should use actual integrator filters to allow solid triggering on the vertical sync and VCR servo pulses. The Sencore SC60 dual-trace scope will meet all of these VCR service requirements. However, if you want to quickly troubleshoot a VCR and perform the required alignment adjustments, the Sencore SC61 with the built-in frequency and digital voltmeter would be an excellent choice.

Frequency counter

If you don't have an SC61 scope, you'll need a stable frequency counter to set the reference oscillators found in the VCR color circuits and for adjusting the servo circuits. The Sencore FC45 counter has a 25 mV sensitivity for measuring low-level signals for direct readings of oscillator frequencies down to 1-Hz resolution. Using the PR50 audio prescaler with this counter allows the adjustment of the 30-Hz control track signals to an accuracy of .01 Hz. This unit also provides a direct readout of the frequency, which is updated every second. There's also a built-in filtering system that prevents false double-counting due to signal noise that's often present in these low-frequency signals.

Digital voltmeter (dvm)

A good digital readout voltmeter is very important for VCR servicing because most of the circuitry is contained on ICs, and voltages to these chips must be correct for proper operation. The dvm is used for adjusting the regulated power supplies to the correct output voltages and for measuring the low-level signals found in some of the chip circuits. There are several Sencore dvms and DMMs for you to choose from.

Transistor-diode tester

You'll find there are several VCR circuits that are controlled by discrete transistors and protected by various diodes. A VCR machine can contain more than 100 transistors, plus ICs that are used for various control and voltage regulation functions. Thus, you need a good in-circuit tester or a go/no-go quick-check transistor checker for VCR troubleshooting. The Sencore TF46 SuperCricket is an in-circuit transistor checker that you might want to consider. The TF46 also includes an automatic gain test to help in grading or matching transistors used in crucial circuits.

Special VCR adjustment tools and test tapes

Figure 10-1 shows the Sony Beta Max VCR tool kit on the left and a Nortronics VCR maintenance cleaning kit. The tool kit contains tension gauges, fixtures, eccentricity gauge, reel table tension fixture, alignment and lapping tapes, metric feeler gauges, head demagnetizer, and a set of metric Allen wrenches. The clean-

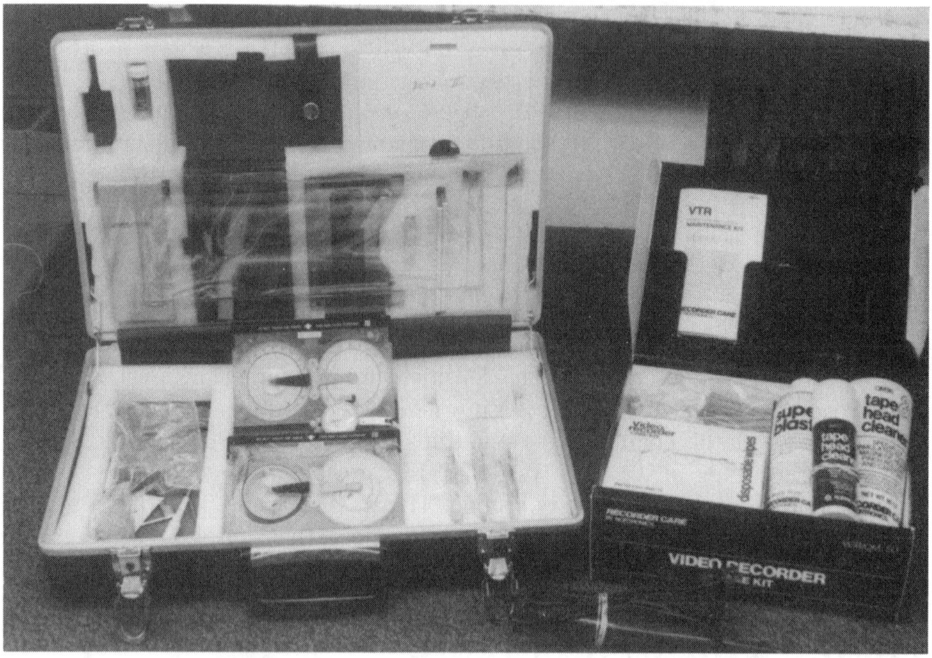

10-1 Sony Beta Max tool kit and Nortronics cleaning kit.

ing kit has head cleaners, swabs, canned air, cassette head cleaner tape, and antistatic cleaning cloth.

VCR magnetic recording fundamentals

Any tape recorder, be it audio or video, must be able to convert an electronic signal to a magnetic signal that's used to magnetize a layer of oxide that coats the magnetic tape. The magnetic tape is pulled past a recording head that is actually an electromagnet. The recording head consists of a U-shaped piece of metal that has a coil wrapped around it. The coil (which receives a signal from some type of amplifier system) causes a magnetic field to build and collapse within the metal in step with the signal applied to its coil. The open end of the U-shaped metal head is a very narrow slit called the *head-gap*. This head gap is the portion of the head that makes contact with the tape. The magnetic flux lines passing between the two parts of the head-gap also pass through the tape that's in direct contact with this area of the head, causing the magnetic particles on the tape to be magnetized. The amount of magnetization is directly related to the amount of current passing through the head coil at the instant it's in contact with the tape.

In audio recording, you find signals in a frequency range of about 20 Hz to 20,000 Hz, while in video, the frequency range is from 30 Hz to 4 MHz, as illustrated in Fig. 10-2. This difference presents several problems when you consider the characteristics of a recording head and the associated system. You'll find that the head gap imposes the greatest limitation (influence) on a system for recording the video spectrum.

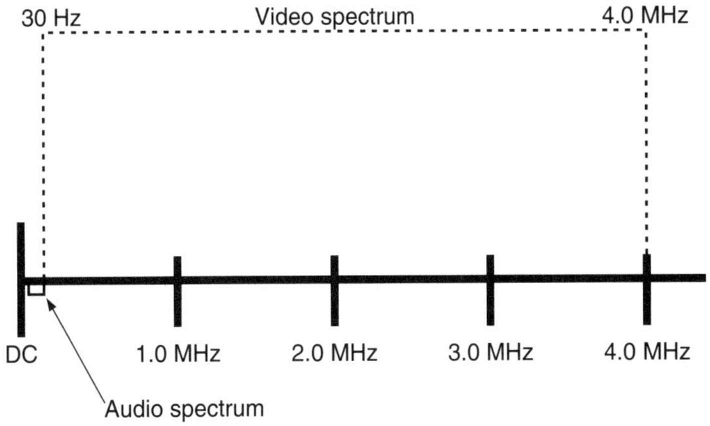

10-2 Video frequency spectrum chart.

The relationship between the width of the gap in the record head and the length of the wave on the tape is reflected in the voltage output and indicates that output is maximum when the gap is one-half of the wavelength and there's a point beyond which the system is ineffective.

Now, if you apply audio recording principles to a video signal of 3.0 MHz and a record head with a gap width of .0025 mil, this will provide maximum output at 3.0 MHz. However, should you use a .3-mil head and increase the tape speed sufficiently to record 3 MHz (which is possible), you would use 1800 inches of tape per second, which is just not acceptable.

A better way to increase the head-to-tape speed is to move both the head and tape, rather than just moving the tape past a stationary head. A rotary head system is used to increase the relative speed between head and tape. In conjunction with a selected head gap (narrowed), it's possible to record and play back the high video frequencies.

This technique can thus provide the required head-to-tape speed and still provide an acceptable length of recording time without consuming a large amount of tape. Not only must the higher frequencies be recorded, but remember that a wide range of video frequencies must be recorded.

The technique usually used (currently) is to impress a carrier on the tape and modulate the frequency of the carrier with the video signal, as illustrated in Fig. 10-3. Carriers are usually found in the 2-MHz to 5-MHz range. The one shown in Fig. 10-3 uses a carrier frequency of 3.4 MHz, with sync tips appearing at 3.4 MHz and peak whites at 4.4 MHz. Also shown in Fig. 10-3 is a fairly common color record system. Note that, before recording, the chroma information is down-converted to a band in the 629-kHz area of the spectrum. This basic system solves the problem of output differential for frequency response in the 4-MHz video range.

Mechanical tape recorder system

The first working video tape recorder was developed for the TV broadcast industry. It had very tight specifications and, after a few modifications, developed into a video

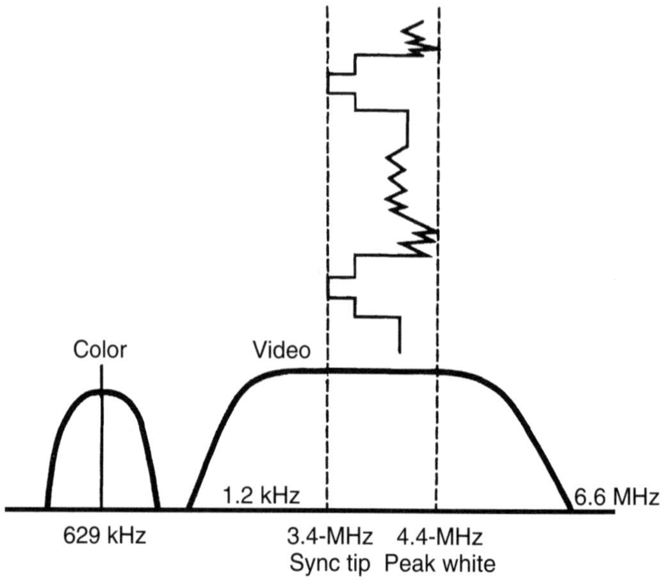

10-3 Typical video spectrum recorded.

taping system that's in use today. Much of the TV programs today are coming from this type of machine system. The system uses four record heads mounted on a wheel that rotates so that the heads scan tracks across the tape, similar to the simplified paths shown in Fig. 10-4. The system uses a two-inch-wide tape traveling at a speed of 15 inches per second. While this system does an excellent job, it's much too complex and expensive for home use.

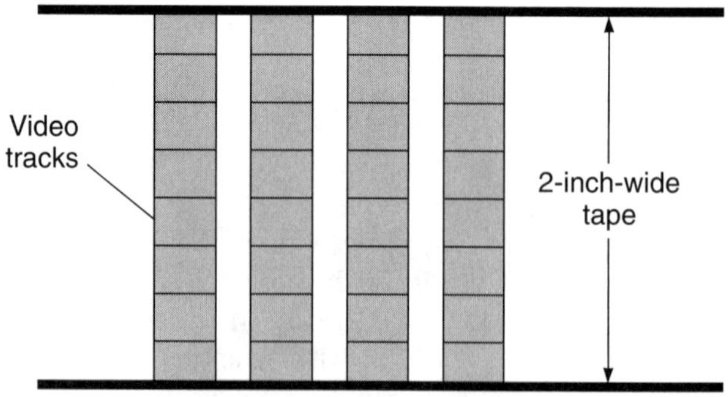

10-4 Track path of Quad-Head recorder.

The quad-head machine records about 16 lines of video on each track, and 16 tracks are required for one complete field of a TV picture (262 plus lines). For the consumer market, it would be more desirable to have a way to record one complete field on one track. Slowing the headwheel down would not work, as the head to tape speed would be too slow to provide adequate frequency response.

Thus, the helical scan system was developed for consumer home use. Instead of four heads traveling across the tape at almost right angles, the tape is wrapped around a rotating drum containing two heads and traveling at an angle to the plane of head rotation as shown in Fig. 10-5. By selecting the proper video head diameter, degree of tape wrap, tape speed, and head speed, it's possible to record a long-enough track (across the tape at an angle) to hold a complete TV field and still have proper head-to-tape speed. One helical scanning concept used a "full" tape wrap around the drum (see Fig. 10-6) and used one record head. Later-model machines use a half-wrap with two heads. Several different techniques are used to form the head-to-drum tape path and/or tracking.

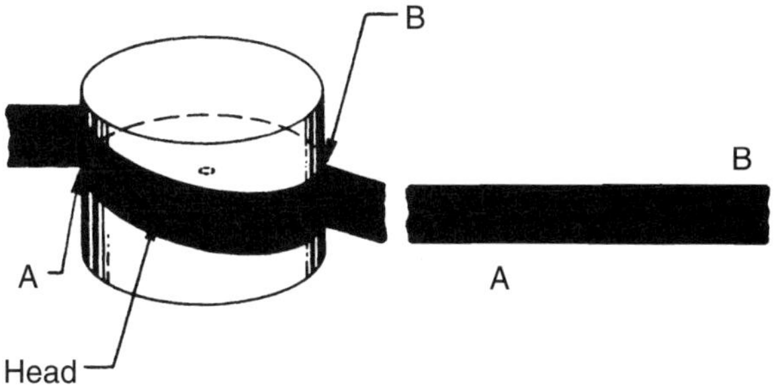

10-5 Helical scan recording track.

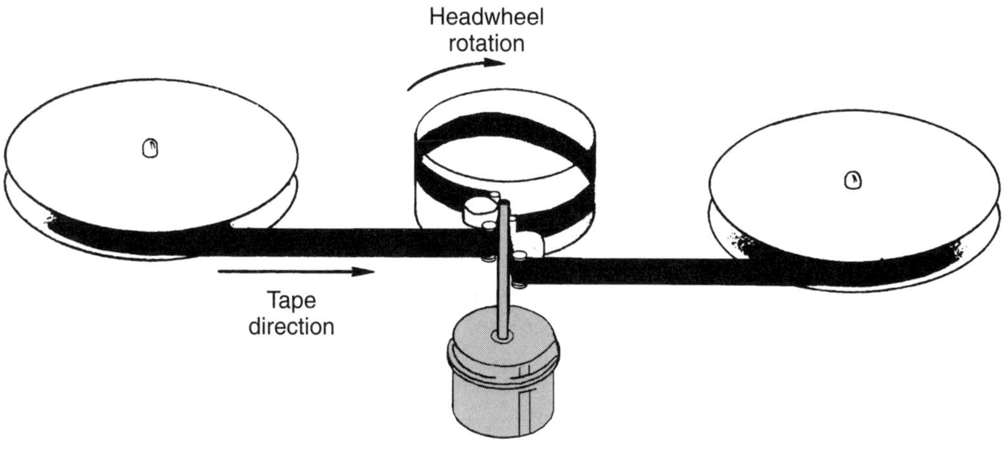

10-6 Full-wrap helical scanning concept.

A typical head drum arrangement is shown in Fig. 10-7. The record heads are positioned 180 degrees apart and are made to protrude slightly beyond the surface of the drum. The energy is transmitted to (or from) the heads through the use of a rotary transformer as an integral part of the head drum assembly. The drum rotates at 30 RPS so that each head records a complete field, and a complete rotation of the head drum records a full frame (2 fields) of TV information (525 lines).

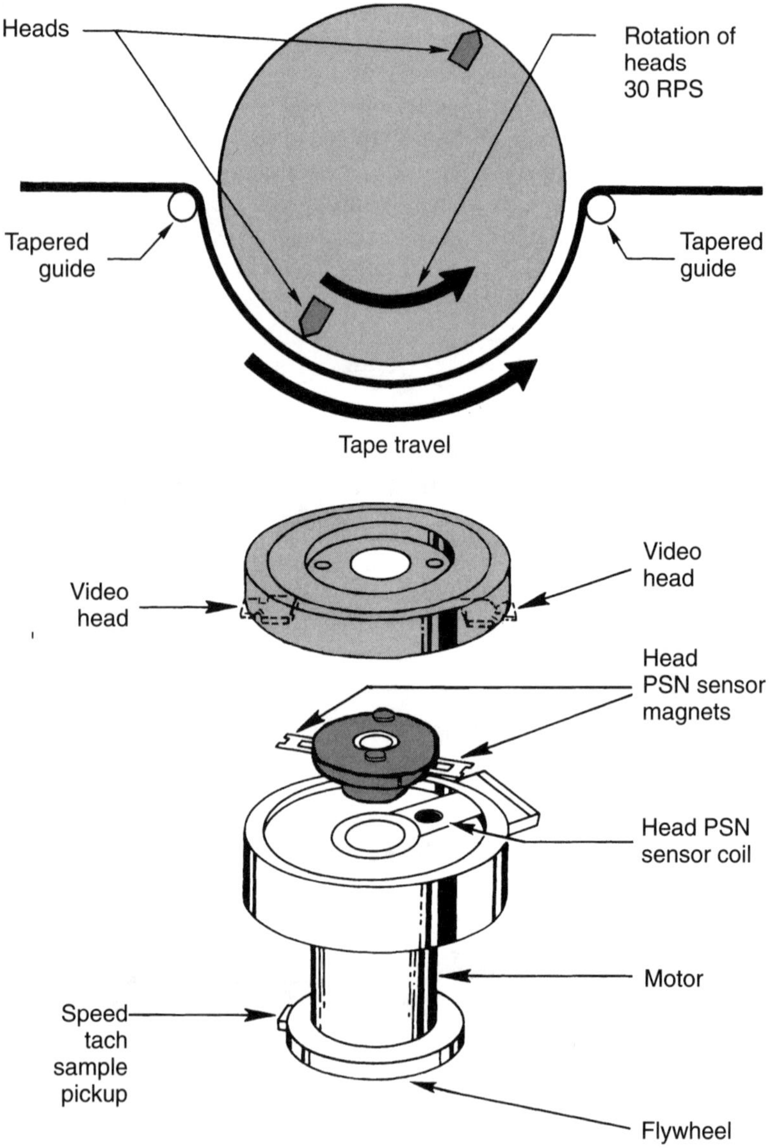

10-7 Exploded view of head drum assembly.

Now this recording is put down on the tape, as seen in Fig. 10-8A. Note that each track has one TV field that consists of 262½ horizontal lines with all other information such as luminance signals, chroma signals, burst pulses, sync pulses, etc. The sound carrier is demodulated in the system and is recorded on the tape as audio, which is shown in Fig. 10-8B. The control track, also shown in Fig. 10-8B, consists of pulses generated and recorded by the VCR system and used to keep the pick-up heads properly aligned with the recording on the tape during playback. The various heads required to do all of this are positioned as shown in Fig. 10-9. As with audio

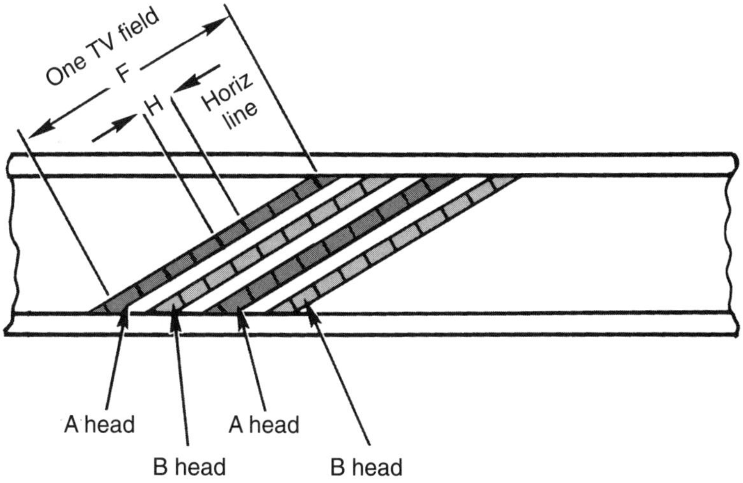

Each head records one field ($262^1/_2$ lines)

(A)

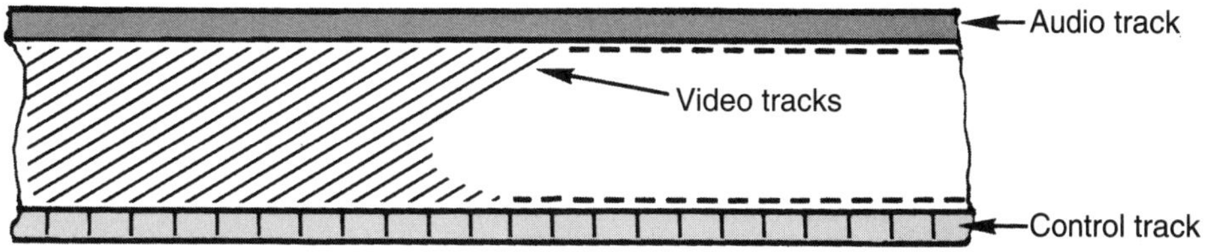

Simplified track(s) placement on helical scan system

(B)

10-8 A&B The track locations on helical scan systems.

tape recording, the tape passes over a full erase head to remove any residual magnetism that might be on the tape, passes the rotating video heads where the picture information is recorded, then moves to the other head unit (two) where the audio and control pulses are recorded. In the playback mode, the same video, audio, and control track heads are used to recover the signals from the tape.

Some of these helical VCRs use three motors—one to drive the video recording head, one to provide the capstan drive, and another to load and unload the tape into and out of the mechanism.

The video head is coupled directly to the motor, and this is called *direct drive* (DD). (Refer to Fig. 10-7.) Since both the speed of rotation and the head position (with relation to the playback tracks on the tape) are crucial, the video head motor is controlled by a servo system.

The tape drive motor is also controlled by a servo system, driving the tape via a capstan mechanism. This motor also provides drive for the take-up reel through a

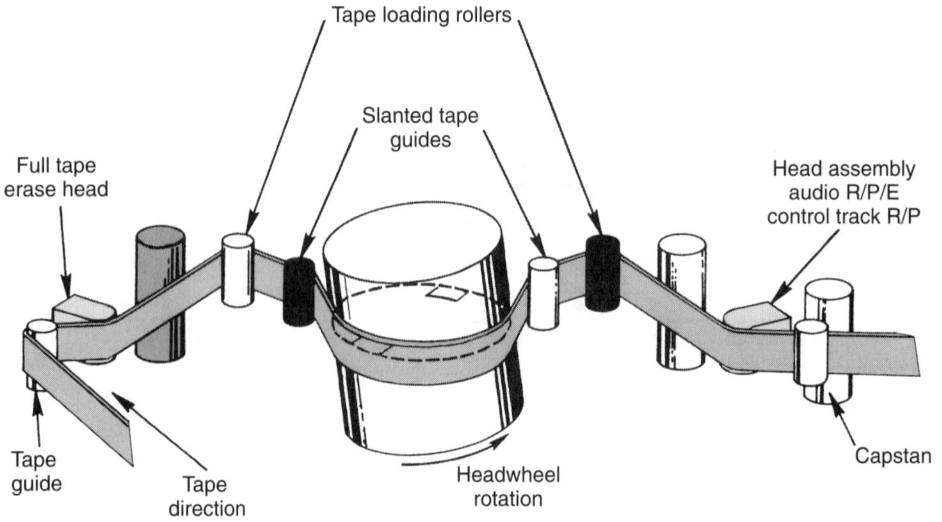

10-9 Head layout for a helical scan VCR.

slip-clutch mechanism similar to that found in an audio tape recorder. The load/unload motor is used to operate the cassette tape load/unload mechanism and, of course, requires no servo control.

A simplified block diagram of the electronic system in a two-head VCR is shown in Fig. 10-10 and indicates some of the complexity involved in a VCR system. As noted previously, the heads are used for both record and playback.

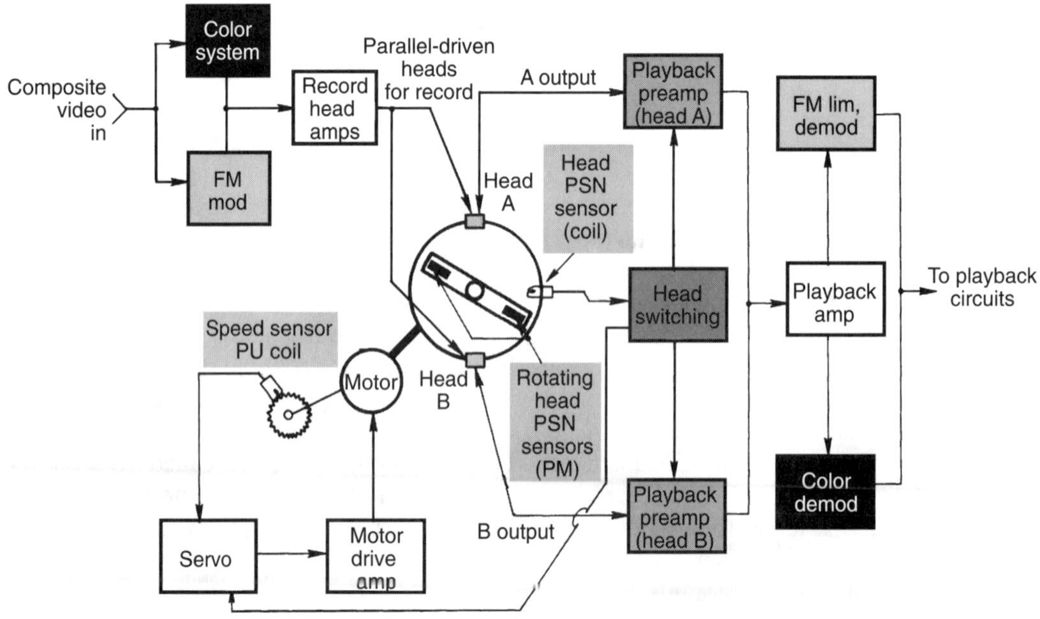

10-10 Interface of VCR play/record electronics.

The video head motor (which in this machine is called the direct drive, D-D, cylinder head) is shown in Fig. 10-7. The video head drum (two heads) mounts on top of the cylinder, and the flywheel is on the bottom. In this machine the head assembly can be easily removed/installed by taking out two retaining screws and disconnecting four wires.

The speed sensor system is located on the bottom of the head drum. On top of the video heads you'll find the head position sensors—consisting of the pickup coil and polarized magnets (on the rotating arm). This transducer gives a negative or positive output pulse for head identification and position control of the video heads. (The recorded track is traced and played back by the same video head that made the original recording.) Some videotape recorders also use a servo for the capstan drive system.

VCR signal crosstalk

Proper video playback is possible only when a single recorded track is played back at one time. Engineers must be careful in the design of a videotape format to prevent the video heads from playing more than one signal track at a time. If part of a second signal path is picked up during the tape playback, a condition known as *crosstalk* will be present. Crosstalk will show up in an audio recorder that has an improperly adjusted head that plays back the desired signal and part of another signal on the tape. Video crosstalk will cause a noisy picture.

Two systems are used to prevent video crosstalk. The first system uses "guard bands," which are blank areas between each track of video information. This system is used in the Sony U-Matic and other reel-to-reel formats. The disadvantage for the use of guard bands is that no information is recorded on the guard bands, which of course uses more tape per minute of operation. Eliminating these guard bands increases tape efficiency by "squeezing" the individual stripes of video information closer together.

Guard band elimination

With audio tape recorders, you have probably encountered a loss of high-frequency response due to improperly aligned playback heads. A tape head will provide its highest signal output if the angle of the gap of the playback head is exactly the same as the recording head. An audio head is aligned by adjusting the position of the head with the head "azimuth" adjustment, which tilts the head position slightly. Both the Beta and VHS VCR formats use two video heads located 180 degrees apart on the head drum assembly. Some of the newer machines with special effects have three and four heads. Two heads are used to allow one head to be in contact with the tape, while the other head is located on the back side of the video head. The first head records one field of information, and the second head records the next field of information as the heads rotate. Thus, the individual stripes of video information are recorded alternately by the two heads.

The gap in each head is set at a 7 degree (6 degree for VHS) slant. The slant of one head is exactly opposite the other. This means that the azimuth angle of each stripe of information differs by 14 degrees from the adjacent stripe. The result of this 14-degree azimuth shift is that each head will pick up the signal that was recorded at the correct angle but reject the crosstalk information recorded at the other angle. Thus, the need for extra tape for guard bands is eliminated.

Video head alignment

There are several reasons that the playback head must exactly follow the same path on the tape as was followed during recording. The first one is that the best signal pickup (resulting in a better signal-to-noise ratio) depends on the playback head following the center of the recorded path. Another important reason is that many VCR users want to be able to use their VCR to play back a tape that was recorded on another VCR.

The head must trace the recorded tracks at exactly the same angle during playback and recording. Should the head take a slightly different angle, the result is an increase in playback noise in either the top or bottom of the picture, as the head begins to lose contact with the recorded signal path. The angle of the recording and playback path is determined by the diameter of the video drum, the path around the drum, the angle that the tape approaches the drum, and the speed of the tape as it passes the spinning video heads. Each of these variables is related to the tape transport mechanism. Proper tape angles require very accurate tape guides. Proper tape speed requires synchronous motors and precise tension adjustments on both the supply reel and take-up reel to prevent the tape from stretching a different amount during recording and playback.

VCR recording system

The heart of the VCR is the video record/playback head. The head must be capable of recording and reproducing signals in the 4-MHz region. Even with the increased head-to-tape speed that the helical scan system provides, the ability to reproduce those frequencies requires a rather small head gap.

The current model video heads are made of ferrite material formed somewhat like that shown in Fig. 10-11. Typical heads are made of material about 1.5-mils thick and have a gap of 13 to 15 microinches.

The video information is impressed on the tape in the form of a frequency modulated (FM) squarewave carrier. Figure 10-12 shows how the magnetic impression would look when the carrier is modulated by a horizontal sync pulse. The frequency at which the pulses occur is a function of the intensity of the video information. Whites are at the highest frequency, 4.4 MHz, and the sync tips (blacks) are at the lowest frequency, 3.4 MHz, as shown in Fig. 10-3. The period of time a given repetitive rate lasts is a function of the horizontal width (or frequency) of the video information.

The block diagram in Fig. 10-13 illustrates the way in which the video information is broken down before it's recorded. A standard video signal is fed to the input amplifier. The gain of this stage is controlled in order to have a stable output. The signal then passes through a low-pass filter to remove any information above 3.8

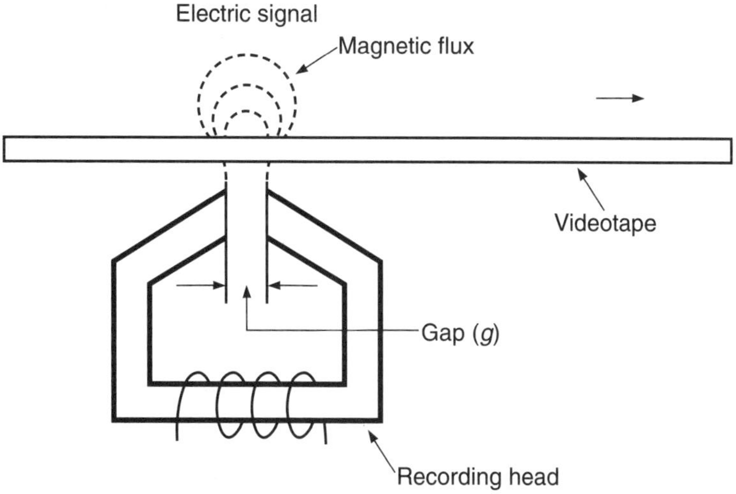

10-11 A typical video recording head.

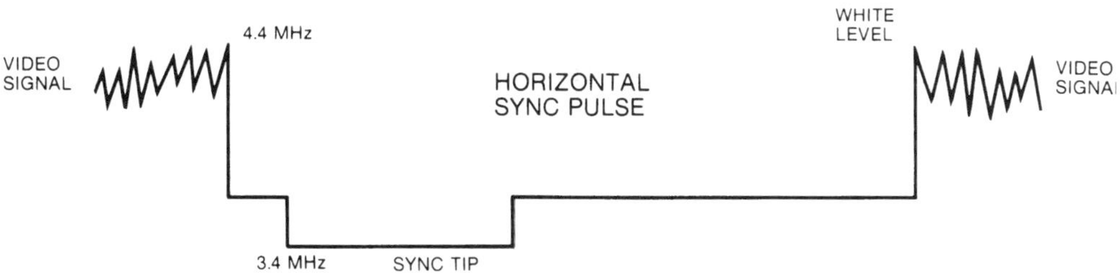

10-12 FM video information recorded on the tape.

MHz, and the clamp circuit restores the dc level of the signal, which is referenced to the sync tips. A preemphasis circuit then emphasizes the high-frequency response and a clipping circuit limits the amount of overshoot developed at both extremes of the intensity spectrum. The signal is then fed to an FM modulator, then to a record amplifier, which drives (in parallel) the record heads through rotary transformers.

VCR playback system

In the playback mode, the process is the reverse of record in that the heads now pick up the information recorded on the tape and it's fed to the playback circuits. One difference is that the heads aren't in parallel. Each head is connected through its own rotary transformer to its own video preamplifier. These amplifiers are switched on

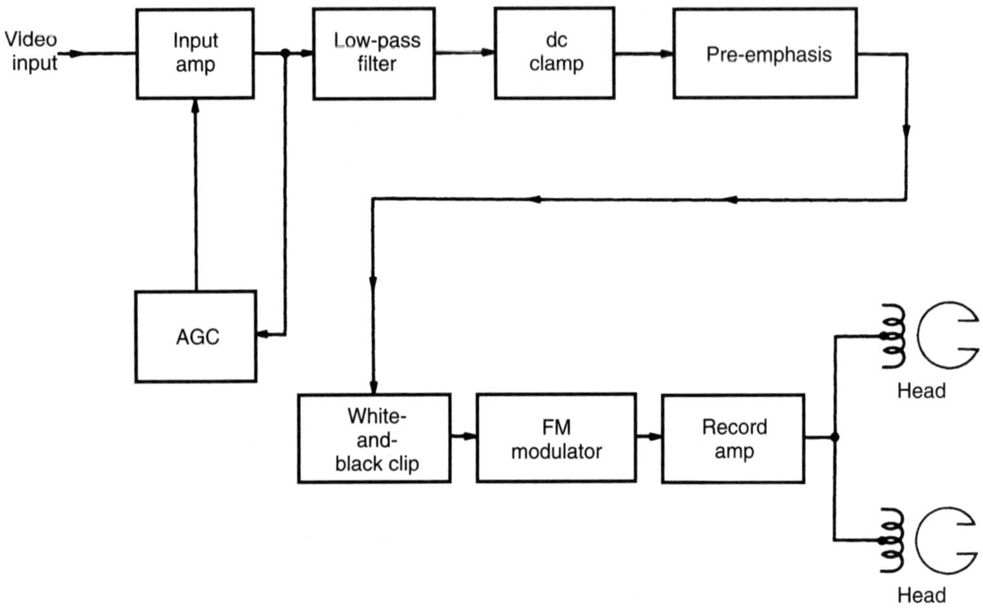

10-13 Block diagram of a luminance recording chain.

and off at a 60-Hz rate in such a manner that only the head that's in contact with the tape is in the circuit at any given instant of time. The reason this is done is because the head not contacting the tape will introduce undesirable noise into the picture. The outputs of the preamplifiers are connected together, and from that point on, the two signals merge into one, as shown in the Fig. 10-14 block diagram.

The combined output from the two record heads is fed to a series of limiters where any amplitude variations caused by noise, extraneous signals or changes in head-to-tape contact are removed. The clean signal is then fed to the demodulation circuitry, where it's converted back to video information.

One of the problems encountered in video playback occurs when, for various reasons, the head breaks contact with the metallic coating on the tape. This can be due to dirt on the tape or from minute defects in the coating. The resultant phenomenon is called a *dropout*, and, if ignored, can create unacceptable flashing in the picture. To eliminate this phenomenon, a circuit called a *dropout compensator* is used.

The block diagram in Fig. 10-15 shows a typical dropout compensator. During playback, the video information follows the path marked direct video. This signal is also impressed on a delay circuit that stores one complete line of video. However, the delayed information doesn't enter the signal path because its input to the selector is switched off by the FM rectifier and its switching circuit.

In the event that the FM carrier drops below a predetermined level, the voltage developed by the FM rectifier will open the switch and permit the information stored in the delay circuit to enter the signal path. The result is that if there's a dropout of

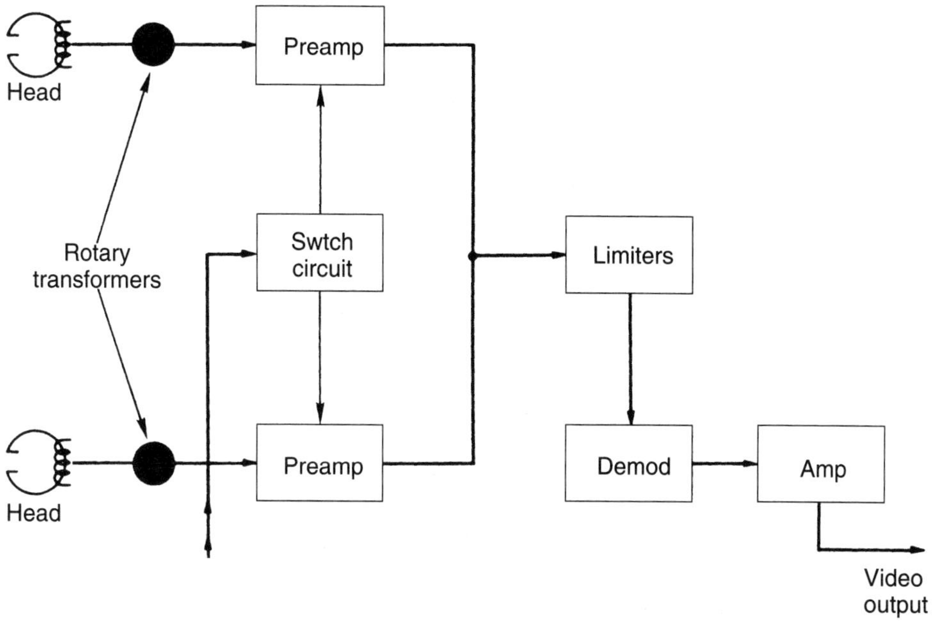

10-14 Block diagram of a luminance playback system.

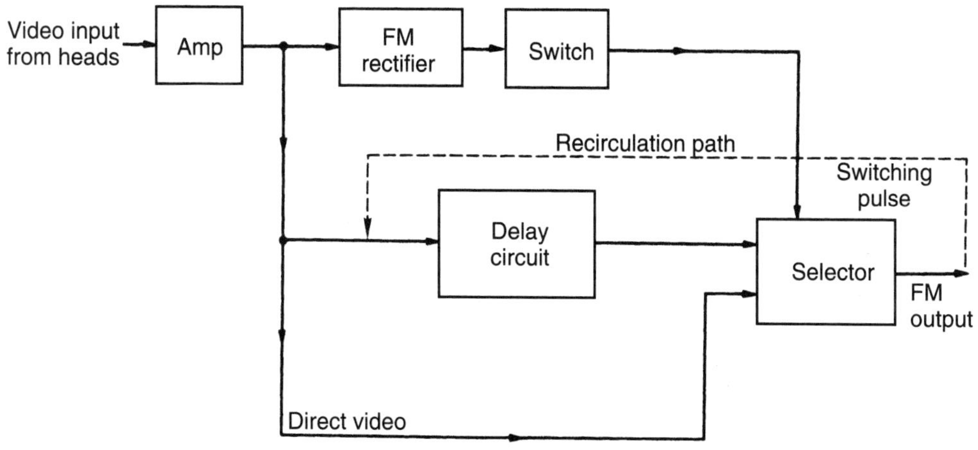

10-15 Dropout compensator block diagram.

carrier, the compensator will enter the information that appeared on the previous line and the picture goes on unnoticed by the viewer. In actual operation, this compensation can go on for 3 or 4 lines, due to the recirculation path, but at that point there's a noticeable deterioration of picture quality.

VHS color system

Generally, helical scan video recording systems use some method of separating color information from the luminance information before the signal is recorded on the tape. A common method is to convert the chroma to some other frequency. In Fig. 10-16, this system converts the 3.58-MHz subcarrier to 629 kHz, with the color sidebands appearing around that point. In this system, steps must be taken to limit the low-frequency excursion of the luminance FM signal to prevent interference patterns in the color portion of the picture.

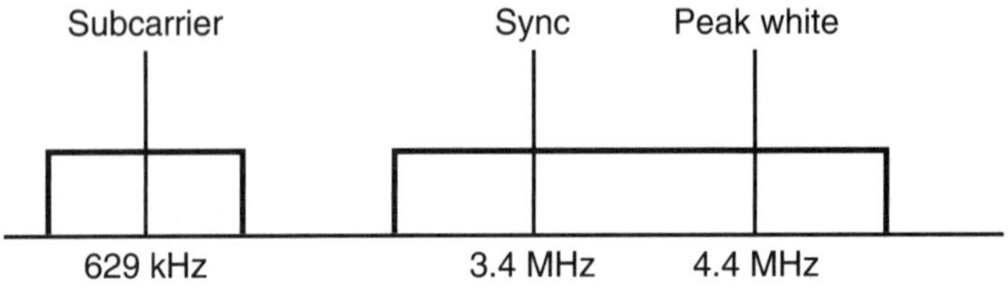

10-16 Color carrier frequencies on the tape.

The procedure for converting the 3.58-MHz signal to 629 kHz is shown in the Fig. 10-17 block diagram. The chroma is separated from the composite video signal before it enters the luminance recording chain. The 3.58-MHz signal is then beat against a 4.20-MHz oscillator, and the resultant 629-kHz frequency, containing all the chroma information, is then recombined with the luminance FM signal and recorded on the tape.

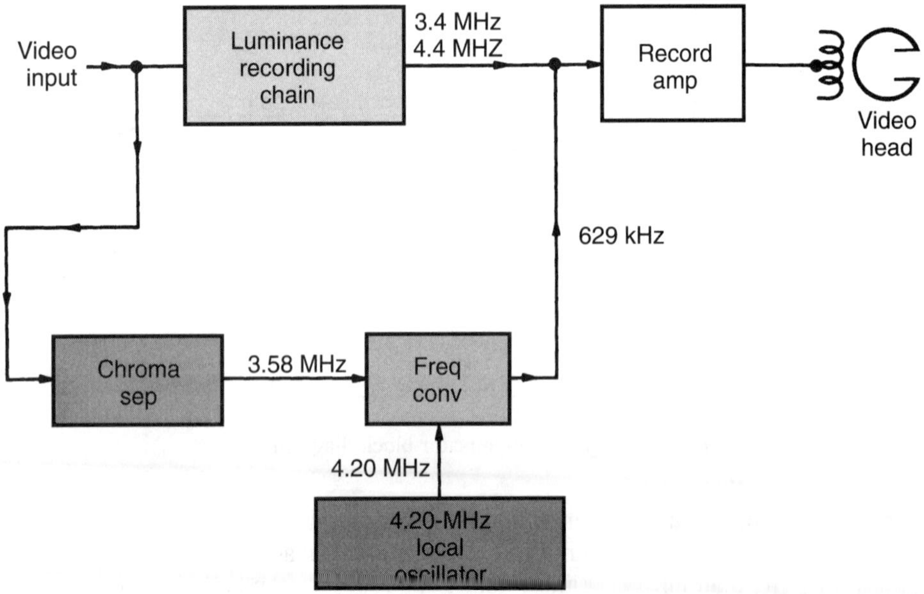

10-17 VCR recording with the converted subcarrier method.

VCR control system

With this brief look at tape recording, it's obvious that in order to reproduce a picture from magnetic videotape recording, it's essential that the playback head scan the recorded track with great accuracy. As an example, in the two-head helical scan system, each line of recorded information contains one complete field of the TV signal and the head switching occurs just before vertical blanking (so it will not appear on the screen). It's therefore necessary that the playback head contact the tape (track) in exactly the same spot that the record head did at the beginning of each track. Figure 10-18 illustrates what might happen should the drum head speed and tape speed get out of time with each other. For this reason, the drum wheel drive and the tape drive must be controlled both when recording and during playback. The necessary control of timing and positioning is accomplished by use of a servo control system.

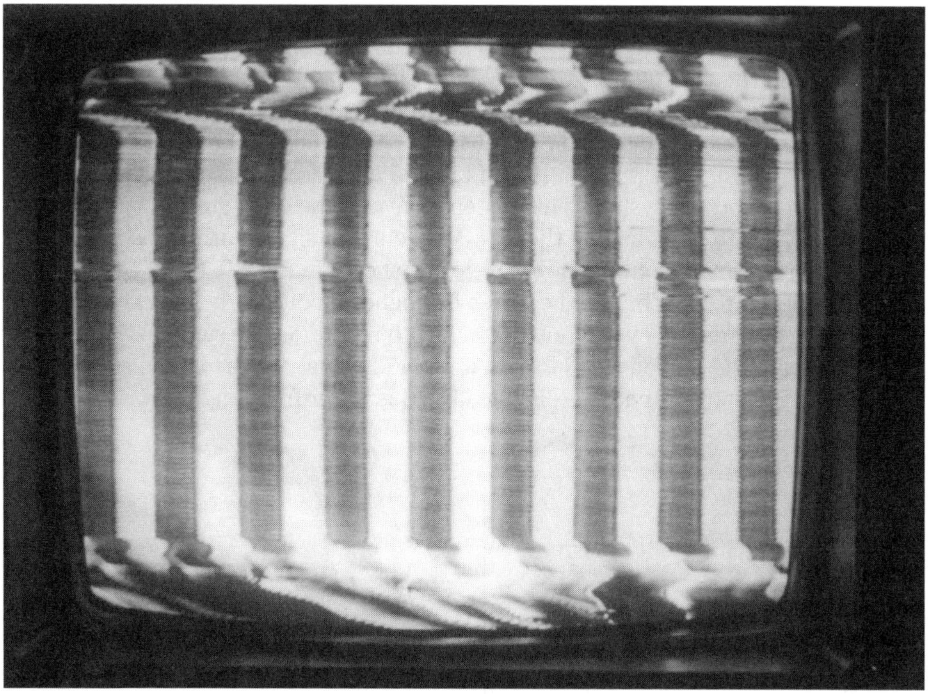

10-18 Results of a tape and head drum speed that's out of sync.

A *servomechanism* is an automatic device for controlling and correcting the performance of a mechanism. To perform its function, a servo needs at least two bits of information:

1. An input from the machine to determine how it's performing (feedback).
2. A constant (reference) source to determine the degree to which the machine deviates from the desired performance.

The color-under system

It's possible to record the color information at the same time as the luminance signal. However, you'll find that the phase relationships of the chroma signal are what determines that the color is reproduced correctly by the TV. Thus, these phase relationships are very important.

Phase errors are caused by several mechanical factors. First, the speed of the video head is not always constant. The head is controlled by a servo system that adjusts the speed slightly to make sure the head traces the same path on playback as when the tape was recorded. These slight variations would cause the colors to shift if the chroma were recorded directly with the video. This type of phase error (distortion) is called *phase jitter* or *servo error*.

The next cause of phase error is tape tension. The tension of the tape is constantly changing as the amount of tape on the takeup and supply reels changes. Tension differences are also present when a tape is played back on a different machine than the one used for recording. As the tension changes, the tape stretches different amounts. If the tape stretches, as an example, the horizontal sync pulses will be farther apart and the color subcarrier frequency slightly lower than desired. If the tension is decreased, the opposite occurs. This type of color subcarrier frequency error is known as *time-base error* and can cause various changes on the color picture.

The effects of both phase-jitter and time-base error are minimized with the *color-under* recording system. This system converts the color signal to a frequency of about 700 kHz by just mixing the 3.58-MHz chroma signal with an oscillator during recording. (Note the frequency waveforms shown in Fig. 10-19.) The chroma is then recorded directly (but at the lower frequency) along with the FM luminance signal. Also, this frequency will not be too low to record on the same heads used for the FM signal. The reason for this is that the FM signal acts as a bias signal to increase the efficiency of the recording head at this lower frequency.

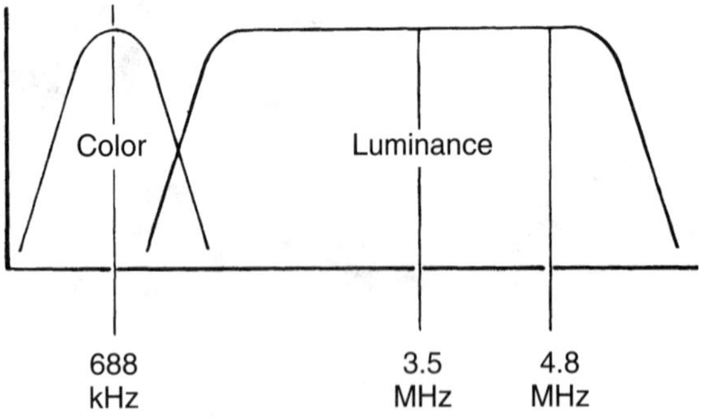

10-19 Frequency of chroma information converted to a lower frequency by heterodyning.

During playback, the low-frequency color information is converted back to the NTSC color frequency (3.58 MHz) by again mixing it with an oscillator. During playback, however, this oscillator doesn't operate at a fixed frequency. Phase correction circuits are used to compensate for the slight frequency changes that might affect the color playback. The horizontal sync pulses (recovered from the FM signal) are used to indicate the amount of phase jitter or time-base error and are used to correct the chroma frequency. The result is a close duplication of the original chroma information.

Color crosstalk elimination

Let's now delve into the VCR crosstalk problem a little bit more. The FM signal used to record the black-and-white detail is recorded at two different azimuth angles to allow "zero guard band" recording. Azimuth recording only attenuates the crosstalk of the higher recorded frequencies. Since the chroma information occupies a frequency range of 188 kHz through 1.19 MHz (500 kHz above and below 688 kHz), you don't have enough cancellation of adjacent tracks on playback. Thus, additional crosstalk rejection to provide a noise-free color picture is required.

Both the Beta and the VHS formats use a system of phase shifts to prevent color crosstalk. The phase of each line of color information is changed compared to the previous line. The Beta format just inverts the phase of every other horizontal line when one (of the two) video heads is recording a field of information. The other head records the chroma information without this line-by-line inversion. During playback, the phase correction circuits reinvert the line-by-line chroma information for every other field to restore the original phase relationships. A "comb filter" is used to cancel out any crosstalk.

Comb filter operation

As previously discussed, the comb filter is a device that can separate two signals that have a fixed phase relationship. The comb filter will provide two outputs, one that passes only those signals that are in phase with each other, and the other that passes only those signals that are out of phase with each other. Let's see how these in- and out-of-phase applications can be used in the VCR system. The first application allows the rejection of color crosstalk using the phase processing techniques just described. The second allows the separation of chroma from the luminance information TV signal with better effectiveness than the bandpass amplifier found in a color TV receiver. Both applications depend on the fact that the phase of the chroma information changes each horizontal line.

The chroma subcarrier of the NTSC signal is inverted every horizontal line. This means that the phase of one line is shifted 180 degrees from the phase of the previous (and following) line. This line-by-line phase inversion results in *frequency interleaving*. Thus, frequency interleaving means that the frequency of the color subcarrier (and all of its harmonics) falls between the harmonics of the luminance information. This frequency interleaving is determined by the color subcarrier frequency (3.579545 MHz) and the horizontal sweep frequency (15,734.26 Hz). Actually, all this means is that the two frequencies are related to each other in such a way that the required phase-inversion takes place when the two signals are phase-locked together, as they are at the TV station. With this phase inversion in mind, let's first look at how the comb filter is used to eliminate color crosstalk.

The phase shifting that's produced by the VCR during recording and playback in both the Beta and VHS format results in crosstalk information that's a continuous phase for every line. The two formats produce this continuous phase crosstalk information with slightly different signal processing as explained previously, but the results are basically the same. Since the unwanted crosstalk information has a continuous phase, and the desired chroma signal is phase-inverted line-by-line, the comb filter will separate the signal from the crosstalk.

The comb filter consists of a delay line (designed to delay the signal by one horizontal line) and a resistor bridge. The bridge has three input points and two output points. The first input is for the nondelayed signal, and the other two are for the delayed signal. One of the output terminals will provide a signal only if the delayed and nondelayed signal are both the same phase, and the other will provide an output if the two signals are 180 degrees out of phase. The second output provides the desired chroma signal without the crosstalk. The reason that the two signals are separated is that the delayed signal is taking a different path around the resistive bridge than the nondelayed signal, as shown in Fig. 10-20. The left-hand side of the resistor bridge adds the two signals if they're the same phase. The signals are canceled if they're out-of-phase. The right-hand side of the bridge has the delayed signal flowing the opposite direction through the resistors and is inverted by 180 degrees. Now, the signals will add only when they're of opposite polarities.

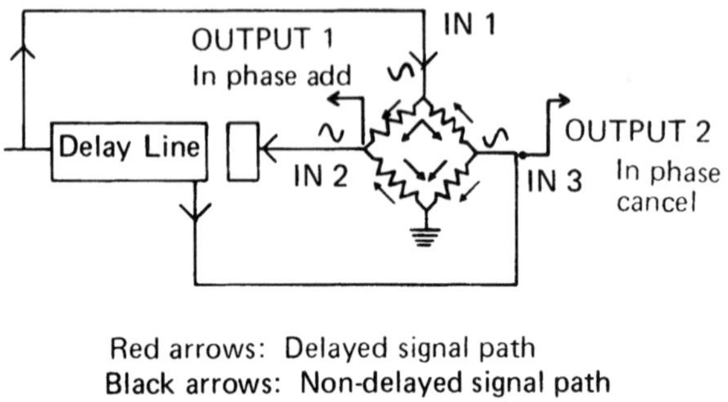

10-20 Block drawing of how a comb filter separates signals.

The second application of the comb filter is used by Sony in two-speed machines to separate the luminance from the chroma information during recording. Previous videotape recorders used a low-pass filter to separate the luminance signal, which was then used to produce the FM signal that's recorded. The chroma signal was separated with a bandpass filter similar to that used in a TV receiver. This system produced good results in the fast, one-speed Beta format machines. However, the low-pass filter restricted the high-frequency portion of the video signal. When the two-speed machines were introduced, Sony added a separate video equalization circuit for the 1-hour and 2-hour recording speeds. Since the high video frequencies suffer more from a poor signal-to-noise ratio than the lower frequencies, it's desirable to have more high-frequency information at the output of the stages that separate the luminance and chroma information. The comb filter is once again used to obtain the higher video frequencies that were lost in the low-pass filter.

This application (see Fig. 10-21) takes advantage of the interleaving that's part of the NTSC color signal to allow signal separation by phase detection. The low-pass filter used in the 1-speed machines restricted the video frequency response in the areas where the luminance and chroma information share the same frequency band, and thus restrict the highest video frequencies to about 3 MHz. The luminance information of a composite color signal is not phase-inverted every other line, but the chroma information is. This means that the comb filter's in-phase output node provides the high-frequency luminance, and the out-of-phase output provides the chroma information. The comb filter used for this separating is the same one that's used for crosstalk rejection during playback. Switching of the input and output connections results in the dual use of the same filter.

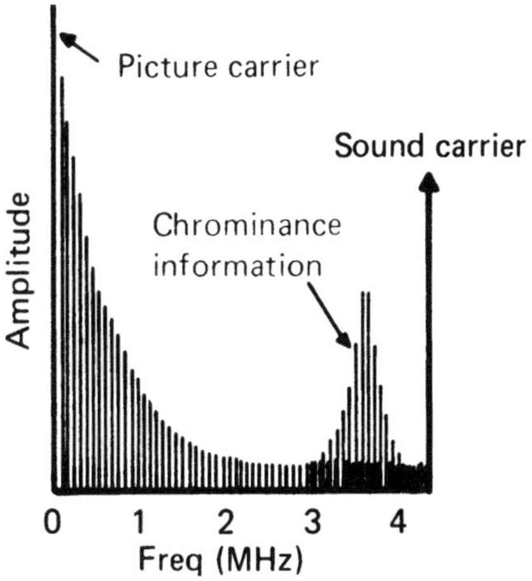

10-21 The line-by-line phase inversion of the chroma signal causes frequency interleaving, which allows the comb filter to be used to separate the chroma information.

Test tapes

Test and alignment tapes are required for verifying proper VCR playback and circuit alignment operations. Test and alignment tapes are available from the VCR manufacturer's service or parts department. Many of the tests required for playback servicing can be performed with a test tape that you can record on a known-good machine, using the signals from the VA62 Sencore Analyzer. This tape should have about 50 counts of the staircase pattern, bar sweep pattern, chroma sweep pattern, chroma bar sweep pattern, color bar pattern, and the crosshatch pattern.

This tape will not replace the need for an alignment tape, but supplements the alignment tape. The advantage of having your own recorded test tape is that it will save much wear and prevent the possibility of damage to your expensive alignment tape. Now, should a defective machine damage your own test tape, you'll be able to replace it for the cost of a blank tape rather than having to purchase another new alignment tape. Your own recorded test tape will let you confirm operations of the color circuits as well as the performance of the luminance circuits.

Any adjustments that affect compatibility with other VCR machines should be made with the manufacturer's alignment tape. This tape has been carefully prepared to provide a standard from one machine to another so that a tape recorded on one machine will play properly on another machine. This is very important if your customer uses commercially prepared prerecorded tape programs or if your customer has more than one machine and plans to play tapes on a different machine than used for recording.

When you record your own test tape, be sure to use a machine that's operating properly. New machines will generally be set up to record properly. Be sure to feed the signals directly into the "camera" input jack of the VCR rather than connecting them through the antenna input and onto the tuners. This will assure you that an improperly adjusted IF stage or tuner fine-tuning adjustment will not cause the quality of your tape recording to be less than ideal. Use the VCR tape counter to let you easily locate specific portions of the tape so you can quickly find the pattern that's needed. It would also be desirable to include an audio signal for at least a portion of the tape so that you'll be able to check out the audio playback circuits with the same tape. This audio signal is available from the "audio switch position" and the drive signal output jack of the VA62A. You can select audio frequencies of 333 Hz, 1 kHz, 5 kHz and 7 kHz.

Recording your own VCR test tape

- Connect the VTR standard output from the VA62A Analyzer to the camera input of a known-good VCR recorder. Now flip the input switch to the "camera" position.
- Set the tape counter to "000."
- Select the fastest tape-recording speed on machines that have more than one speed. This might be called "SP" or "X1" on some units and Beta II or Beta III on others.
- Now select from the Analyzer the video pattern that you want to record and use for playback testing. The counter mark on the VCR can be used as an aid for locating the desired pattern for playback. For example, the first pattern can be recorded from 000 to 050, the second from 050 to 100, and so on.

VCR tuner and RF/IF circuit checks

You can use the VA62A Video Analyzer to check out and align the RF/IF stages in the VCR in the same way you would for a TV receiver. If you suspect a tuner or IF problem in the VCR, make the following preliminary checks to help you isolate the defective stage:

The first step is to use the 1-volt VTR standard test signal from the VA62A and feed it into the camera input jack of the VCR. With the VCR switched into the camera mode, you should now receive a good test signal on the TV monitor. If you can feed a good signal through the VCR in this test, then all of the machine's E-to-E circuits are good and the problem is in the tuner, IF stages, or coax (TV/VCR) changeover switch.

Subbing the VCR tuner

The IF test signal from the VA62A can be used to check for a faulty head-end tuner of the VCR machine. You can easily substitute for the electronic tuner or a solid-state band-switch type of tuner.

The first step is to disconnect the coax IF cable plug from the tuner and inject the IF test signal at the input to the IF amplifiers. Set the Analyzer to the 45.75-MHz video IF position and the RF level to the (LO) position. Start with the RF-IF level vernier control at mid-range. If the tuner is faulty, you should now see a good test pattern on the TV monitor screen that's connected to the VCR. If not, then suspect IF, video detector, or AGC problems. IF alignment checks and adjustments can also be performed with this same test setup. You can select any of the video patterns to modulate this IF test signal. The block diagram in Fig. 10-22 shows how to connect the VA62A for use as a tuner subber.

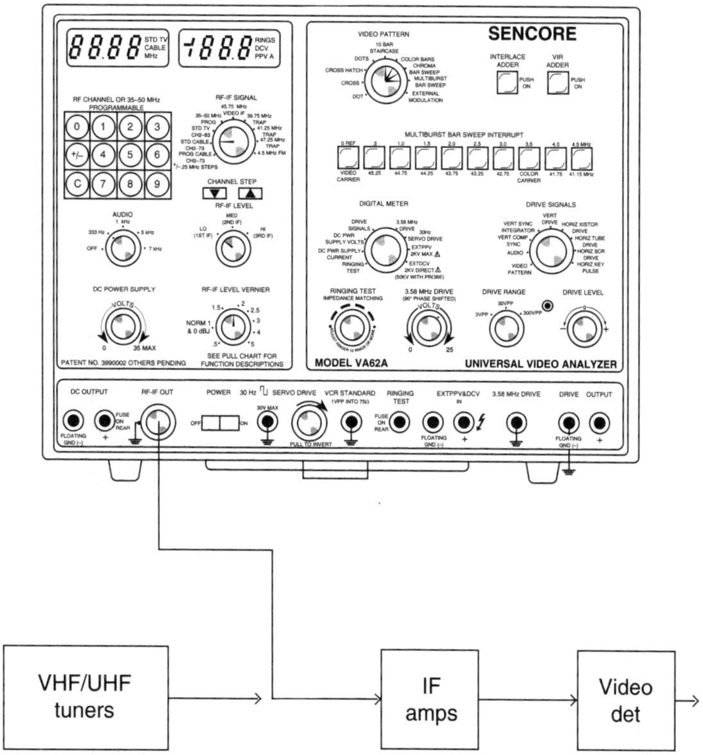

10-22 The VA62A Analyzer being used as a VCR tuner test subber.

Coax switch and splitter checks

Should you still not be able to obtain a good TV station signal through the VCR, such as a total loss or a snowy picture, you might now suspect the antenna input splitter or the coax changeover (VCR/TV) switch. You can quickly check these by viewing the TV monitor and injecting the RF signal from the Analyzer into the various coax cable connections.

Program the VA62A to the desired RF channel and feed the signal into the splitter. If you can't feed a signal through or have a loss of signal (snowy), then feed the signal into the two output cable connections from the splitter. If the picture is now good, then the splitter is defective. If not, go on to the coax changeover switch and perform the same input/output signal test. Also, don't overlook a faulty coax cable or connector during these various checks.

The block diagram in Fig. 10-23 illustrates the test setup for checking out the coax switch and splitter by signal injection with the VA62 Analyzer. Figure 10-24 shows the VA62 being used to check out the coax changeover switch.

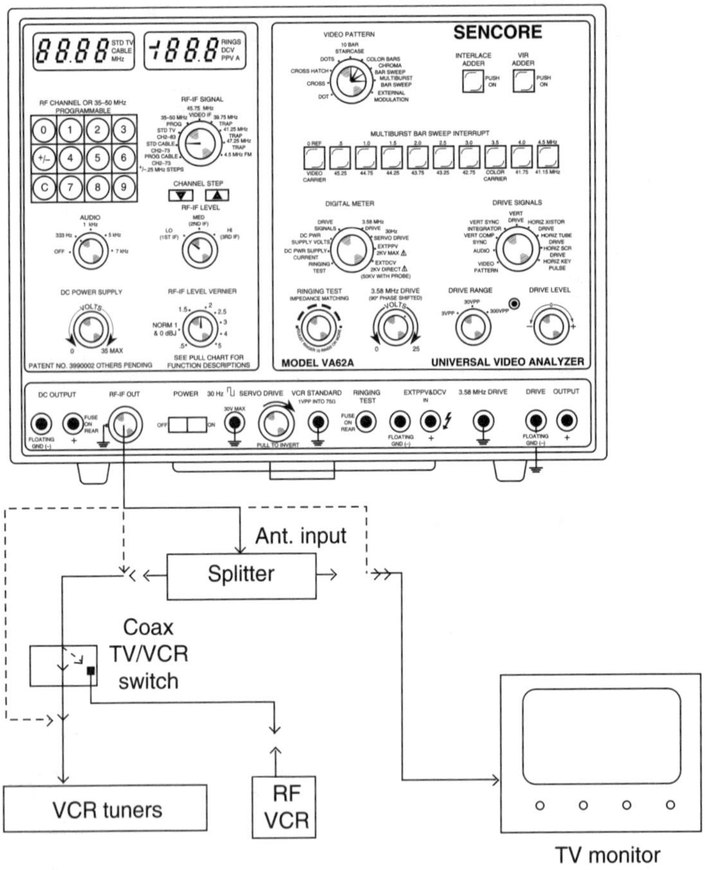

10-23 Signal injection setup to check out the splitter.

Troubleshooting the VCR IF stages

If you suspect trouble in the VCR IF stages, you can use the VA62 to inject the 45.75-MHz IF signal into any IF stage to quickly isolate the fault. The IF carrier frequency

10-24 The VA62A being used to check out the coax switch and splitter.

is microprocessor controlled, and the levels of the output signals are set for a normal voltage gain of 10 per stage. Thus, this makes the testing of stage gain as easy as moving the RF-IF level switch to the position for that IF stage under test with the injected test signal. The block diagram in Fig. 10-25 shows the VA62 Analyzer control settings and the various IF circuit injection points.

Use the following steps to quickly isolate a defective IF stage in the VCR:

1. Set the Analyzer to the desired video pattern.
2. For the first check, turn the RF-IF level signal's switch to the third IF position.
3. Inject the IF signal at the third IF input (usually base of the transistor) and adjust the RF-IF level vernier for a smooth, snow-free picture.
4. If you obtain a good picture in the preceding step, then move the injection probe to the input of the second IF amplifier stage. Change the RF-IF level switch to the second IF position.
5. If you find that the signal is still being passed at this point, then repeat the test at the input of the first IF stage. Switch the RF-IF level control to the first IF position. You should now have isolated any defective stage in the video IF system.

IF stage gain checks

You can determine the amount of IF stage gain with the calibrated VA62 RF-IF level vernier. Inject the 45.75-MHz IF signal at the output of the stage and adjust until a

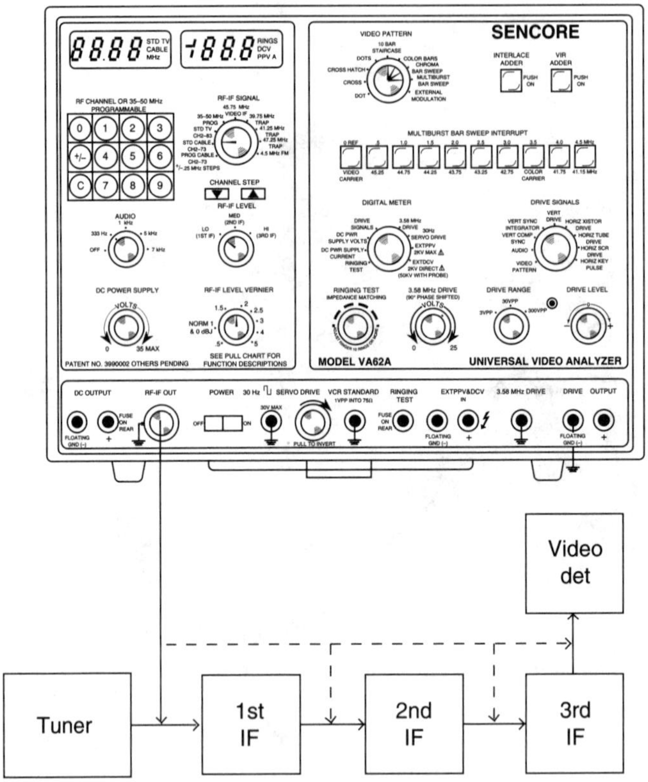

10-25 VCR IF block diagram and test signal injection points.

snowy picture is produced. Then move the injection probe to the stage input and set the control again for a snowy picture. Then find the voltage gain of the stage by dividing the output signal by the input signal.

Keep in mind that the IF signal level with the RF-IF level switch in the third IF position is ten times that in the second IF position, which in turn is ten times that found in the first IF position. For example, if the output of the second IF has a signal level of two with the RF-IF level switch in the third IF position, and the input requires a setting of five in the second IF position, just divide the output signal level by the input signal level. In this case the output signal is two times 10 (since there's a voltage difference of 10 between the second and third IF signals), or 20. This number is divided by five (the input voltage) for a stage gain of four. If the gain is desired in dB, use a calculator or dB table to calculate the gain with this formula: dB (gain) equals 20 times (log of the voltage gain). In this example, the gain would be 12 dB.

You can also use the built-in digital voltmeter in the VA62A to check IF and other stage gains in the various VCR circuits. Set the digital meter to EXT. PPV and the RF-IF level control to the third IF position. Connect the external digital meter probes to the output of the third IF video stage and then inject the IF test signal at the same point and note the voltage reading. Now inject the test signal at input to the third IF, note peak-to-peak voltage gain, and then go back to the input of the second IF stage. The last check would be to inject the signal into the IF input to first IF stage.

IF alignment notes

The VA62A Video Analyzer provides a multiburst bar-sweep pattern that's designed for both IF alignment and IF circuit troubleshooting. This pattern provides reference frequencies that relate directly to the key frequencies that are found on sweep/marker generators used for video IF alignment.

The multiburst bar sweep pattern is part of a composite video waveform. This technique lets you perform a complete IF alignment without the need to bias the AGC circuit. This bar sweep pattern is available with an RF or IF carrier. This allows the signal to be fed into the antenna terminals or the IF input of the VCR to be tested or aligned. The results of feeding the multiburst bar sweep signal through the VCR receiver can be evaluated on the TV monitor screen or with an oscilloscope connected to the video detector output in the VCR.

Programmable IF generator

The VA62A will produce any frequency from 35 to 50 MHz by using the "35–50 MHz PROG" position of the RF-IF signal switch. You just enter the desired frequency at the keyboard by entering four numbers. All four numbers must be entered for the microprocessor to recognize the desired frequency. The microprocessor automatically places the decimal point as the desired frequency is entered from the keyboard.

The programmable IF function has many applications. It can be used to rough-in the alignment of the IF coils before performing a full alignment with the multiburst bar sweep. Also, it can also be used to set any trap in the VCR IF system that might have a nonstandard frequency. The programmable generator is only used to set special traps that aren't covered with the other trap functions. Any of the four audio frequencies can be used to modulate the programmable IF signals.

Troubleshooting the VCR record circuits

Let's now lead off with the VCR troubleshooting by checking out the record circuitry operation and service procedures. Use the numbered blocks in Fig. 10-26 to follow along with these various steps. This should help you see how each of the individual circuits fits into the total recording picture. Most all brands of VCRs will follow these basic block diagrams.

The first block (1) is the low-pass filter and AGC amplifier stages. The low-pass filter circuit reduces the high-frequency components of the composite video signal to reduce noise in the picture. These stages can be tested with a reference input level of 1 volt peak-to-peak from the VA62A to make sure that the output is of the proper amplitude. For this test, inject the output of the VTR standard jack directly to the camera input of the VCR and adjust the AGC control for the proper output level using the multiburst bar sweep video pattern. The circuit's checked out further by using the adjustable video output supplied from the drive signal's output jack instead of the VTR standard jack. This adjustable output should be varied from 0.5 to 2 volts (negative polarity) while the video signal is viewed on the scope for this AGC action.

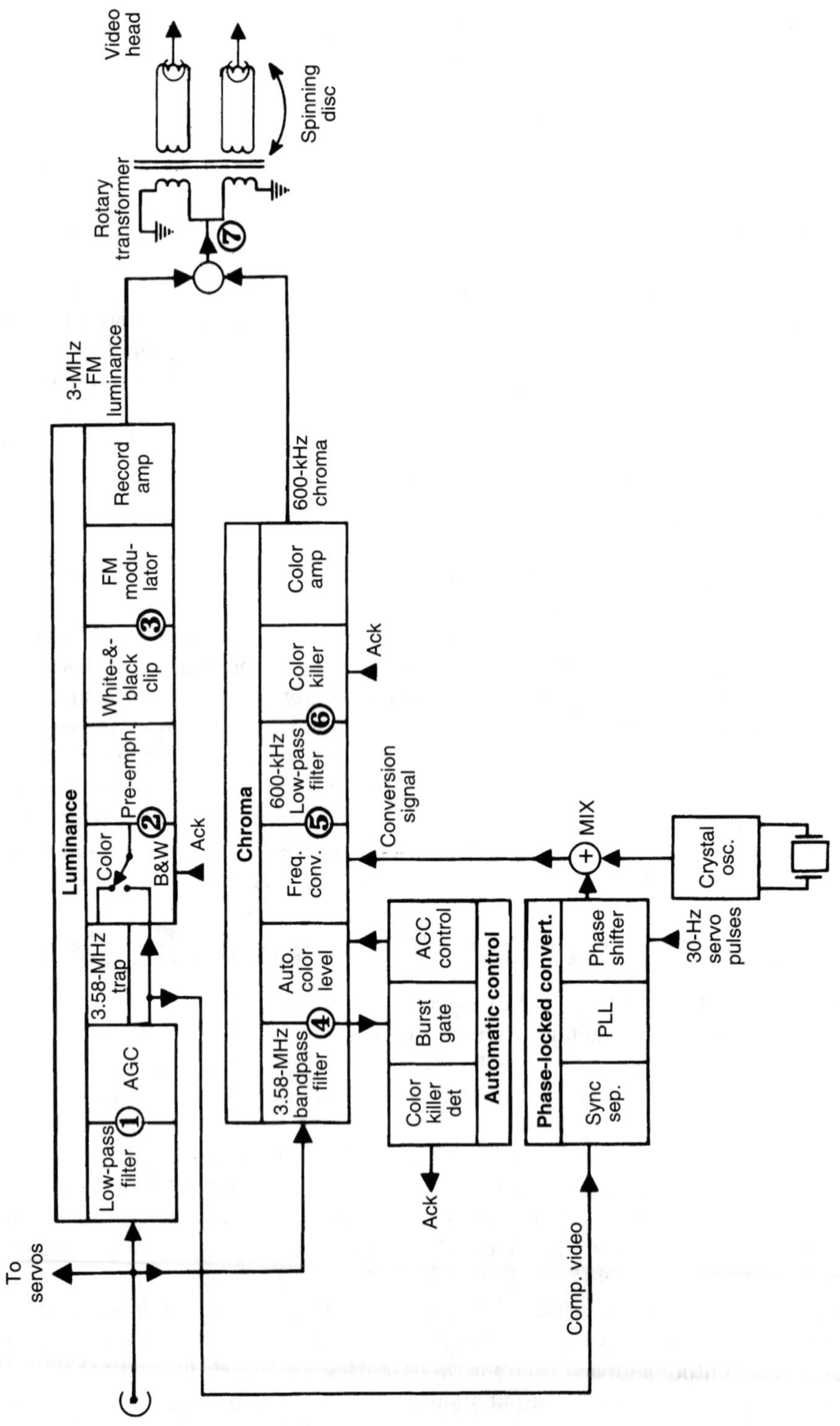

10-26 A block diagram of record circuits.

If the AGC system is not working, you must now determine if the AGC stage, the dc amplifier, or the AGC detector is at fault. To do this, connect a bias voltage (using the dc power supply on the VA62) to the AGC line and vary the voltage when the standard 1-volt signal from the analyzer is fed to the VCR camera input. The signal level at the output of the ACC stage should change as you vary this voltage. If not, you know that the trouble is in the gain-controlled stage. If the voltage does produce a change, then go back to the ACC detector and inject the dc voltage at its output. Note the ACC bias voltage and other check points shown in the Fig. 10-27 block diagram.

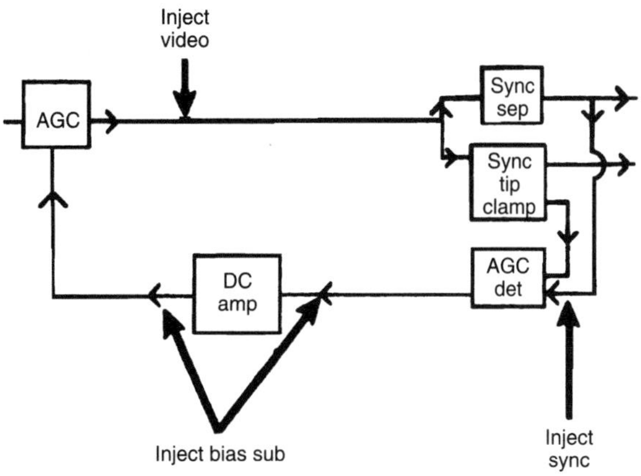

10-27 Drive signals for the VA62A can be used to check each stage of the AGC circuit.

For the last check in this stage, you can provide substitutes for both of the signals feeding the AGC detector itself by injecting the composite video signal at the camera input and then feeding the composite sync pulses from the VA62A drive signal's output in place of the machine's sync separator output. This check allows you to locate an AGC defect to a single stage, which is very important when troubleshooting defects caused by defective ICs or poor solder connections in the signal path.

Block (2) contains the 3.58-MHz trap, color killer (ACK), and the preemphasis circuits. The automatic color killer (ACK) circuit switches the trap into the luminance path when a color burst is detected, preventing the chroma signal from interfering with the luminance FM carrier. The ACK circuit switches the trap out of the system for black-and-white signals for increased high-frequency picture detail.

You can use the dc power supply voltage as a bias to see if the color and black-and-white filters are switched in and out of the circuit's signal path. Set the bias to about 4 volts and inject into the chip that controls these filters. Normal operation of this circuit should allow more high-frequency response during black-and-white reception, compared to a color program. Thus, you have proof positive that the circuits are operating properly.

The multiburst bar sweep pattern is thus used as a reference for checking the video frequency response of the various circuits that are processing the video signals. This multibar pattern gives a total frequency response check of the video circuits. For proper operation and check-out of the preemphasis circuits, use the multiburst bar sweep pattern and the scope. You'll find that the multispeed VCR machines require different amounts of preemphasis for each tape speed.

Let's now move on to block (3), which contains the black-and-white clipper, FM modulator, and the record amplifier stages. The black-and-white clipper, directly in front of the FM modulator stage, prevents excessive modulation of the FM carrier by limiting the positive and negative overshoots on the leading edges of the sync pulses and on fast transitions in the video signal. On a scope waveform you'll notice that the overshoots have the same amplitude at the sync pulse and the white bar of the bar sweep pattern. This balance is lost when the clippers have problems. Only use the multiburst bar sweep white bar (zero reference) and sync pulse as your reference when testing this point because the close spacing of the other test bars will make the rest of the waveform on the scope difficult to interpret. The low frequencies represented by the sync pulse and white bar are good enough, as the stage is amplitude related, not frequency related.

If the clipper circuits aren't properly set, they'll cause the picture to tear out during playback. The different frequency bars in the multiburst mode thus give you a dynamic check of the clipping circuits at different video frequencies.

The chroma recording circuits

Referring again to the block diagram in Fig. 10-27, you'll find that the first stage of the chroma circuits isolates the chroma signal from the composite video signal that was fed into the VCR. The chroma circuits then convert the 3.58-MHz color signal to a lower frequency (629 kHz for VHS or 688 kHz for Beta machines) before it's recorded onto the tape.

Block number (4) represents the 3.58-MHz bandpass amplifier stage. This stage separates the chroma signal from the composite video signal. This stage can be checked by injecting the chroma bar sweep signal and viewing the output of the stage on the scope.

In block number (5) you'll find the chroma frequency converter stage. This stage mixes the 3.58-MHz composite chroma signal with a 4.2-MHz local oscillator signal to convert the chroma down to about 600 kHz for recording onto the tape. The important thing to watch for in this stage is that the frequencies are being properly mixed and converted.

Block (6) contains the filtered 600-kHz chroma circuitry. The 600-kHz filter removes the mixing products from the previous signal. The only signal remaining is the down-converted chroma. The video head amplifier is shown as block (7) and will be covered in detail a little later in this chapter.

Analyzing VCR color record circuits

The VA62A Analyzer test signals are all phase-locked, which lets you quickly troubleshoot all of the VCR chroma circuits. The Analyzer has a separate output jack and level control for the 3.58-MHz color oscillator signal. This will give you greater versatility and efficiency for isolating VCR color problems. The 3.58-MHz signal is phase-locked to the video pattern signal and the RF signal for positive indications when troubleshooting "tough-dog" color problems.

Let's now see how signal injection can help you analyze these VCR color circuit stages. Let's consider a VCR that has changing color levels when a tape is played back. First, you must know if the color levels are changing during the recording or playback of the videotape. This is easy to confirm by playing back a tape recorded on the suspected faulty machine on a known-good VCR unit. If the color levels remain the same, you know that the defect is in the playback circuits. In this case, however, you find that the levels are changing when the signal has been recorded on the suspected machine and played back on the known-good machine. Thus, you now are certain that the trouble is in the recording circuits of the suspected VCR unit.

Changing color levels could be caused by a defect in any of the circuits shown in the Fig. 10-28 block diagram. These blocks make up the automatic chroma control (ACC) system in the VCR. A defect (such as a faulty IC, transistor, or poor solder connection) anywhere in these stages would produce almost the same symptom-changing color levels with different input signals. Of course, the scope will allow you to find a missing or low-level signal, but signal injection lets you have a more positive check because you can actually duplicate the signals that should be found at the output of each stage.

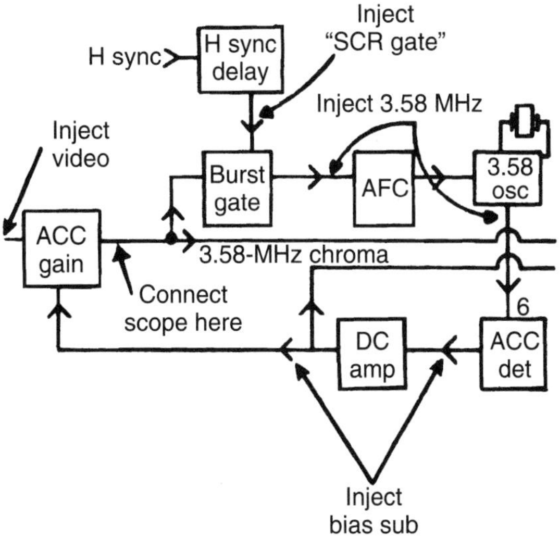

10-28 Points where the drive signal can be used to check out the ACC circuit.

To make these checks, you must provide a reference signal at the input to the ACC stages. This is accomplished by connecting the VTR standard output of VA62A to the camera input jack of the VCR. Now select the chroma bar sweep pattern to provide a reference color pattern for analyzing these color circuits.

Let's now check the circuits that produce the burst flag signal. The input to this stage is the composite sync pulses that have been separated from the luminance signal. For this substitute signal, use the V&H composite sync signal from the drive output of the VA62A. Keep this output low so as not to damage a solid-state circuit device. You can monitor this peak-to-peak voltage with the digital meter switch set to the drive signal position.

Start off this check by injecting the composite sync signal into the horizontal sync delay circuit and note if proper operation can be obtained. Use a scope to monitor the operation of this circuit at output of the ACC-controlled stage.

If the injected test signal doesn't return the proper signal amplitude at the output of the ACC circuit, then move one stage and sub in the burst flag pulse. For this check, change the drive signal's switch from the composite sync position to the SCR gate drive signal. This signal provides the correct test pulse for the burst flag because the pulse produced by the SCR gate signal is stretched the same amount as the burst flag. The pulse is present during the color burst and will operate the burst gate the same as a signal produced by the VCR's chroma processing stages. If operation of the ACC circuit returns to normal, you know that the trouble is in the horizontal sync delay stage.

If the SCR gate signal doesn't return to its proper operation, you can continue the stage-by-stage injection at the output of the burst gate. For this check, use the separate 3.58-MHz (phase-locked) signal from the 3.58-MHz drive jack on the VA62A. You can also check the dynamic operation of the ACC circuits with this test signal by just varying the amplitude of the 3.58-MHz signal by its separate drive control. For example, if you increase the amplitude slightly, the ACC output signal should reduce in amplitude. If this change is noted, you know that the ACC circuit is working and the trouble is in the burst gate stage. Conversely, if changing the amplitude of the injected signal doesn't change the ACC output, you know that the problem is in a later stage.

The 3.58-MHz drive signal can also be injected at the output of the 3.58-MHz oscillator. Varying the amplitude of this injected signal should again produce a change in the ACC output level. If it will not, you'll need to step forward one more stage.

The output of the ACC detector is a dc voltage whose amplitude is related to the amplitude of the burst signal. You can use the adjustable dc voltage from the VA62 to check the operation of both the dc amplifier and the ACC-controlled stage. Start this check by setting the output of the dc voltage for bias to the level indicated on the VCR service data. You should now raise and lower this voltage about 10 percent. The result should be a change in the ACC output signal level. If you still don't have proper operation, inject the dc test voltage at the output of the dc amplifier (input to the ACC-controlled stage) and again vary the voltage. If a changing output level is still not obtainable, you know the problem is in the ACC-controlled stage itself. With this technique you'll be able to tie down all of these circuits in a somewhat complicated feedback system to locate a fault in any single stage or in some cases the defective component.

This might seem like lots of checks, but in the real world you wouldn't have to sub in each injection point. You could, for example, start at the output of the ACC detector and feed in the substitute dc voltage. This divides the circuit in half. If proper operation returns, then the defect is somewhere in front of this stage. Should machine operation still not be correct, the trouble is either in the dc amplifier or the ACC-controlled stage. Thus, you now know which direction to go in order to further analyze the stages completely. This direct injection technique is a real time-saver once you get used to applying it to various types of circuit defects that always crop up. The combination of signal tracing and signal injection provides the best of both worlds for speed and accuracy in circuit trouble analysis.

VCR playback circuits (luminance)

Refer to the VCR playback circuit block diagram shown in Fig. 10-29 as you read through these various stages. The action begins when the machine picks up the signal recorded on the videotape, amplifies this minute signal, and splits it into separate luminance and chroma components. You'll usually find that the luminance circuits aren't perfect. The high-frequency detail is reduced to about half of that recorded on the tape. However, properly operating and aligned luminance circuits will keep these losses to a minimum.

In block (8) you'll find the FM detector circuits and high-pass filters. In the FM detector block you should find the 4.3-MHz FM carrier passing through this stage. Later filter stages remove the carrier, leaving the composite luminance signal. The luminance signal should be observable on the scope in this section of the VCR.

Block (9) contains the de-emphasis/filter network circuitry. This stage removes the unwanted FM carrier and compensates for high-frequency losses in the video-to-FM-to-video conversion process. The *de-emphasis* portion of this circuit compensates for extra gain added to the high-frequency video information during recording to keep the amount of high-frequency noise at a minimum. The key point to look at on the scope trace is the overall waveshape, especially the relationship of the sync-to-white levels and the relationship of the multiburst frequency bars to each other. The sync-to-white levels should match closely to the waveshape you would view coming directly out of the VA62's "VTR standard" output jack. The high-frequency losses shouldn't be greater than 50 percent.

Block (10) contains the luminance amplifier and output stage. The luminance signal is completely recovered by the time the signal arrives at this point. On a scope you'll notice that only the luminance signal is present at point (10), but at test point (14), both the luminance and chroma signals will be found in a properly operating machine in the playback mode.

VCR (chroma) playback circuits

Again referring to the block diagram in Fig. 10-29, you'll note that the VCR separates the color signal from the other video signals picked up by the video heads. The machine then routes these signals to special circuits that convert the color signal back to 3.58 MHz. If you look closely at a scope waveform, you'll notice some losses of the high-frequency detail in the color circuits, just the same as in the luminance circuits. As with the luminance circuits, however, these losses are small enough to produce a good, viewable color picture.

A good way to check out these VCR color circuits is to play back your own test tape on which you've recorded the color-bar and chroma bar sweep patterns from the VA62 Analyzer on a known-good machine. For some circuit problems, this tech-

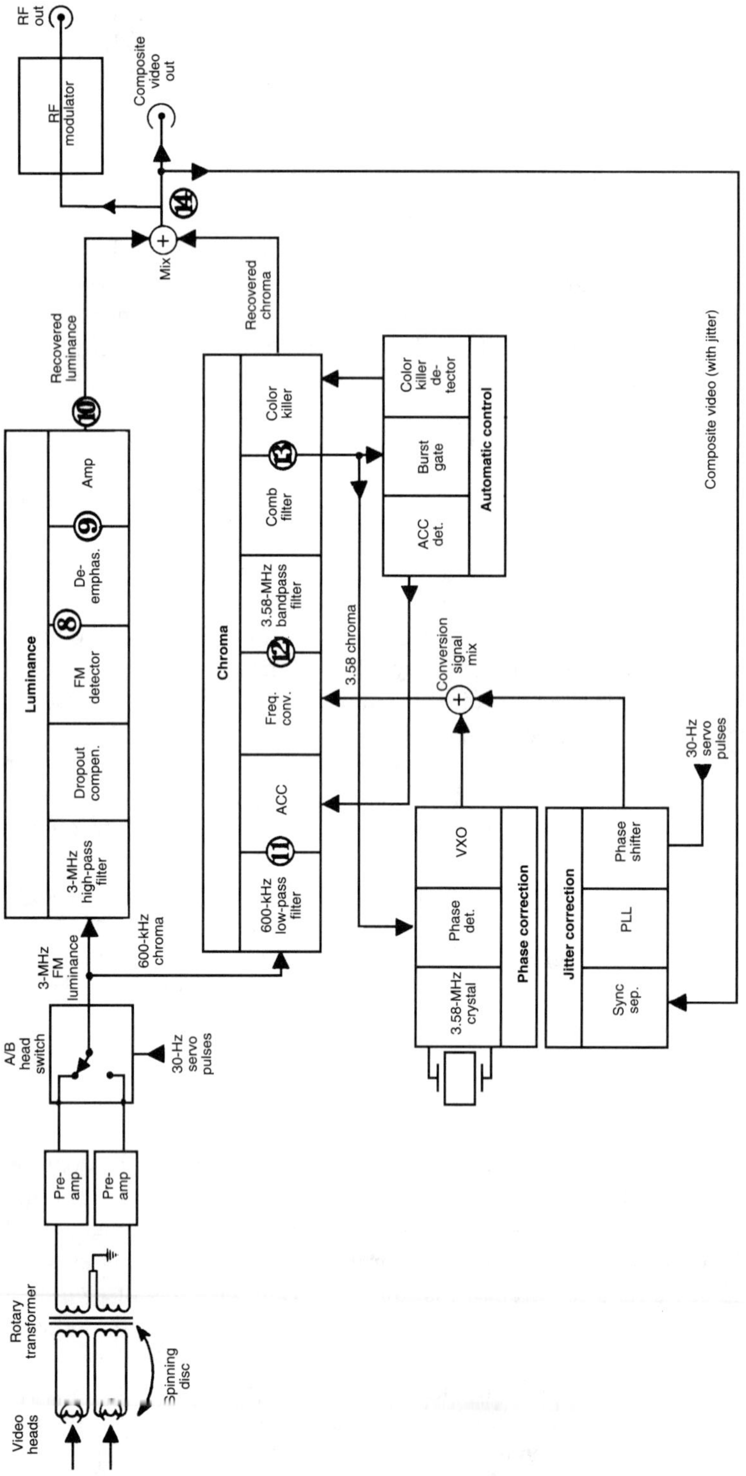

10-29 Block diagram of playback VCR circuits.

nique of signal tracing with the scope will find the trouble quicker than by using signal injection.

The video signal splits after leaving the preamplifier, and one branch goes into block number (11), the 600-kHz low-pass filter stage. When playing back the chroma bar sweep pattern from the test tape, you should notice on the scope pattern waveform a loss of the highest frequencies. This loss is normal and due to the fact that the 4-MHz bar beats down to about 100-kHz in the down-conversion process. This frequency is too low for efficient recording by the high-frequency video heads. In theory, this much signal loss should cause the amount of color in small areas of the picture to not be seen. However, in actual operation, the 3-MHz bar carries enough detail for a good, acceptable color picture.

Block (12) is where the 3.58-MHz converted chroma circuitry is located. This stage heterodynes the 600-kHz chroma back up to 3.58 MHz by mixing it with a 4.2-MHz local oscillator. The same mixer is usually used for the down-conversion process in the recording mode and the up-conversion process in the playback mode. The mixer stage contains all of the following frequencies: 600 kHz, 3.58 MHz, 4.2 MHz, and 4.8 MHz.

The comb-filtered 3.58-MHz chroma and color killer circuits are referenced as block number (13). The 3.58-MHz bandpass filter removes the unwanted mixing products from the converted chroma stage. The comb filter cancels "crosstalk" from adjacent video stripes on the tape to prevent interference to the chroma signals. The main points to look for on a scope trace at this point are the amplitudes of the bars in the chroma bar sweep pattern. The center (3.58-MHz) bar should always be larger than the other two. The right-hand bar is generally reduced, but shouldn't be much lower in amplitude than shown in the typical scope waveform of Fig. 10-30. More details on VCR comb filter operation and alignment will follow later in this chapter.

The video output and mixer stage is shown as block (14). In this block, the recovered luminance and chroma signals mix to produce the composite video signal sent on to the RF modulator and video output. In this stage you can expect to find high-frequency information losses in both the bar sweep and the chroma bar sweep scope waveforms. Proper alignment of the various VCR-tuned circuits will keep these losses to a minimum.

Frequency response within the VCR

An important test of the VCR playback system is to see if the machine is providing the frequency response it was designed for. The multiburst bar sweep pattern from the VA62 can easily be used for a dynamic test of the entire system's video response. For this complete dynamic test, just connect a scope to the video output jack of the VCR. Terminate this jack with a 75-ohm resistor, connect the scope, and view the waveform as you play back the multiburst bar sweep pattern portion of your test tape. A properly operating VCR should produce the scope pattern shown in Fig. 10-31. The waveform in Fig. 10-32 indicates a need for alignment.

If you find a frequency response problem, the first place to suspect is the adjustment of the video head equalization circuits that are used to compensate for the nonlinear output of the video heads. The best method for checking this equalization

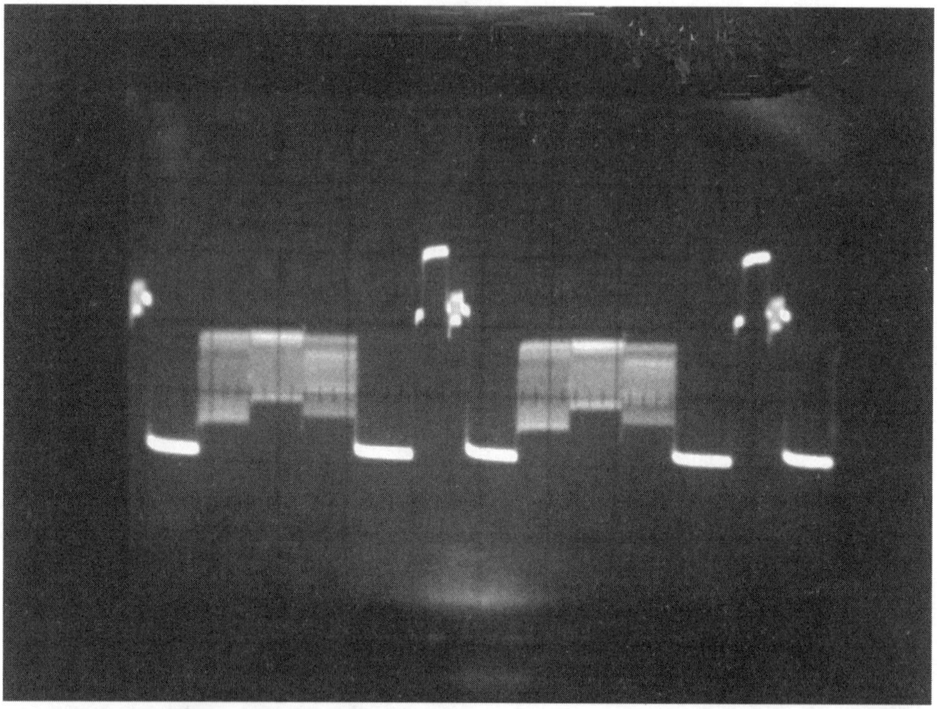

10-30 Waveform used to check out comb filter adjustments.

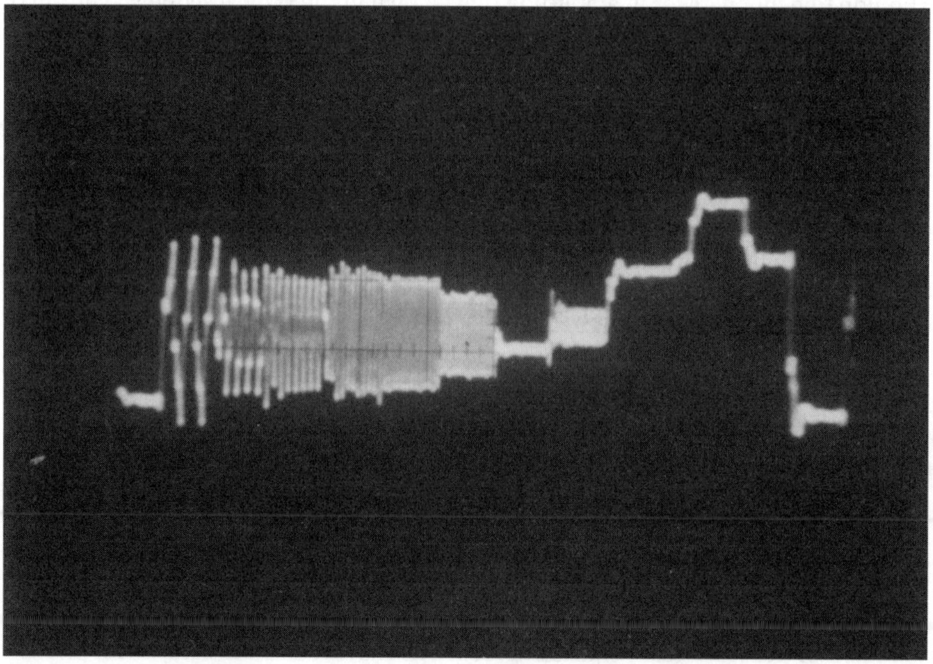

10-31 A properly operating VCR should produce this scope waveform.

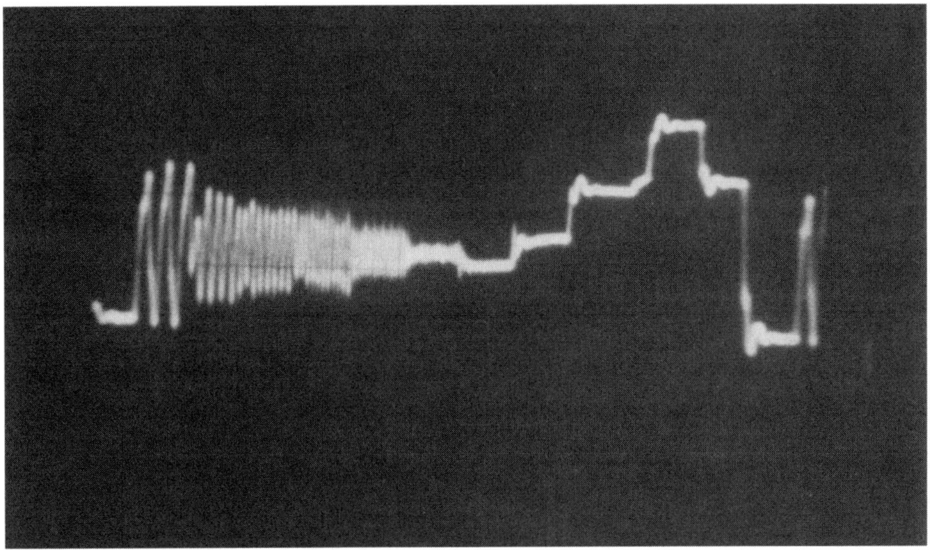

10-32 This scope waveform indicates that the VCR needs alignment.

is to use the VCR manufacturer's alignment tape and follow the procedures recommended in the service information.

The chroma bar sweep provides a check of the chroma frequency response and even circuit troubles. This test is one of the many reasons for using the VA62 patterns to record your own test tape because there's no other complete test for all of the VCR circuits used to process the color signals. The chroma bar sweep checks the color circuits at both the upper and lower frequency limits that are required for good color detail. The center (3.58-MHz) bar provides a reference level for a comparison of the frequency detail 500 kHz above and below the subcarrier. A key feature of the three test bars produced by the chroma bar sweep is that they're phase-locked to the horizontal sync pulse and have a 180-degree phase shift every horizontal line. This means that the bar signals will be properly separated by the comb filters used to cancel color crosstalk during the playback mode.

Comb filter checks and adjustments

Setting the comb filter used in the playback portion of the VCR color circuits is easy to do as long as the scope used has good vertical channel sensitivity, at least 5 mV. Start by using a direct probe to connect the scope to the output of the comb filter bridge that's not connected to the chroma amplifiers. This output point will show the crosstalk rather than chroma output. The signal at this point doesn't have sync pulses, so the external trigger input of the scope should be connected to the video output jack for triggering. Now, set the scope to trigger at the horizontal sweep rate. Scopes with built-in TV sync separators will provide a stable trace with the composite video signal used as the reference.

Now, play back the portion of your prerecorded test tape that has the chroma bar sweep pattern recorded on it. Adjust the comb filter's mixer control until the amplitude of the signal for the color carrier bar (3.58 MHz) has the minimum amount of crosstalk. Be sure to use a direct scope probe and have the scope set for maximum sensitivity, as the signal level is very low at this point. The two phasing coils used in the comb filter on some VHS-system VCRs should be left in the factory-aligned positions, as a high-cost TV broadcast type vectorscope is required to set them properly. The RCA service information, for example, doesn't give a procedure for setting these coils.

Servicing VCR servos

The test signals from the VA62A analyzer are ideal for making VCR servo checks and alignment adjustments. The phase-locked broadcast-quality sync makes it desirable for these applications. For best results, inject the signal from the VTR standard jack on the VA62A directly into the video or camera input jack on the VCR. This procedure will eliminate the fine-tuning error that might occur when feeding an RF signal into the VCR tuner input terminals.

On some machines there are several adjustments that call for the counting of the horizontal sync pulses just before the vertical sync pulse. When using a TV-station signal, the equalizing pulses are present and can cause an error in counting, resulting in the wrong setting of a crucial control. The phase-locked sync signals don't have the equalizing pulses, the counting of the horizontal sync can be done correctly without the equalizing pulses causing confusion, and this makes the proper setting easier to obtain. Equalizing pulses aren't used in the VCR circuits and aren't important for the servo alignment procedures.

The servo control circuits in a VCR compare two signals to indicate if the heads are in the correct position during both record and playback modes. During recording, the two signals are the vertical sync pulses and the head position signal (PG pulses). The two signals should arrive at the servo comparison circuits at the same time. If they're not properly timed, the servo circuits adjust the rotation speed of the video heads until the two signals are in sync. A block diagram of the servo system for a Beta machine is shown in Fig. 10-33.

During playback, the same head reference pulse (PG pulse) is compared with the output of the control track head. Again, the two signals are compared and the speed of the rotating cylinder head is adjusted until the two pulses are properly timed. The servo drive signal from the VA62A can be used to analyze many sections of the servo circuits where a scope or other techniques will not work.

Servo troubleshooting examples

Beta VCRs use the servo control pulses as a protection device as well as a control system. The servo logic circuits "look for" a servo pulse coming from the spinning heads. If this pulse is missing, it might indicate that the video cylinder head is not spinning due to a jammed tape, faulty belt, or defective head drive motor. Any time the PG pulses are missing, the tape deck automatically shuts down (goes into the stop mode) to prevent the possibility of more machine damage.

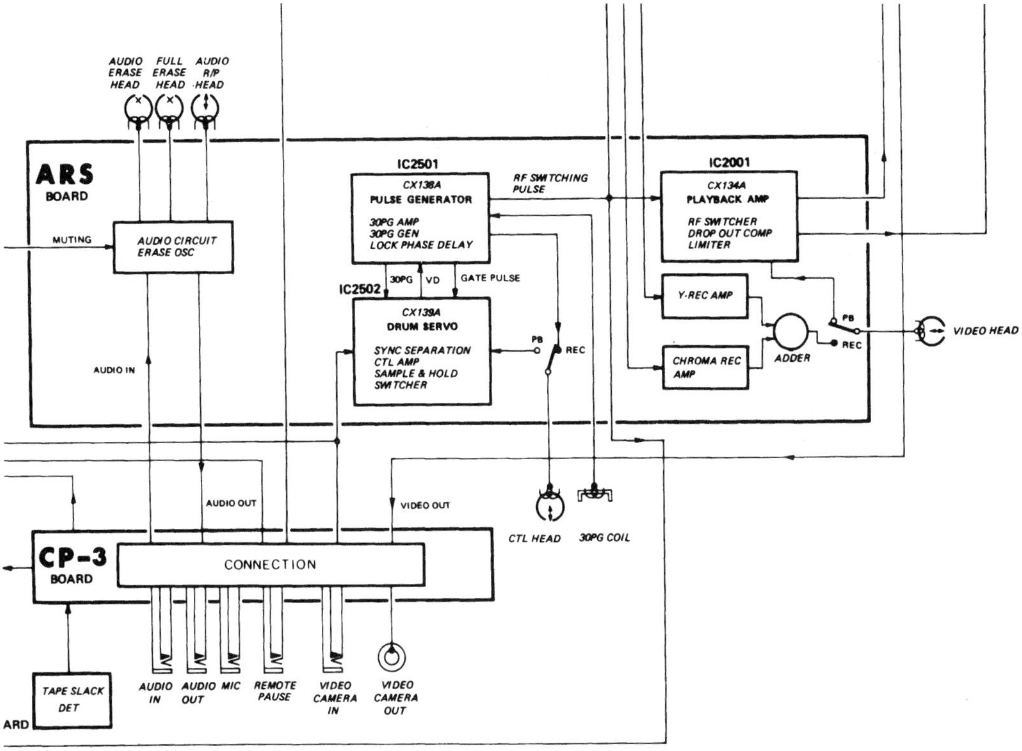

10-33 Block diagram for a Beta Max servo system.

As an example, let's consider a VCR that will not stay in the play mode. When the play button is pressed, the machine will run for about one second and then automatically shut down. Now, you can't use an oscilloscope to analyze this defect because the pulse produced by the PG head is present only when the video head is spinning. The cylinder head, of course, will spin only when the machine is in the play mode, thus there's no pulse to trace.

At this point you can now use the 30-Hz servo drive signal from the VA62 to substitute the pulse for the PG signal coming from the pickup heads. For this check, adjust the 30-Hz servo drive pulse from the Analyzer for 1 volt peak-to-peak by using the built-in digital voltmeter (the normal signal level at the input to the PG amplifiers) and feeding it to the input of the servo board in place of the PG head output pulses.

Let's assume that you have a machine that will not stay in the play mode. This indicates that the defect is a missing PG pulse caused by a faulty PG head, or trouble in the wiring running from the cylinder head to the servo board. If the substitute signal doesn't correct the defect, just keep injecting the servo pulse and proceed to trace the signals with the scope.

The beauty of this technique is that you don't need to have the VCR running to signal trace through the remainder of the servo circuits. You're simulating the PG pulses that would normally be generated by the spinning head location magnet passing over the PG head.

When troubleshooting playback circuits in the VCR, remember that the correct servo phase is necessary to return color because the machine processes the phases of the color signal differently for the first and second fields. Thus, you should try both phases of the servo signal if color doesn't return, even though the reference signal is fed to the same injection point in each case. This is required because some stages in the color circuits use an inverted servo pulse for some test points.

Introduction to the VCR VC63 test accessory

The VC63 is an accessory to the Sencore VA62A Deluxe Video Analyzer. All of the signals supplied to the VC63 are derived from the signals generated by the VA62A so that all substitute signals are phase-locked to the VA62A reference signals. The VC63 takes these standard signals and converts them to the special signals used in VCR troubleshooting.

You'll find that the VC63 is an entirely new concept for servicing VCRs. This test unit provides signals unique to the Beta and VHS VCR recording and playback systems. For the first time, you can use signal injection into any record or playback stage, including video head amplifiers and equalization circuits, to isolate VCR problems with functional analyzing.

VC63 basic operation

The VA62A Analyzer supplies all reference signals to the VC63. A test reference signal must also be fed directly from the VA62 to the VCR when using the VC63. When working with the recording circuits, the reference signals comes from the VTR standard output jack. When troubleshooting the playback circuits, the 30-Hz servo drive signal forces the electro-mechanical servo circuits into sync with the VA62A Analyzer signals.

All signals supplied by the VC63 are standardized to the video recorder circuits. The signal returns a complete color or black-and-white picture to the screen of a TV monitor connected to the VCR output when injected at the correct test point. This eliminates the need to interpret scope waveforms and voltages to isolate faults in the system.

The test accessory supplies two types of signals for each of the consumer VCR formats. The lum signal is only the frequency-modulated luminance signal used to record the wide-bandwidth luminance signal onto the tape. The lum and chroma signal includes the first (luminance) signal plus the down-converted chroma signal (688 kHz for Beta or 629 kHz for VHS). The chroma signal is not available by itself because the VCR must have a good luminance signal before it can recover the chroma information. A drawing of the VC63 VCR test unit and its control locations is shown in Fig. 10-34.

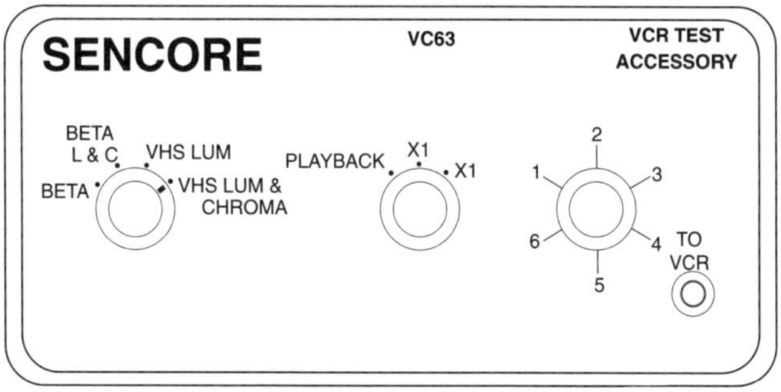

10-34 Drawing of the Sencore VC63 VCR test accessory.

Whenever troubleshooting a complex problem, you should always concentrate on the luminance signals before moving on to the chroma signals. The loss of color, for example, might not involve a problem in the chroma circuits at all, and locating the luminance problem might actually solve all of the VCR symptom troubles.

Checking out the VCR playback circuits

The principal application of the VC63 probably will be for isolating troubles in the playback circuits. There are two reasons for this:

1. Many of the signals in the playback circuits can't be analyzed with an oscilloscope or other techniques because of minor details (such as phase) that affect operation but can't be viewed on the scope.
2. There's no other way (that's known at this time) to inject signals into the playback circuits.

In most cases, the playback circuits should be repaired before attacking a problem in the recording circuits so the VCR can be used to check out its own recording performance. Should you attempt to troubleshoot a recording problem first, you must use another VCR to play back and check the recorded tape. In addition, many circuits are shared between the recording and playback functions. Isolating a playback problem might very well take care of the recording problem also.

The VCR must be locked to the VA62 30-Hz servo drive signal before injecting the substitute test signal. This is especially important when analyzing a color problem because the color recovery circuits must be referenced to vertical via the servo signals. The VA62 Analyzer 30-Hz servo drive signal provides two polarities that are required for VCR troubleshooting. One polarity references the rising edge of the servo square wave to the vertical sync pulse of the first field of information, and the other polarity references it to the second field.

The correct servo phase is necessary to return color because the VCR processes the phases of the color signal differently for the first and second field. When actually troubleshooting, you should try both phases of the servo signal if color doesn't return, even though the reference signal is applied to the same injection point in each case. This is necessary because some stages in the color circuits use an inverted servo pulse for some test points. You don't have to inject the 30-Hz servo drive pulse into the circuit with the inverter (in fact this is not desirable because the other parts of the servos will then not have a reference signal). However, you might invert the reference to reverse the phase of the signal at later points because the servo signal is a square wave. There will be no noticeable change in the other operations of the circuits with the inverted square wave.

The playback head sub

The playback head sub position of the output level control allows injection directly into the first head amplifier of the playback circuits. This is crucial in troubleshooting because the signal normally found at this test point is too small to view with a scope. In addition, the head signals are so special that problems such as improper frequency or phase will result in no output from the circuits that recover the information from the tape.

The playback head sub position duplicates the signal levels found at the head outputs. The range of amplitudes of this function is from zero to only .5 mV. The output connector is floating in reference to ground (differential output) to allow the signal to be injected directly into the different head amplifier found in almost every VCR machine to date. The capacity and drive impedance have been carefully designed to allow injection into these sensitive stages without producing loading or feedback problems.

The VCR test unit's lum signal can be injected into any point before the luminance FM detector. Thus, the VC63 will provide the correct amplitude for signal injection into any of these stages. The tape must be moving (playback button depressed) to inject the signal because most VCRs only connect the playback circuits to the output during playback. Should you experience interference from the moving tape while injecting at a point after the head amplifier, you can jumper the head amplifiers to ground to prevent this second signal from passing through the circuits. This interference will even be present when playing a blank tape (a new one that has never been recorded) because of the random noise a blank tape generates.

The composite lum and chroma signal should only be injected at test points before the point where the chroma signal separates from the luminance signal. This is necessary because the luminance signal must be recovered before the VCR can process the chroma. (The phase-locked loop referenced to the horizontal sync pulses must be locked in order to up-convert the 600-kHz chroma signal.)

Checking out the VCR recording circuits

The VCR test unit supplies enough signal amplitude to inject directly at the video heads during actual recording. The signal connection, of course, must be made back in the circuits because the head itself is spinning at 60 revolutions per second. This is a good check to determine if the video heads are defective or not.

The VA62 Analyzer's VTR standard signal must be injected at the video input of the VCR when injecting the VC63 signals. The VTR standard references the servo circuits so the VC63 signal is properly synchronized on the tape. If you suspect a problem in the servo circuits as well as the video circuits, the 30-Hz servo drive signal can be injected into the servos to provide mechanical synchronization. In most cases, however, servo problems should be corrected before using the VC63, as the servos affect all other functions of the machine.

The X.1 and X1 settings of the output level switch provide the correct amplitude signals for recording checks and tests. The switch and variable control should be adjusted to produce the level of signal shown on the VCR circuit diagrams. The output will relate directly to the peak-to-peak level by multiplying the switch setting times the control setting. Amplitudes from less than .1 volt to 5 volts peak-to-peak are available.

The VC63 test unit must be referenced to circuit ground when injecting signals. Both VC63 output connectors are floating in reference to the common connections of the VA62. Thus, either connector can be connected to the VCR injection test points.

The pattern recorded by the VC63 will be the one selected by the video pattern switch on the Analyzer. There will be no color recorded if the color bars or chroma bar sweep patterns are selected with the special signals switch in the lum position. The patterns will be recorded in color if the lum and chroma position is used. The correct format, of course, must also be selected for color recording. Each format (VHS and Beta) uses completely different frequencies and phase relationships to record the color signal.

The recorded tape must be played back on a VCR deck that's operating properly. This can be the same deck used to record the tape (if the playback circuits are checked with a prerecorded tape to confirm that they operate correctly) or on another VCR machine. You should, however, repair the problems in the playback circuit before delving into the recording problems so that a second VCR machine is not needed.

Some comments

Sencore engineers could have designed the special VCR test signals into the VA62 Analyzer, but they didn't when they realized that VCR systems are still in a constant state of flux and improvement. Also, of course, there's no way to predict what new signals will be used in the future. We've already seen functional improvements that might require additional analyzing signals in the future. Examples include the increased use of the ¼-inch tape format introduced by Technicolor, and more recently, the introduction of stereo hi-fi audio in the Sony Beta machines that are encoded on the tape using the video heads instead of the conventional longitudinal recording techniques that have been used for audio tape recorders for many years. There's nothing to keep the VCR manufactures from making other changes or from introducing other tape formats in the near future. Each of these advances in video technology would make the VA62 obsolete if the test signals were built into the Analyzer. The next video test accessory units will probably be ones that supply special signals required for functional analyzing of TV cameras and video disc machine systems. Other test accessories could be for TV stereo and new types of VCR formats if and when they're developed and introduced on the consumer video market.

Special VCR test signals

Most VCR service technicians concur that one of the toughest amplifier stages to troubleshoot is the low-level input/output circuits associated with the video record/playback heads. The reason for this problem is that the signal levels required to record and play back from the video heads are so low that an oscilloscope is not effective in tracing these signals. The symptom for a defective head is about the same as for a dirty switch contact, a bad rotary transformer, or a faulty head preamplifier. Thus, an injection test signal of the proper phase and amplitude is required to determine which components in the low-level head amplifier circuit is causing the problem.

The signals from the VC63 VCR test unit (see Fig. 10-35) can be injected at any point in this low-level signal path. This test signal is of the proper frequency and phase to pass through the tuned head preamplifier circuits as though it were fed into or picked up by the spinning video heads. The amplitude of this test signal is low enough to duplicate signals normally found in these stages. If the signal is too high, it can cause cross-coupling from one head channel to the other and cause you to make an incorrect trouble diagnosis.

10-35 Photo of the VC63 VCR test unit.

Play and record mode signal circuits

So that you, the technician, will have a better understanding of the complex injection test signal required for VCR troubleshooting, let's now take a brief look at the VHS format videotape recording system. I'll touch on the various signals found in the record and playback modes and show via block diagrams the circuits required to record and play back a videotaped TV signal.

The record mode signal circuit

The simplified signal circuit block diagram shown in Fig. 10-36 is for a model VBT200 RCA video recorder with the VHS format. Note that for recording a video signal (1-volt negative sync) from either the internal TV tuner or an accessory TV camera, the signal is fed to the video input of the machine where it follows separate luminance and chrominance paths.

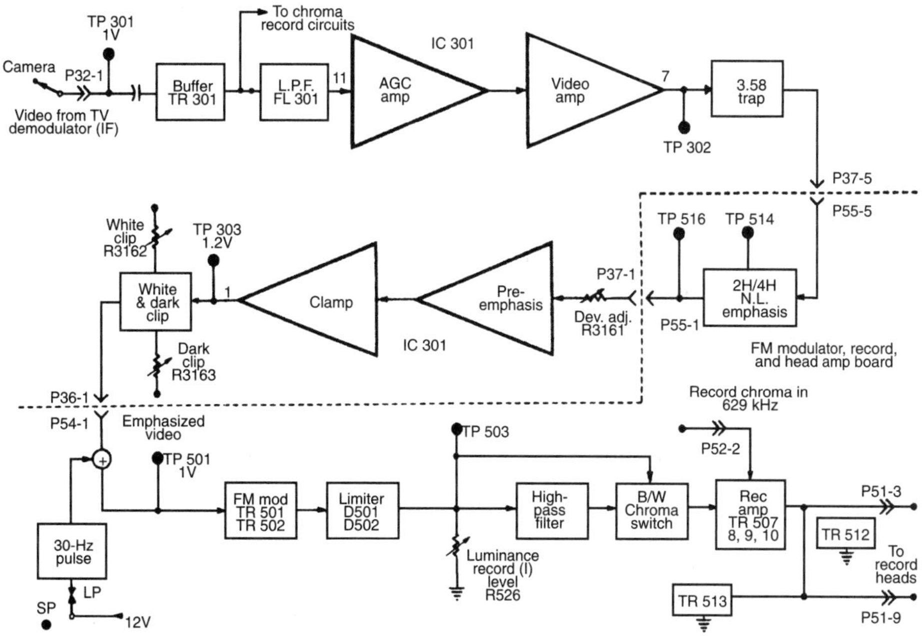

10-36 Simplified VCR record block diagram (RCA).

The luminance, or black-and-white video portion of the signal, enters an AGC amplifier where its level is regulated to remove any variation due to differences between signal inputs. This regulated signal then passes through an electronic-switch-selectable 2-hour or 4-hour emphasis circuit that boosts the high-frequency component of the video signal. The signal emerging from the emphasis network contains overshoots that are clamped to specified levels by black-level and white-level clamps.

The output of the clamping circuits is thus a video signal with specified overshoots. This signal is then applied to the FM modulator, which is basically nothing more than a flip-flop that operates at 3.4 MHz when an input voltage corresponding to sync tip level is present. As the video signal drives the modulator, its frequency approaches 4.4 MHz. The FM modulator output is applied to a record amplifier that boosts the signal level to that required to drive two video heads, which are connected in parallel during the "record" mode of operation.

In the 4-hour operating mode, a 30-Hz squarewave signal is fed to the FM modulator to shift the dc level of the video signal up on alternate fields. This action raises the modulating frequencies by the ½-frequency H (7875 Hz) interval needed to minimize luminance beats caused by nonhorizontal sync pulse alignment.

The chroma signal, after some filtering to remove any luminance signals, is fed to a balanced modulator where it's heterodyned against a 4.2-MHz signal. The difference between these two signals is 629 kHz, which thus becomes down-converted chroma. This signal is mixed with the high-level luminance signal so that the luminance FM is used as bias to record chroma—somewhat like in audio tape recording. During "record," burst is separated from the incoming chroma and fed to an APC comparator. This action phase-locks a 3.58-MHz VXO (variable crystal oscillator) so that it produces a 3.58-MHz cw signal that feeds a second balanced modulator. The other input to this balanced modulator is 629-kHz cw developed by a special VCO (voltage controlled oscillator) and counter circuit. Balanced modulator output is the sum of the 3.58-MHz input and the 629-kHz input, which is about the 4.2-MHz required to drive the second balanced modulator that down-converts the chroma.

During "record," the video signal is fed to a horizontal sync separator to provide horizontal sync that phase locks a 2.517-MHz VCO to a multiple of 160 times the horizontal line frequency. Output from the 2.517 VCO is divided by 4 in the rotary phase switch (divide-by-4-counter). Counter output is 629-kHz cw that phase changes line by line as the video is recorded or played back. One of the 629-kHz signals is divided by 40 and is compared against a horizontal sync AFC, like that used in a TV receiver. The control voltage output from the AFC circuit locks the 629-kHz cw signal to the horizontal sync.

The "play" mode signal circuit

In tape playback, the two video heads feed individual preamps that are turned "on" when the particular head is in contact with the videotape. Preamplifier output is then summed together into a continuous RF signal prior to entering separate luminance and chroma circuits.

The luminance (black-and-white) signal is 3.4- to 4.4-MHz FM, which is fed to limiter stages that remove all amplitude variations including the chroma signal content. Limiter output drives an FM demodulator that recovers the original black-and-white video signal complete with the overshoots introduced by the record emphasis. In order to correct frequency response of the video signal (and attenuate high-frequency noise), the signal is passed through appropriate de-emphasis networks that are switchable for 2- or 4-hour operation. After de-emphasis, the video signal passes through some noise cancellation circuitry and emerges as a 1-volt peak-to-peak negative sync video to drive the RF converter in the VCR machine to provide channels 3 or 4 RF output to a TV receiver.

A dc compensation stage is switched into the circuit when in the 4-hour operating mode to remove the dc level shifts that were introduced in "RECORD" to produce the half line offset. This dc compensation is a 30-Hz squarewave that corrects the dc component of the video signal in the shifted fields to agree with the dc level of the video in the nonshifted fields. Note the "playback" signal block diagram shown in Fig. 10-37.

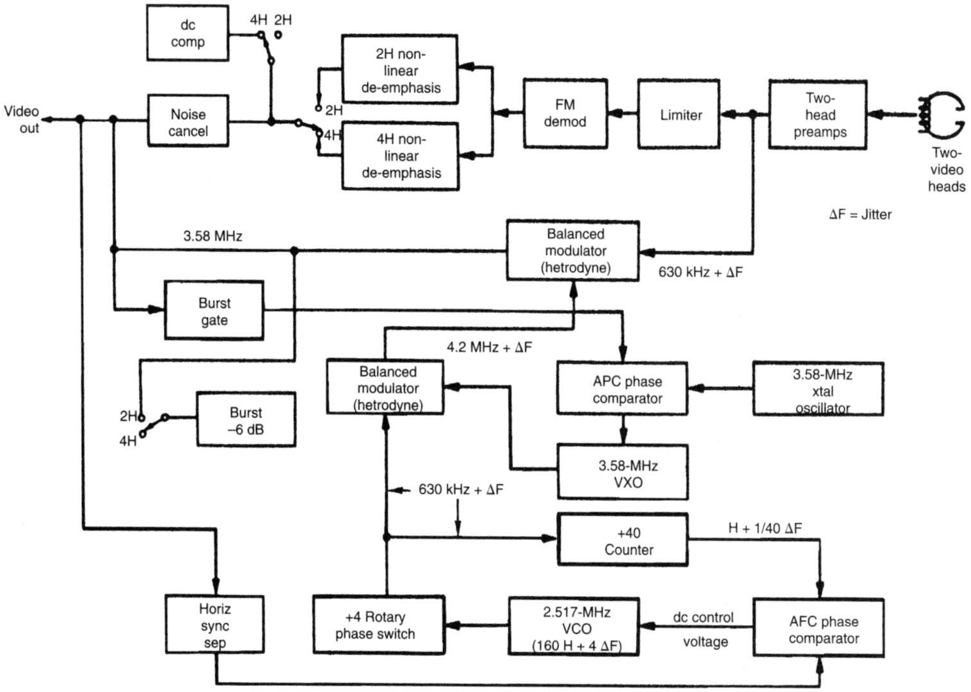

10-37 Block diagram of a VCR playback circuit (RCA).

Jitter-frequency component

The chroma signal consists of 629-kHz RF plus a *jitter-frequency* component. Let's take a look at this "jitter frequency" component. Generally, the video-head speed and the tape-transportation speed are held to very tight tolerances in actual operation. However, an extremely small change in video head velocity, which would produce nothing noticeable in a black-and-white picture, can produce phase errors so bad as to make the picture not viewable. Thus, this effect of minute speed variations, which are called the jitter component, must be removed in order to produce proper color. Note that if the 629-kHz down-converted chroma signal containing jitter is beat against a 4.2-MHz signal that contains the same jitter component, the difference will be a constant 3.58-MHz signal without the jitter component. This is accomplished by passing the 629-kHz signal into a balanced modulator where it's beat against 4.2-MHz signal with jitter, thereby producing a constant output of 3.58 MHz chroma. In playback, this 4.2-MHz signal plus jitter is produced by the second balanced modulator.

During playback, the 3.58-MHz VXO is locked to a locally generated constant 3.58-MHz signal derived from a crystal oscillator. The other input to the balanced modulator is 629 kHz plus the jitter factor. The 629-kHz signal is provided by the same 2.517-MHz VCO used in "record." Recall that the VCO locks to horizontal sync because in playback, the horizontal sync pulse contains the small time-base errors caused by minute variations in head cylinder speed. Now, the playback horizontal sync pulse can be fed to the AFC phase comparator to produce dc control voltage with jitter modulation. Consequently, this jitter is introduced into the 2.517-MHz signal whose output is divided to 629 kHz. In this way, the 629-kHz output contains the jitter component and thus cancels the jitter in the 629-kHz incoming chroma.

VCR servo comments

Basically, the VCR has two servo systems, which are the capstan servo and the video cylinder head servo system. In the "record" mode, TV vertical sync is the reference for the cylinder servo. (Note the block diagram of the cylinder servo in Fig. 10-38). The feedback signal is a pulse signal taken from the cylinder P.G. (pulse generator), which provides an accurate indication of where the video heads are on the tape and which head is in contact with the tape. During "play," it's only necessary that the cylinder head rotate at a constant 1800 rpm. Thus, a 60-Hz reference signal is supplied by a counter circuit that operates from an internal 3.58-MHz crystal oscillator. The capstan servo controls the actual movement of the tape. During "record," the capstan servo moves the tape at a constant speed. In the "playback" mode, a reference signal pulse (30 Hz) is used to lock in cylinder head rotation. When using the VC63 test unit you'll need, at times, to use the 30-Hz servo test pulse from the VA62 Analyzer in order to lock in the video pattern for proper trouble diagnosis.

Video heads and head amplifier circuit comments

The symptom for a defective playback head amplifier circuit's usually easy to spot, but might also turn out to be a faulty video cylinder head. The playback picture (for a faulty amp) will have a severe flicker and will be very noisy, as shown in Fig. 10-39. The reason for this symptom is that only every other video field is being played back to the video monitor. The two-head system uses one of the heads to pick up every odd field and the other head to pick up every even field. When one of the signal paths is defective, you have one complete field followed by noise streaks. The result, for an interlaced picture system, is the picture symptoms that I just described. Should the record-head amplifier be defective, you'd only have noise and streaks on playback and no discernible picture.

Each of the two video heads has its own rotary transformer, which transfers the signal from the video head output to the preamplifier. This rotary transformer is actually made up of two coils, one that's part of the moving head disk, and the other that's part of the stationary portion of the video head assembly. As the video head picks up the signal from the tape, it's inductively coupled to the stationary coil by the moving coil. This eliminates the need for slip rings or brushes that could cause intermittent operation, which would occur after a few hours of operation.

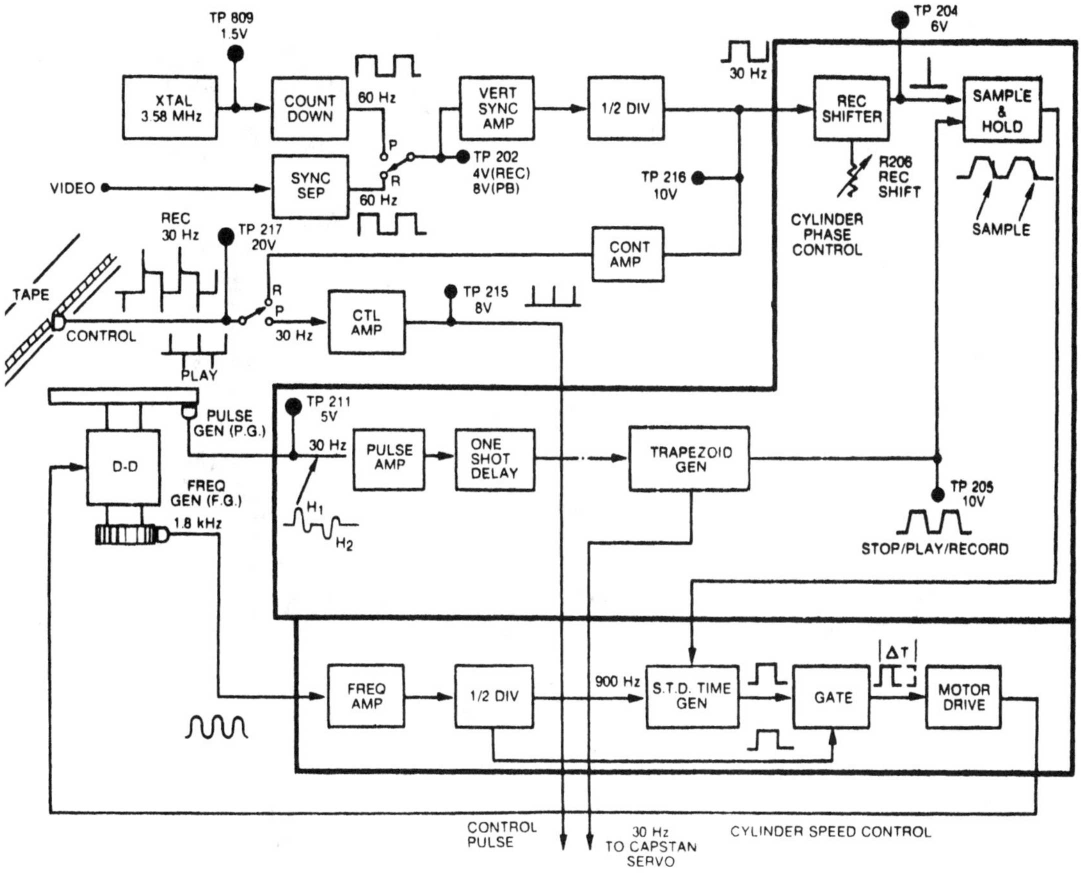

10-38 Block diagram of a cylinder servo system.

The signal that's picked up by the stationary portion of the rotary transformer is passed on to the head preamplifiers. There are two of these preamplifiers, one for each head, and these preamplifiers have been known to be defective. This can be used to locate poor or intermittent contacts on the long slide switches usually found in the VCR record/playback circuit boards. Faulty switch contacts are a common occurrence in older-model VCR machines. Using the signal injection technique along with a scope to signal trace is a good tip to remember when looking for an open or intermittent signal path in VCR or color TV circuits.

Confirming defective video heads

The following two case-history VCR head problems will illustrate how the video analyzer and the VC63 test unit was used to quickly verify defective video heads.

The first example was with a Zenith VR9000 BetaMax VCR that would not play back a good prerecorded tape, but it would record a good tape and produce a fine-

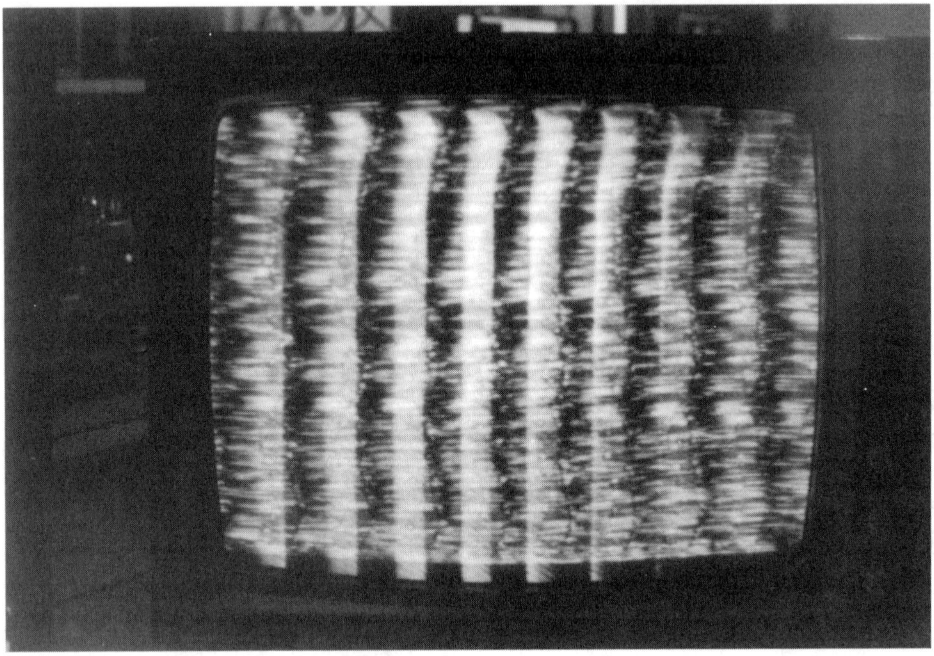

10-39 Symptom for a defective VCR playback amplifier.

quality picture when played back on another machine. You'd almost bet that these symptoms would rule out a defective head. In this case, I injected the test signal from the VC63 (see Fig. 10-40) at the output plug of the VCR heads and thus into the machine's playback head amplifiers. With this check, I could pass a perfect color-bar pattern through all of the machine's playback circuits. When a new head was installed and aligned, the playback picture was excellent.

In another case a RCA VBT200 VHS machine would play back a prerecorded tape, but would only have streaks and lines across the picture when a tape was recorded and played back. For a quick check, I injected the proper test signal from the VC63 into the heads (note Fig. 10-41) when the machine was in the record mode. When the tape was played back, a correct color pattern was shown, which indicated faulty video heads. When new heads were installed in this machine and adjusted, the recorded tape played back perfectly. Thus, another "tough dog" VCR troubleshooting problem was made easy.

Many VCR service technicians have asked and pondered about this question: Are the video heads defective are aren't they? The answer: only the VA62 Analyzer and VC63 VCR test adapter can tell you for certain.

Checking out the record head amplifiers

Let's see how you can quickly check out the record head amplifiers by using signal injection and the oscilloscope. You'll find that the Sencore VC63 VCR test accessory

10-40 Signal being injected into the playback VCR amplifiers.

supplies the proper test signals that can be injected at any point in the low-level record and playback amplifier circuits. This VCR test unit shown in Fig. 10-42 is being used for signal injection in a VHS system head amplifier board. This test signal is of the correct amplitude and phase to pass through the tuned-head preamp circuits just as though it were being picked-up or recorded on the tape by the spinning video heads. Make sure you adjust the substitute test signal to duplicate the signals usually found in these stages. If you inject too much signal, a cross-coupling might occur from one head channel to the other, and you can't really tell if you've located the defective stage.

Generally, the scope can't be used very effectively for tracing a signal through the various head amplifier stages because of the many different frequencies and phases found in these circuits. A signal found in a properly operating VCR head amp will just show up as a fuzzy-looking, thick line of grass because the scope can't lock

10-41 Test signal being injected directly into the video heads.

10-42 VHS video head amplifier being checked with the Sencore VC63.

onto any one frequency. (Note the scope trace in Fig. 10-43.) This, of course, can make scope signal tracing very confusing. However, you'll learn a little later how signal injection and the scope can be used to help you speed up troubleshooting the record amplifier stages.

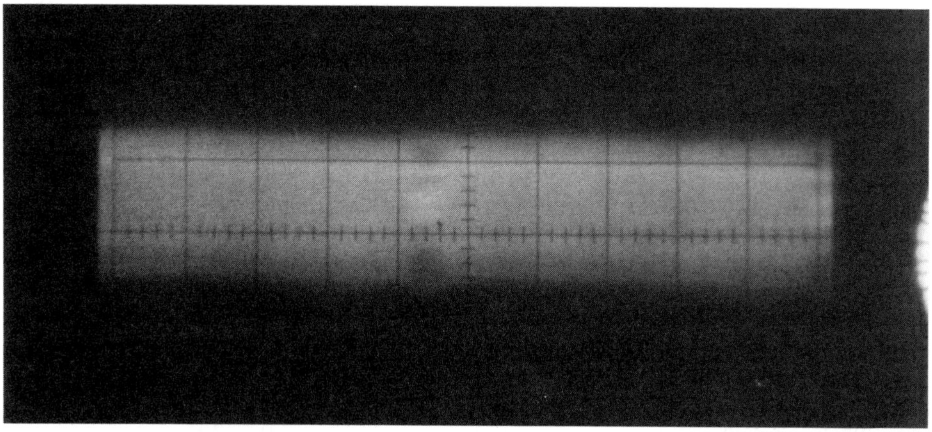

10-43 Scope grass signal found in the VCR head amplifier system.

Analyzing the record head amplifier circuits

Let's now see how you would analyze a record head amplifier problem in an RCA model VBT200 VCR. As you go through these checks, refer to the marked triangle signal injection points on the circuit diagram in Fig. 10-44. Note: Before troubleshooting the record mode, the machine must be able to play back a prerecorded tape with a good picture.

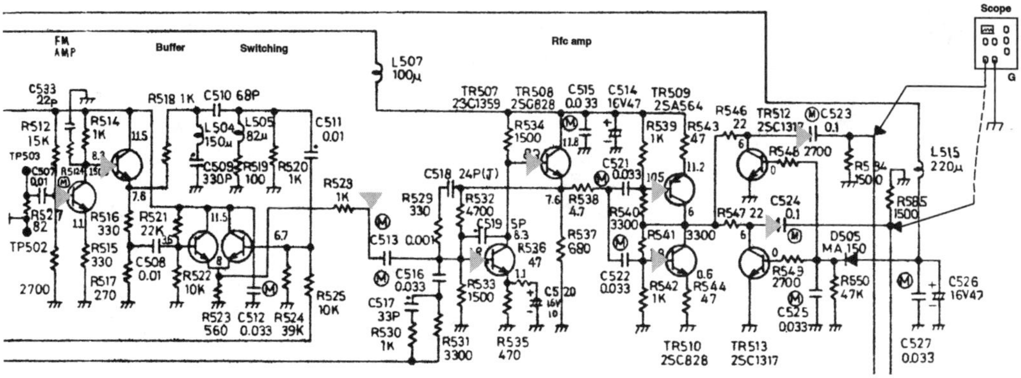

10-44 Head amplifier circuit for an RCA model VBT200 VCR.

The first step to preparing the machine for these signal injection tests is to connect the standard 1-volt video test signal from the VA62 into the video input jack of the VCR and switch to the camera mode. Now, with the machine in the record mode but with no video input signal, check with the scope for a 0.3- to 0.7-volt peak-to-peak pulse at approximately 3 MHz at test point TP502. If no pulses are present, check out the FM modulator stages TR501 and TR502.

Before any signal injection is performed, ground test point TP502 to eliminate any stray signals. With the VCR in the record mode, inject a signal from the VC63 as shown in Fig. 10-44. Set VC63 to VHS lum & chroma and the analyzer to a color bar pattern. A good place to start signal injection is at the base of TR508. Now, play back the tape and, if it has recorded the test bar pattern, move the signal injection probe back to base of the TR504 buffer and recheck the tape on playback. If it didn't record a signal on the tape, the fault is between these two injection points.

To help you speed up the record head amplifier circuit analysis, you can use the following procedure.

Connect the scope as shown in the record head amplifier circuits (input to heads). (See Fig. 10-45.) Again, to eliminate any unwanted signals, ground test point TP502. Now, with the VC63 set to VHS lum & chroma at approximately 0.1 to 0.3 volt output, inject the signal at the base of the TR503 FM amp and then to the TR504 buffer amp base. Then see if there's any signal amplitude on the scope screen. If no signal is indicated, then move on to the base of the TR507 and the TR508 record am-

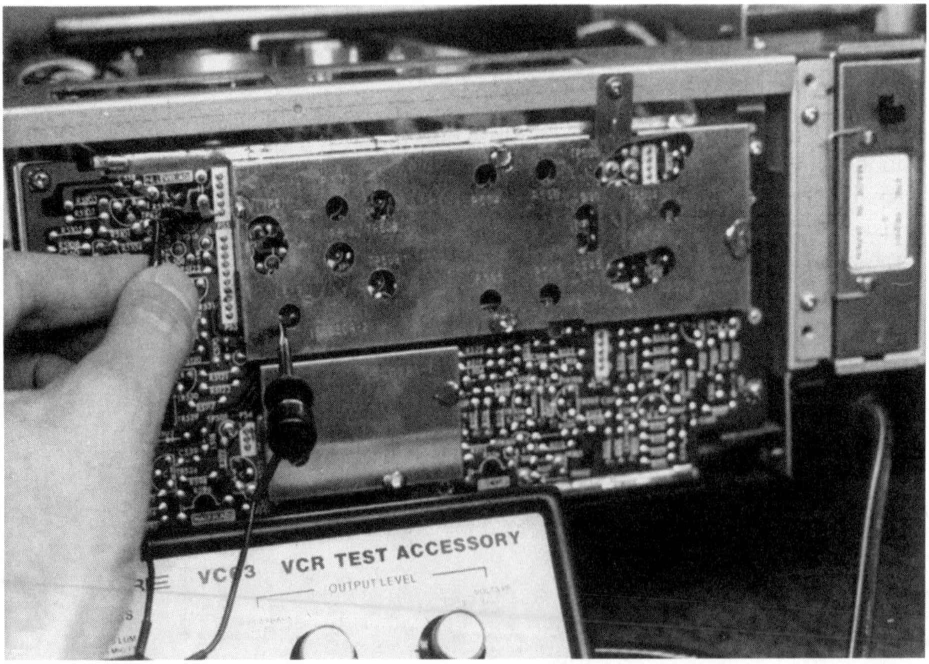

10-45 VCR record circuits being checked with the VC63.

plifiers and note any scope waveform action. When the scope indicates some signal waveform, then the stage before that injection point is defective. Use this technique so you won't have to record and then play back the tape after each signal injection is performed in order to find out which stage is defective.

Before looking at troubleshooting the playback head amp circuits in a Beta machine, you should note that one of these VHS machines produced a very poor, noisy picture on playback, with an indication of video output from only one head. A quick signal injection test located IC501, a playback head amplifier chip that was intermittently faulty.

Analyzing the playback head amplifier circuits

For these playback head amplifier checks, I'll be referring to the RP-1 board in a Zenith VR9000 VCR machine. As you make these checks, refer to the injection point triangles shown in Fig. 10-46. Set the VC63 VCR test unit to BETA lum & chroma and start with a 0.5- to 1-volt peak-to-peak signal output. In order to lock the test pattern in while troubleshooting in the playback mode, the 30-Hz servo pulse from the VA62 Analyzer must be connected to the VCR servo control line in the VCR. It's best to override the machine's servo pulse by 2 or 3 volts peak-to-peak, and you'll need to set it for proper polarity (by pushing or pulling the servo control knob in or out) in order to play back the color-bar test pattern in color.

The first step in locating playback trouble is to inject a signal at test point TP4 on the RP-1 board. For these test procedures, put the machine in the playback mode and unplug the video heads at CN5002. If a good color-bar pattern is seen on the monitor with a signal injected at TP4, then the problem is in the RP-1 board playback circuits.

To divide and conquer the fault in this board, you should inject a test signal at pins 24 and 23 of IC1 (CX134A playback amp) to quickly find which section is at fault. If you don't have a test pattern on the monitor, then the trouble is in IC1 or its support components. If you do have a good pattern at this point, then work back towards the video head plug (CN5002) by injecting test signals at the Q7 and Q9 preamps and at the input and output of transformers T1 and T2. Also, should you appear to have a problem in this area (it might look like only one head operating) check out the switch transistors Q6, Q8, and Q10. Note in the Fig. 10-47 that the VC63 is being used to inject a test signal at TP4 on the RP-1 board of a Zenith VR9000 VCR.

Keep in mind that the transistor switch is used to switch between the "A" head output and the "B" head output when the VCR machines are in the playback mode. This switch is usually turned off and on by a pulse that comes from the servo circuits. You can use a scope to see if this pulse is available at these electronic head switches. In this Beta model, transistors Q9 and Q10 are switched when the machine goes from the review to cue modes.

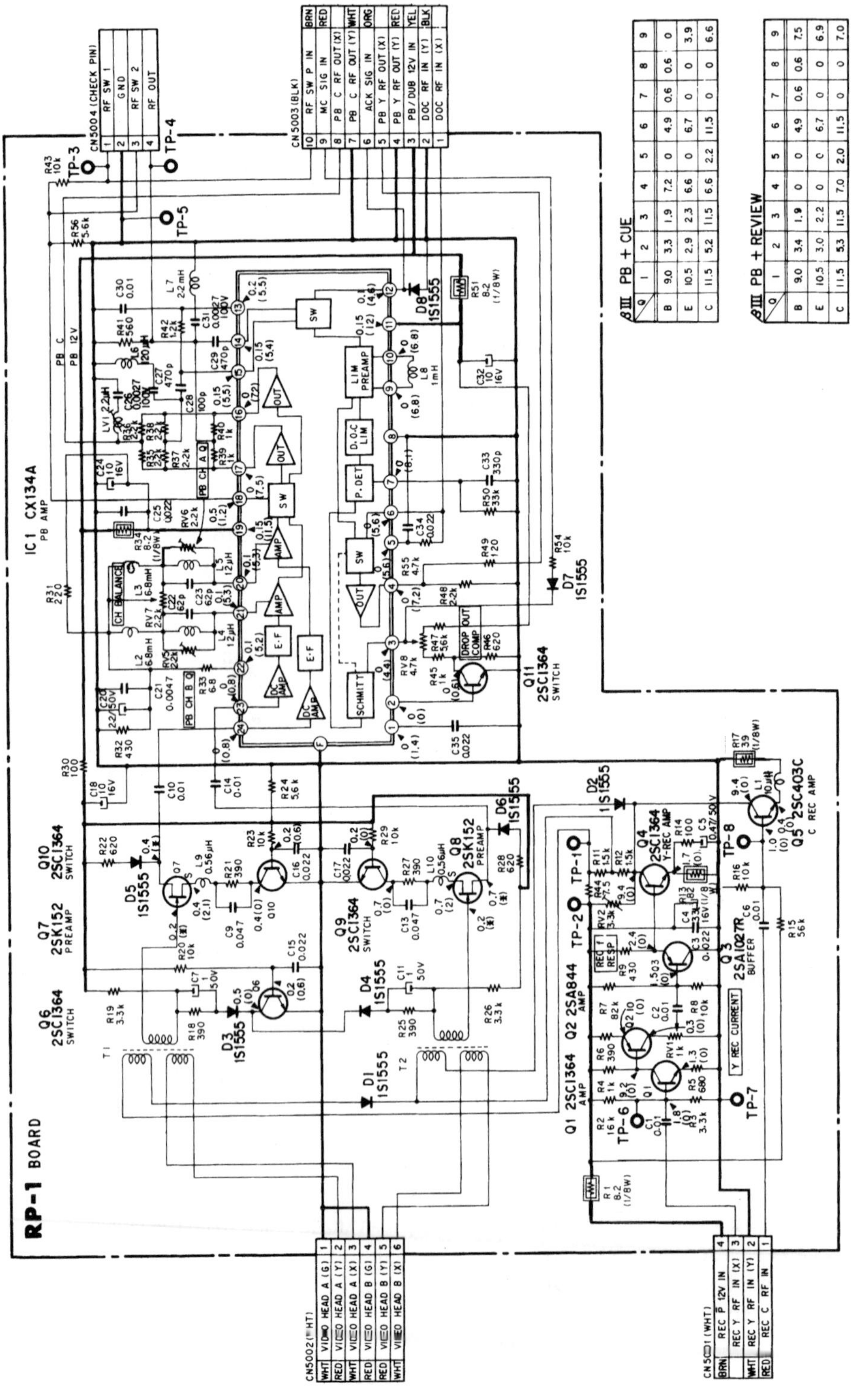

βIII PB + CUE

Q	1	2	3	4	5	6	7	8	9
B	9.0	3.3	1.9	7.2	0	4.9	0.6	0.6	0
E	10.5	2.9	2.3	6.6	0	6.7	0	0	3.9
C	11.5	5.2	11.5	6.6	2.2	11.5	0	0	6.6

βIII PB + REVIEW

Q	1	2	3	4	5	6	7	8	9
B	9.0	3.4	1.9	0	0	4.9	0.6	0.6	7.5
E	10.5	3.0	2.2	0	C	6.7	0	0	6.9
C	11.5	5.3	11.5	7.0	2.0	11.5	0	0	7.0

10-46 Head amplifier playback circuit for a Zenith VR9000 VCR machine.

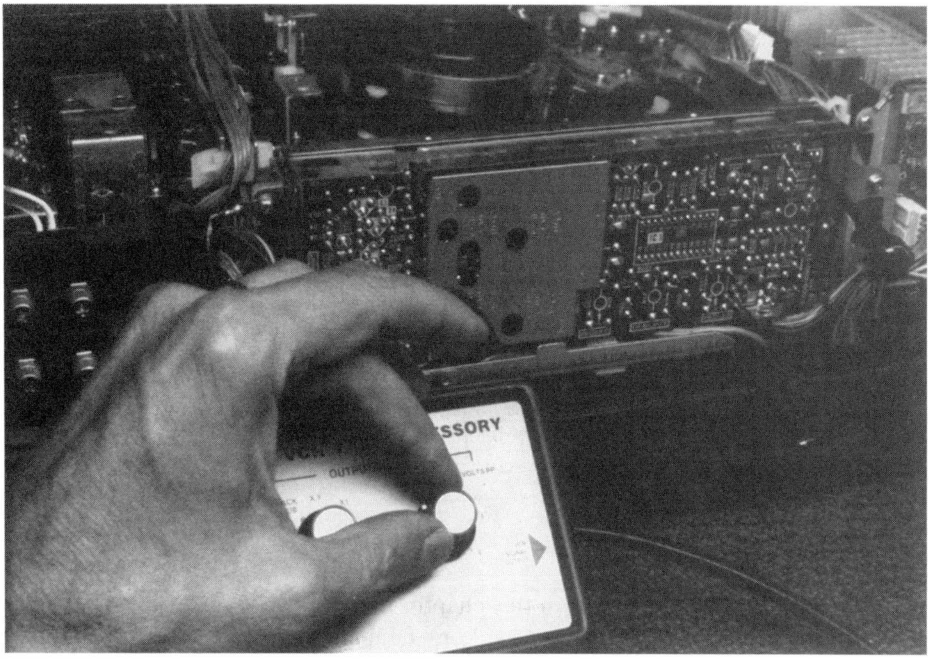

10-47 The VC63 VCR test accessory being used to inject a test signal on the RP-1 circuit board of a Zenith VCR.

The wrap-up

This is just a very tiny sneak preview of what you can do with the new Sencore VA62 and VC63 VCR test adaptor to make VCR troubleshooting much easier. Of course, many more video troubleshooting applications can be performed with the VA62A Analyzer, and more are to come. Other versatile, plug-in adaptors are in the planning stages for testing future electronic products—some not yet even dreamed of.

11 All-format VCR analyzer

IN THIS CHAPTER YOU'LL FIND OUT HOW TO USE THE SENCORE VC93 VCR ANALYZER to its fullest capabilities. The chapter is divided into four sections that cover the major functional portions of a VCR or camcorder:

1. Luminance.
2. Chroma.
3. Audio.
4. Servo.

You can use the information in this chapter in two ways. First, you can use it as a reference for how to do a specific test. For example, details on how to test a rotary transformer are in the section entitled "Testing rotary transformers."

You can also use this information as a step-by-step troubleshooting guide. You might want to consult the "trouble tree" in Fig. 11-1 to help you decide what tests and signals to use for a given symptom. The figure outlines a logical troubleshooting sequence that will lead you to the defective stage in the least amount of time. This chapter provides you with specific details and procedures for troubleshooting defective machines. After you've become familiar with the test procedures, you troubleshoot by following the steps outlined in the "trouble tree." Begin your troubleshooting by selecting the "trouble tree" path that best fits the symptom you observe.

Some starting tips

The VC93 improves troubleshooting effectiveness through a technique called *functional analyzing*. With this technique, you inject known-good signals, supplied by the VC93, into the functional blocks. If the output returns to normal, you're injecting after the defective stage; if the output remains the same, your injection is before the defective stage.

Begin your troubleshooting by checking the machine for obvious operator problems. Check the position of the controls for a clue to the problem. A "tracking" control that's all the way to one end of its range, for example, is a clue to a possible tracking problem.

After checking for obvious operator problems, place a work tape into the machine and observe the playback audio and video on a monitor connected to the machine's output. Always troubleshoot a VCR or camcorder in the playback mode first.

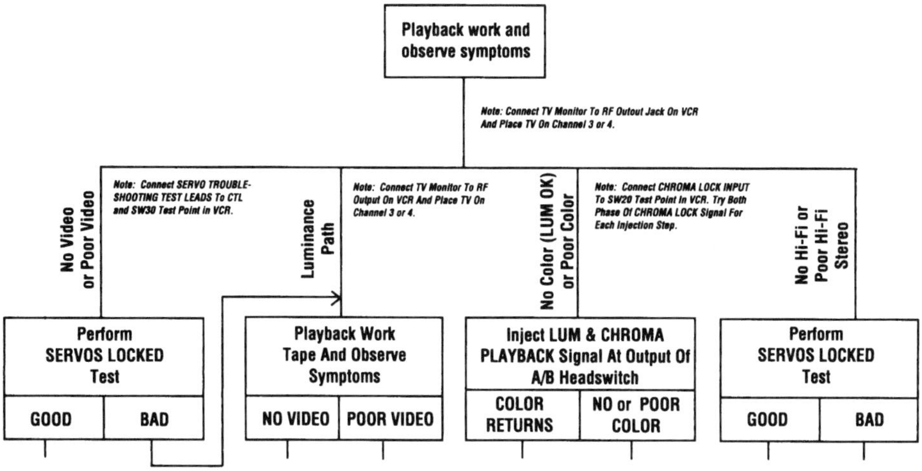

11-1 Select the path on the "trouble tree" that most closely matches the machine's symptoms.

Many circuits are common to both playback and record. Ensuring that the playback circuits work eliminates most of the potential circuits that could be at fault. To check the record circuits, you must also record a signal and play it back.

Caution

Do not use the servo performance test tape or factory alignment tape for the initial checkout. Doing so could ruin a valuable tape. Always check the machine using a work tape first to ensure that the machine will load and play the tape without disastrous results.

VCR overview

A VCR contains five major sections: luminance, chroma, audio, servo, and system control. The following is a brief description of each. (See Fig. 11-2).

The luminance section receives the recorded information from the tape and processes it into a form that can be used by a TV or monitor. Inject the VC93 playback signals in the circuits between the video heads and the FM detector. Use the drive signals between the FM detector and the video output.

The chroma section processes the down-converted chroma signal so it can be combined with the luminance signal to create a color playback signal. The VC93 playback signal contains a down-converted chroma signal and other key signals to troubleshoot the chroma conversion circuits.

The audio section converts the linearly recorded audio and FM modulated stereo hi-fi signals to baseband audio. Use the VC93 playback "stereo" signals to troubleshoot the hi-fi stereo section. Use the audio drive signals to troubleshoot the linear audio section.

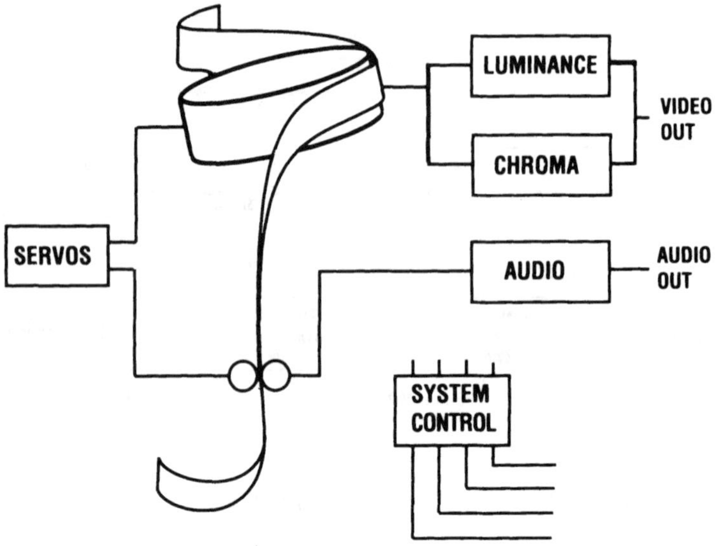

11-2 The five overall functional sections of a VCR.

The servo section controls the tape movement and adjusts the speed and phase relationship of the heads with respect to information recorded on the tape. The VC93 servo analyzer tests evaluate the servo operation, and the servo sub bias supply is used to further troubleshoot the servos.

The last section is the system control. This section has one or more microprocessors that monitor the entire operation of the machine and tell the various circuits what to do at a certain time. The microprocessor program is unique to a particular model. This, combined with the digital signals involved, means signal substitution is not the best method for troubleshooting the system control section. Instead, troubleshoot this section with an oscilloscope and dvm.

Luminance troubleshooting

Use the VC93 playback and drive signals to troubleshoot the luminance section. Use the playback signals to substitute in any circuit from the video heads to the FM detector. Use the "composite video" and "lum" drive signals in the luminance circuits between the FM detector output and the video output jack/input or the RF modulator. (Note the setups in Fig. 11-3.)

Video head troubles

There are two symptoms of a bad video head, depending on whether one or both heads are bad. The first symptom, caused by two bad heads, is complete loss of video. In older VCRs, the entire monitor screen is snow, as shown in Fig. 11-4. Newer machines often mute the video to produce a blank or solid color raster, as shown in Fig. 11-5.

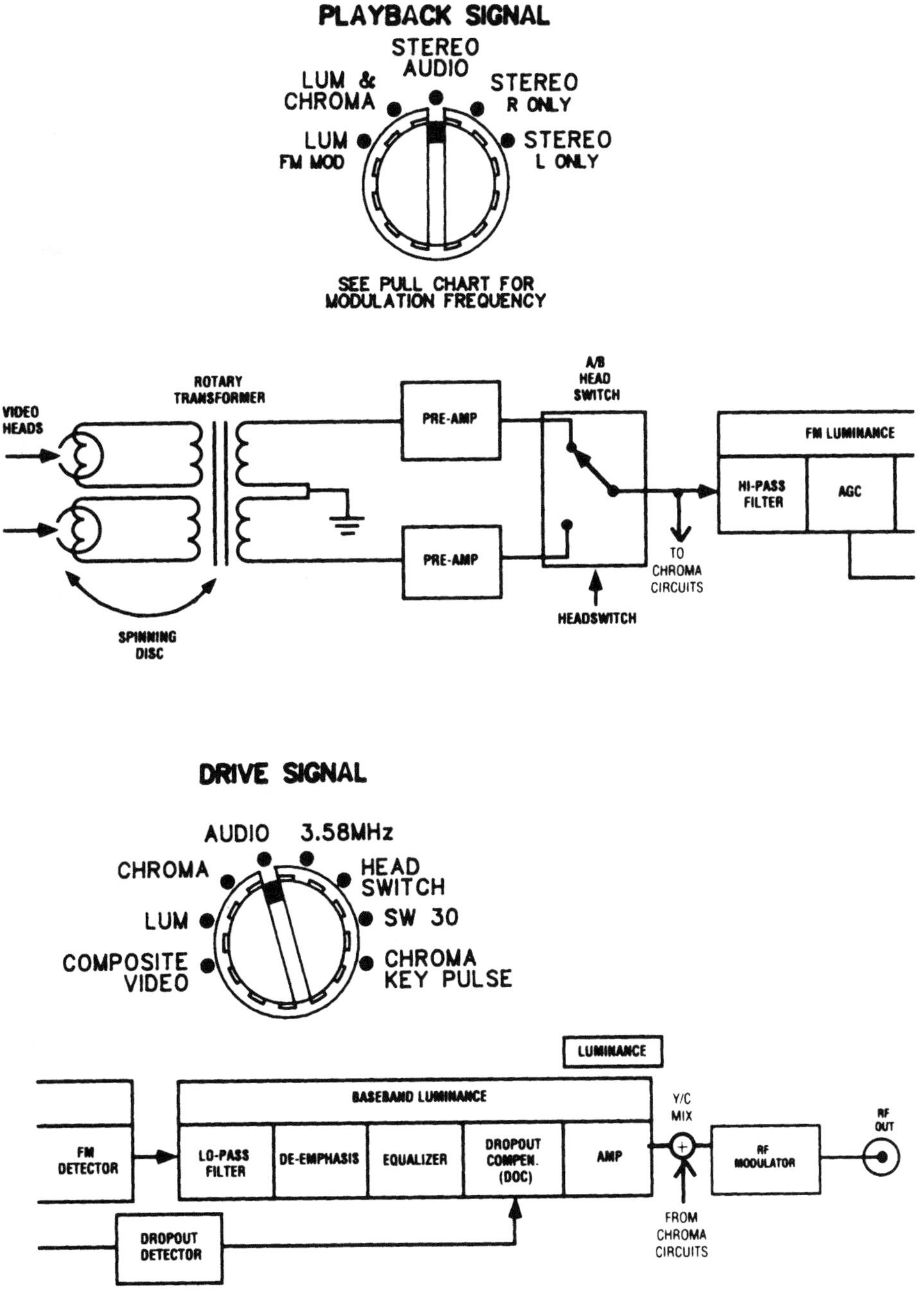

11-3 Use the VC93 playback signals to inject before the FM detectors, and use the drive signals to inject after the detector.

The second symptom appears when one head fails. In this case, the machine produces a noisy playback picture. An important part of interpreting this symptom is that the noise must cover every part of the picture. If any section of the picture is

11-4 Snow on the monitor screen indicates a bad video head.

11-5 On newer machines, the screen is blanked out when you have a loss of video. This could indicate a bad video head.

clear (even if it's only a few inches somewhere on the screen), you don't have a defective head. The problem is likely servo or tape path related.

Note: TV sets that use vertical countdown circuits will show several inches of clear picture at the bottom of the screen, even if one video head is bad. These sets blank the noise from one field of video. For best results, use a monitor that doesn't use a vertical countdown circuit.

Symptoms alone don't prove when video heads are bad. Defects in other luminance circuits can produce identical symptoms. If the symptoms suggest bad video heads, use the VC93 playback signal to substitute for the head signal. Because the heads rotate, you cannot inject the signal directly into them. Instead, inject at the output of the rotary transformer. This will determine if the problem is in the video heads/rotary transformer or if the problem is after the injection point.

To determine if the circuits after the rotary transformer are working:

1. Insert a blank tape into the machine and press "play."
2. Set the VC93 as follows: (See Fig. 11-6.)
   ~ VCR format to match format being serviced.
   ~ Modulation to "color bars" or "external."
   ~ Playback range to "playback head sub."
   ~ Playback signal to "lum."
3. Connect the head substitution test lead to the playback output jack.
4. Connect the head substitution test lead to the ch. A and ch. B head preamp inputs of the VCR.
5. Observe the playback monitor for an improved picture.
6. Adjust the playback level control for the best picture.

A good picture shows that all the circuits after the injection point are good. This leaves either the rotary transformer or the video heads to be defective. Test the rotary transformer following the procedures in the test section. If the rotary transformer tests out okay, go to the servo analyzer tests. If they don't indicate a problem, then replace the video heads.

If you don't obtain a good picture, then a stage after the injection point is bad. Refer to the "Testing video stages" section for information on troubleshooting those stages.

Testing rotary transformers

Rotary transformers fail when either the rotating or stationary windings develop an open or short. Test the transformer by injecting the "lum" playback signal. If it passes the signal, the rotary transformer is good.

To test the rotary transformer:

1. Turn the VCR off.
2. Set the VC93 as follows:
   ~ VCR format to format being serviced.
   ~ Modulation to "color bars" or "external."

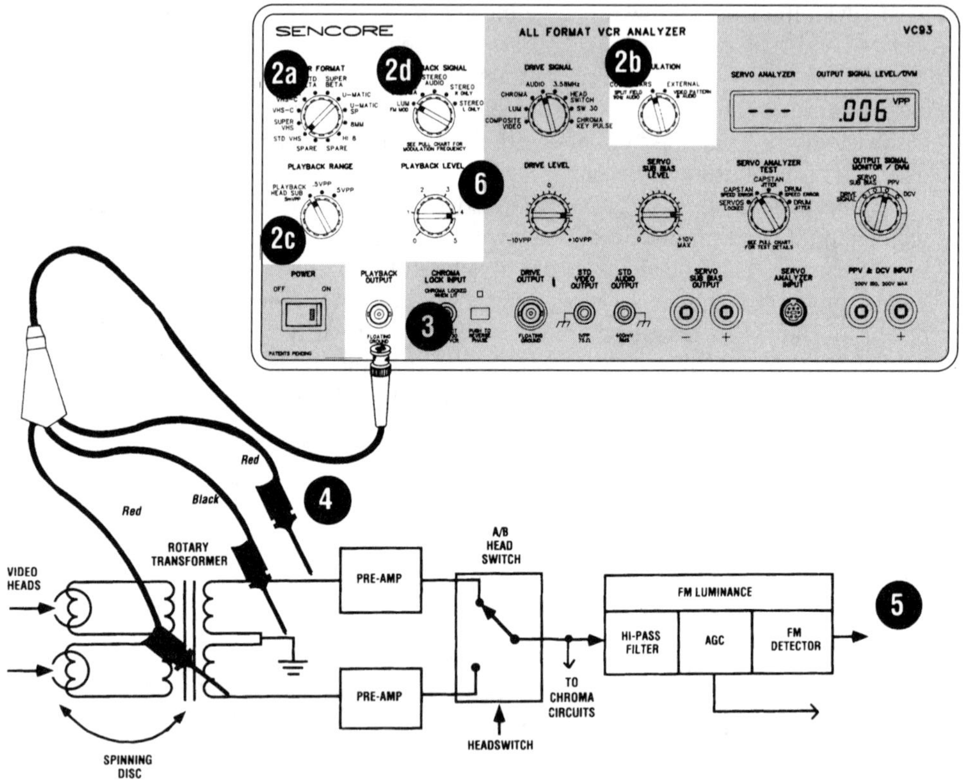

11-6 Setup for injecting the playback signal into the luminance sections.

- Playback range to "5VPP."
- Playback signal to "lum." (Note the setup connections in Fig. 11-7.)

3. Connect the direct test lead to the playback signal output jack.
4. Set the playback output level:
   - Set the output signal monitor/dvm to "PPV."
   - Connect the dvm test leads to the PPV & DCV input.
   - Connect the dvm test leads to the direct test leads.
   - Adjust the playback level control for an output of 2 VPP.
5. Perform the rotary transformer test:
   - Connect the direct test lead to the rotary transformer primary.
   - Connect the dvm test leads to the rotary transformer secondary.
   - Read the results on the output signal level/dvm LCD readout.
   - Repeat the previous three steps for the remaining windings.

If the readings are similar for each winding, the rotary transformer is good. If any winding reads lower than the rest, or gives no reading at all, replace the rotary transformer. Note: Some rotary transformer shorts occur only at certain spots in the rotation. This typically produces a streak in the playback picture. If you suspect this problem, connect an oscilloscope in place of the VC93 PPV meter. Then slowly turn

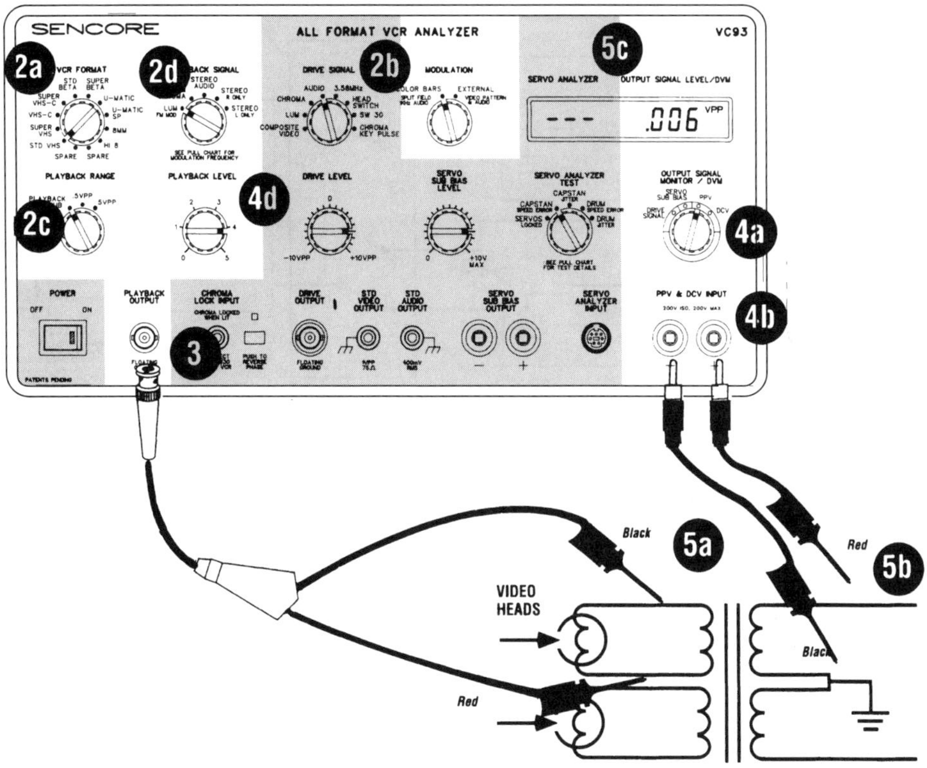

11-7 Setup for checking out the rotary transformer.

the transformer while pressing down on the upper drum assembly. If the output signal disappears at any point in the rotation, the transformer has an intermittent short and must be replaced.

Testing four-head VCRs

Some VCRs use one set of video heads for the fastest tape speed (SP for VHS and Beta I for Beta VCRs) and another set for the other speeds. Testing these four head-head machines requires a work tape recorded at the different tape speeds. (Note the four-head circuit shown in Fig. 11-8.)

Play back the tape and observe the picture at each speed segment. If you see symptoms that suggest a bad head, note at what speed the tape is playing. The noise can be caused by a bad head, but it also can be caused by the video head selection circuitry.

To determine if the problem is in the head selector or in the heads, locate the rotary transformer windings that correspond to the video heads used for the speed in question. Inject the VC93 "lum" playback signal before the video head selector for that winding. The picture will either stay bad or clear up. Note the test setup in Fig. 11-9.

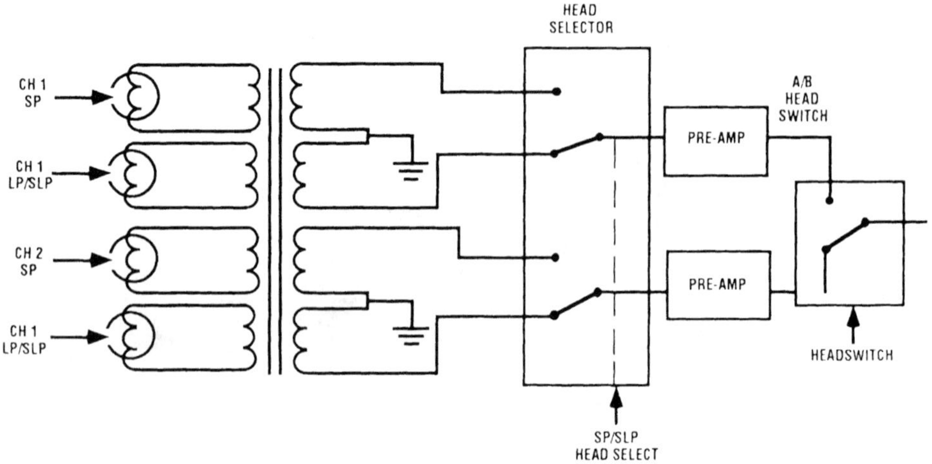

11-8 A four-head VCR has additional rotary transformer windings and switching circuits.

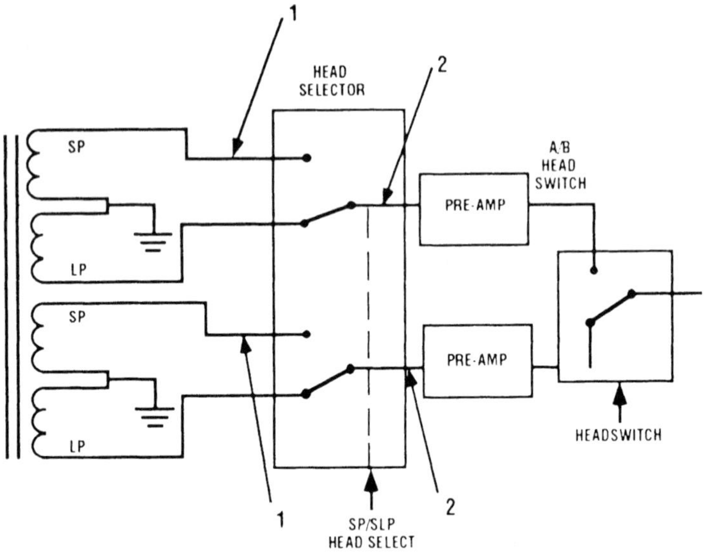

11-9 Injecting the VC93 playback signal both before (1) and after (2) the video head selector switch should produce a good picture. The example shows injection into the SP windings.

If the playback picture stays bad, either the head selector or the circuits after it are defective. To determine which, inject the playback signal after the head selector. If the picture clears up, the video head selector is bad. If the picture stays bad, the problem is after the head selector.

If injecting into the head selector produces a good picture, either the rotary transformer or the video heads are at fault. Follow the "Testing rotary transformers" procedure to determine which component is defective.

"Trick" VCR heads

Some VCRs use a "special effects" or "trick" head to minimize playback noise when the machine is placed in the pause, fast forward, or reverse modes. A head select circuit similar to that used in four-head machines selects the "trick" head.

A problem in the "trick" head or in its selector circuitry causes a noisy picture when the machine is placed in pause. To determine if the problem is in the selector or in the "trick" head, determine which rotary transformer winding corresponds to the "trick" head. Also, identify the winding that's used for both the "trick" and the normal mode. Connect one of the "red" head substitution test leads to the rotary transformer "trick" head output and the other red lead to the common output. Place the VCR in "pause" and observe the playback monitor. Note the connections shown in Fig. 11-10.

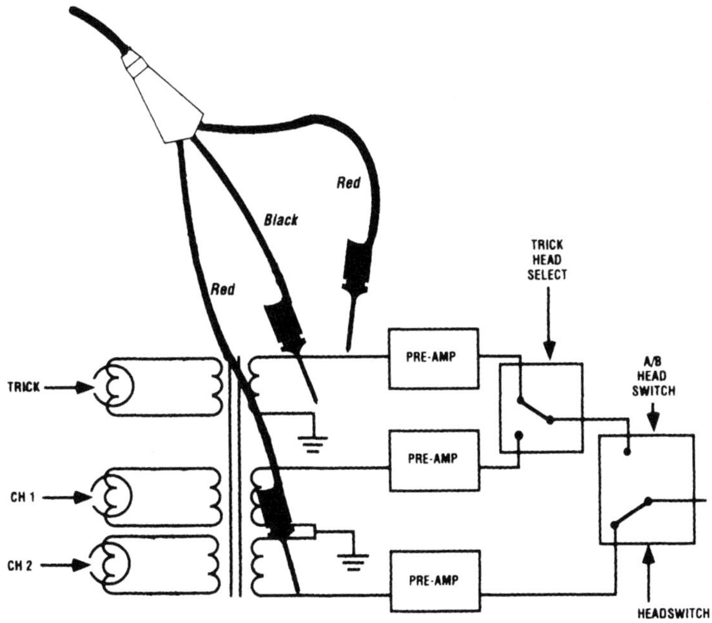

11-10 Place the VCR in "pause" and inject the playback signal into the "trick" head selector input to determine if the switching circuits are working properly.

A good picture indicates that all the circuitry after the injection point is good and the problem is either the rotary transformer or the "trick" head. Do the "rotary transformer test" to check the rotary transformer.

If the picture is bad, suspect the "trick" head switcher. Test the switcher by injecting the VC93 signal at the output of the "trick" head switcher. If the picture returns, the "trick" head switcher is defective.

Checking video-head preamps

The video head preamplifiers amplify the weak signals from the rotary transformer. One preamp is used for each head. A defective preamp produces a snowy picture that can easily be misdiagnosed as a bad video head. Use the VC93 "lum" playback signal to positively identify and locate head preamp problems.

To check video head preamps:

1. Insert a blank tape in the VCR and press "play." (Note the test setup in Fig. 11-11.)

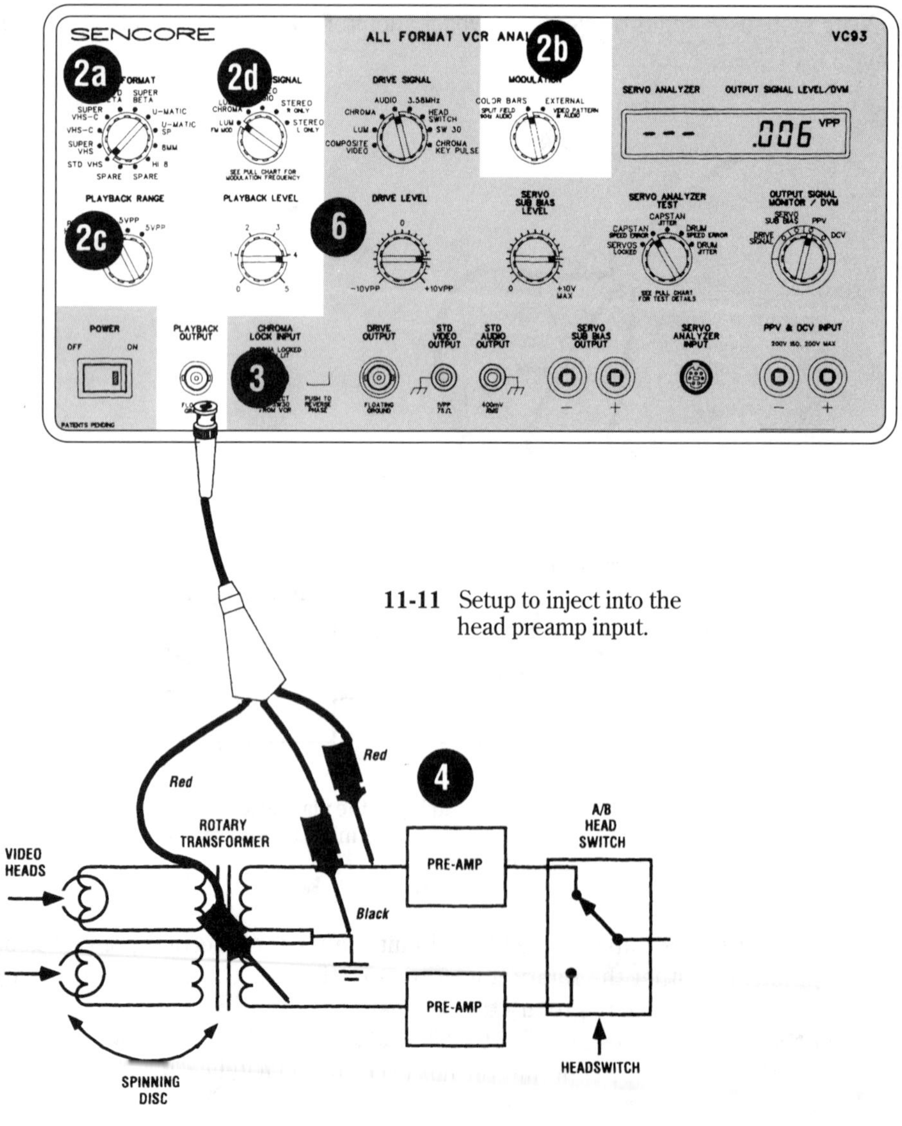

11-11 Setup to inject into the head preamp input.

2. Set the VC93 as follows:
   ~ VCR format to format being serviced.
   ~ Modulation to "color bars" or "external."
   ~ Playback range to "playback head sub."
   ~ Playback signal to "lum."
3. Connect the head substitution test lead to the playback output jack.
4. Connect the head substitution test lead to the input of both video head preamps.
5. Observe the picture on the playback monitor.
6. Adjust the playback level control for best picture.

If the injection produces a good picture, the circuits after the rotary transformer, including the video head preamps, are good. If the picture is snowy, either a video preamp, or the A/B headswitcher is defective. Note: Use the "playback head sub" position for this test. If you must switch the playback range control to a higher range to get a good picture, one or both preamps are bad.

Most machines incorporate the video head preamps and the A/B headswitcher in the same IC. If so, you don't need to isolate the problem further because both items will be replaced with the IC. Before replacing the IC, check that the A/B headswitcher is receiving a head switching pulse. (See Fig. 11-12.)

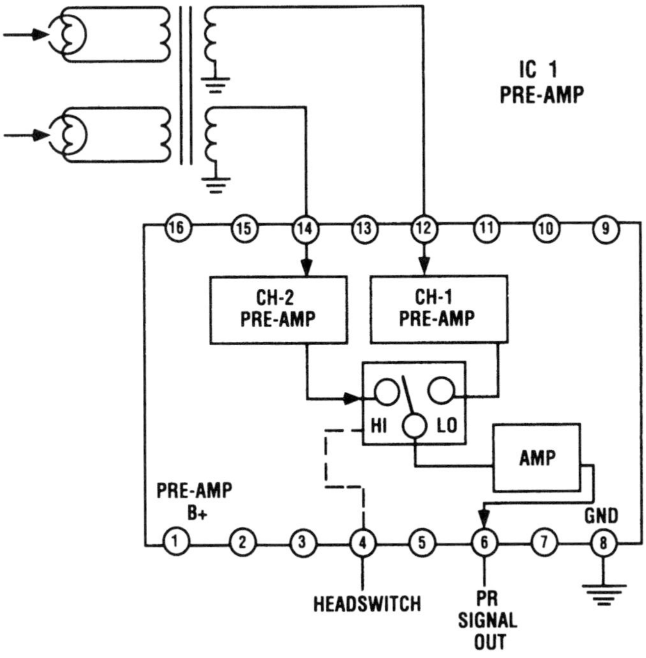

11-12 In many VCRs the video head preamps and the A/B headswitcher are in the same IC.

If the A/B headswitcher and the video head preamps aren't in the same IC, inject the "lum" playback signal at the output of the video head preamps. You'll need to increase the playback range control to the "0.5VPP" or "5VPP" range to obtain a signal level at this point.

Checking A/B headswitchers

The A/B headswitcher selects the output from the video head that's contacting the tape. A defective A/B headswitcher most likely provides a signal from only one head. The playback picture appears snowy, just as it would if one video head is defective or if the rotary transformer or video head preamp is defective.

Headswitcher checks

Here's how to determine if a problem is before or after the headswitcher. (Note the test setup in Fig. 11-13).

1. Insert a work tape in the VCR and press "play."
2. Set up the VC93 as follows:
   - ~ VCR format to format being serviced.
   - ~ Modulation to "color bars" or "external."
   - ~ Playback signal to "lum."
   - ~ Playback range to "5VPP."
3. Connect the head substitution test lead to the playback output jack.
4. Connect the head substitution test lead to the A/B headswitcher.
5. Observe the picture on the playback monitor.
6. Adjust the playback level control for the best picture.

A snowy picture means that the problem is after the A/B headswitcher. (Refer to this chapter's previous section "Injection before the FM detector.") If the test produces a good picture, the problem is before the output of the A/B headswitcher.

Identify the problem

To identify the problem further, do the following;

1. Move the head substitution test lead from the output of the A/B headswitcher to the input of A/B headswitcher.
2. Observe the picture on the playback monitor.

If the picture is still good, refer to this chapter's previous heading, "Checking video-head preamps." If the test produces a bad picture, the problem is associated with the A/B headswitcher. It could be caused by either a defective A/B headswitcher or a missing headswitch pulse.

To prove the operation of the A/B headswitcher:

1. Set the VC93 as follows:
   - Drive signal to "headswitch."
   - ~ Output signal monitor/dvm to "drive signal."

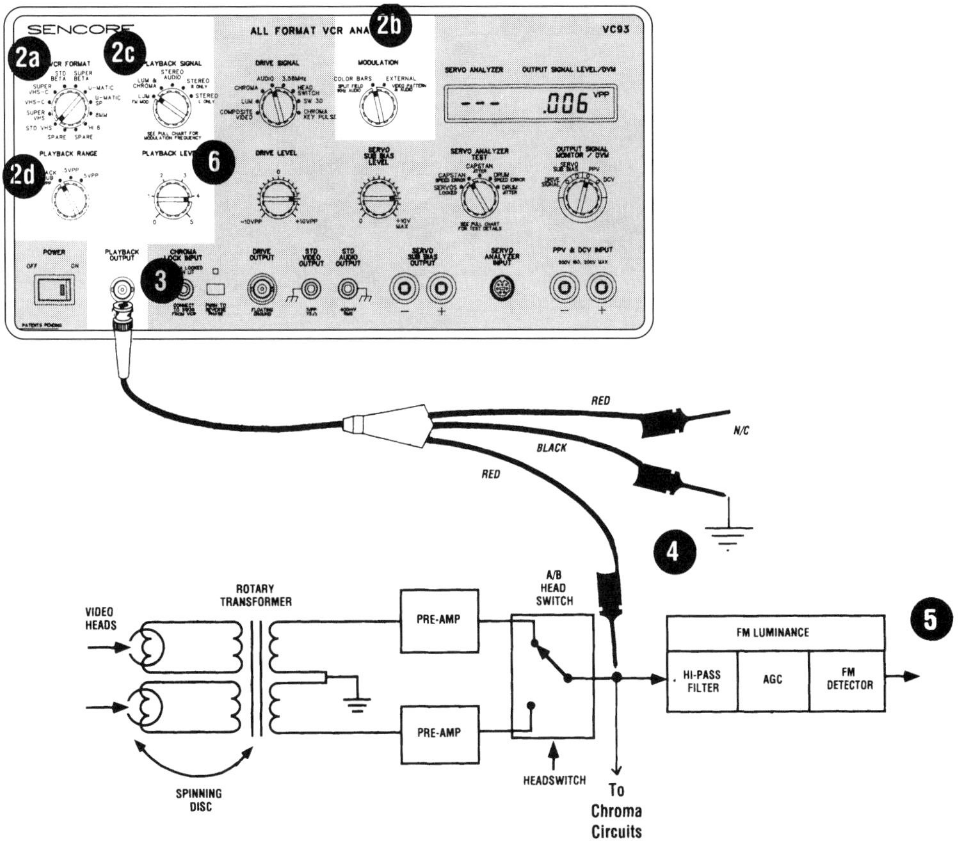

11-13 Test setup for checking the A/B headswitcher.

~ Drive level to "0."

2. Connect the head substitution test lead to the headswitch input.
3. Increase the drive level control while monitoring the picture. Note: In many modern VCRs, the A/B headswitcher, video head preamps, and other circuits are contained in the same IC. If so, set the playback range switch to to "Playback Head Sub" and inject into the video head preamps. If you don't obtain a good picture, try the opposite polarity of the drive level control. If you still don't obtain a good picture, replace the A/B headswitcher. If you do obtain a good picture, the headswitch pulse is missing.

Injecting before the FM detector

The VC93 playback output jack supplies the FM luminance signals to troubleshoot any problem up to the FM detector. To inject a signal into the FM signal path, refer to the setup diagram in Fig. 11-14.

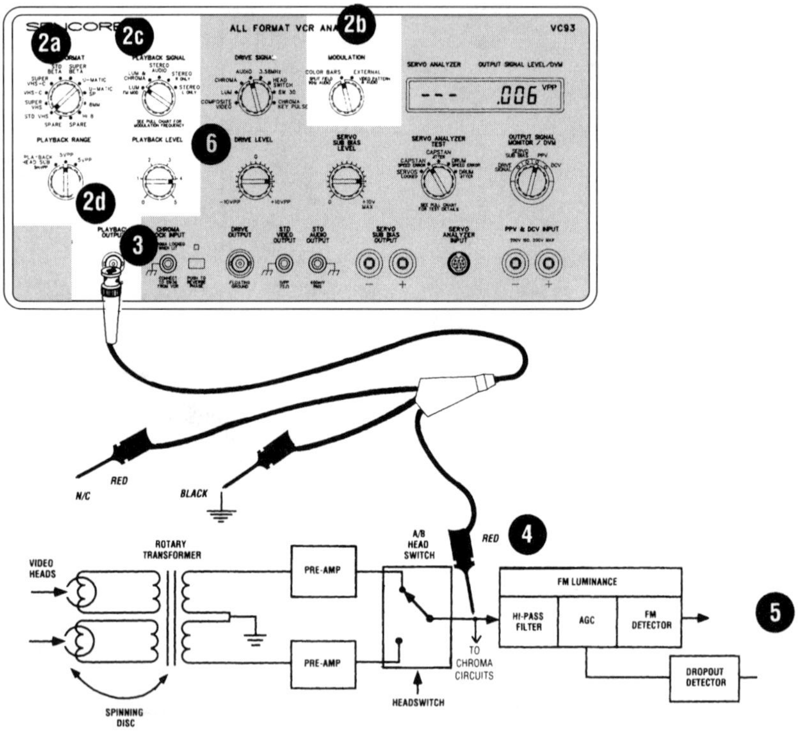

11-14 Setup for injecting into the luminance circuits.

1. Insert a work tape in the VCR and press "play."
2. Set up the VC93 as follows:
   ~ VCR format to format being serviced.
   ~ Modulation to "color bars" or "external."
   ~ Playback signal to "lum."
   ~ Playback level to "0.5 VPP."
3. Connect the head substitution test lead to the playback output jack.
4. Connect the head substitution test lead to the injection point.
5. Observe the picture on the playback monitor.
6. Adjust the playback level control for the best picture.

Note: If the picture remains snowy, switch the playback range to "5 VPP" and adjust the playback level for the best picture. You should obtain a good picture if all the circuits are working properly. If you're using a work tape with a color pattern, it's normal for some color from the test tape to show through.

If the signal injection results in a snowy picture, a circuit after the signal injection point is defective. Repeat the above injection at a later test point that's still before the FM detector. If there are no other injection points available before the FM detector, go on to "Testing video stages."

Testing video stages

The VC93 drive signal produces signals to identify problems in the luminance circuits after the FM detector. To inject a signal in the video stages:

1. Insert a work tape in the VCR and press "play." Refer to the setup connections in Fig. 11-15.

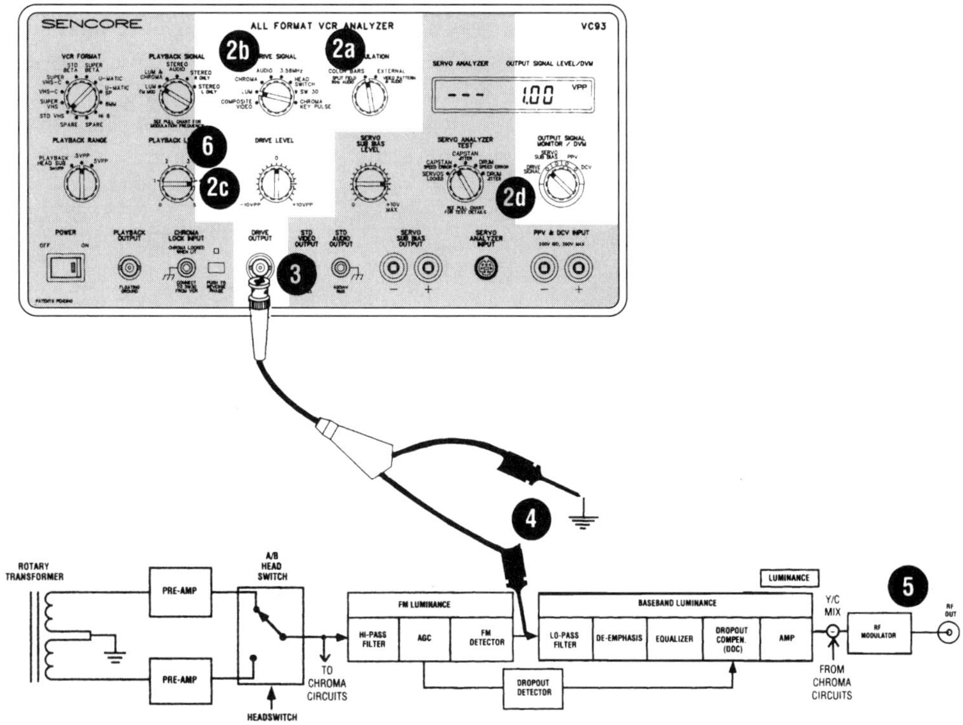

11-15 Setup for injecting into the luminance path after the FM detector.

2. Set the VC93 as follows:
   ~ Modulation to "color bars" or "external."
   ~ Drive signal to "lum."
   ~ Drive level to "0."
   ~ Set output signal monitor/dvm to drive signal.
3. Connect the direct test lead to the drive output jack.
4. Connect the direct test lead to the injection point.
5. Observe the picture on the playback monitor.
6. Adjust the drive level control for the best picture.

This test checks the ability of the video circuits to process a signal and pass it on to the video output jack or RF modulator. If the picture on the TV monitor is good when you make an injection, all the circuits between the injection point and the video output are good. Continue moving your injection point closer to the FM detector until the picture is no longer good. Now you're injecting before the defective stage.

Testing the RF modulator

The RF modulator converts the luminance, chroma, and audio information into an RF signal that can be picked up by the tuner of a TV receiver. A defective RF modulator can result in a snowy or poor-quality picture, or poor or no audio on a TV receiver connected to the RF output of the VCR.

Make a quick check of the RF modulator by injecting a composite video signal or baseband audio into the inputs of the modulator as the symptoms dictate. To check the RF modulator (video portion):

1. Apply power to the VCR.
2. Set up the VC93 as follows. (See Fig. 11-16.)
   - Modulation to "color bars" or "external."
   - Drive signal to "composite video."
   - Output signal monitor/dvm to "drive signal."
   - Adjust drive level for "1 VPP" on output signal level LCD readout.

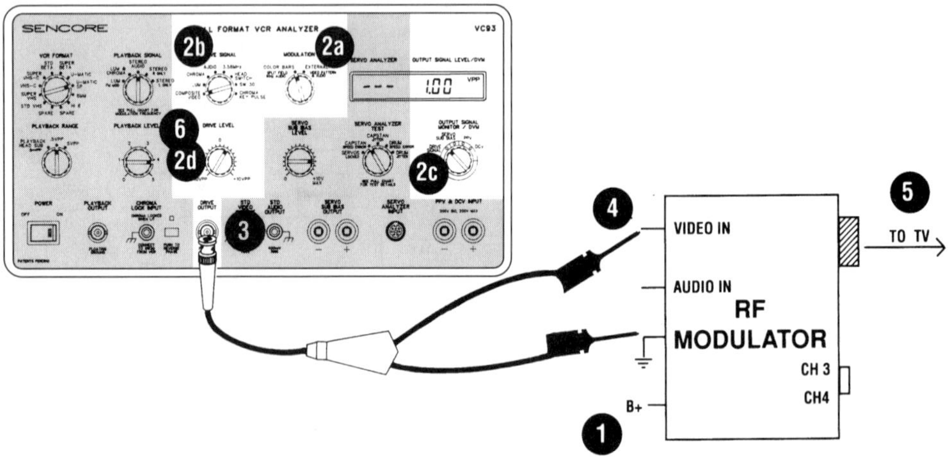

11-16 Setup for checking out the RF modulator (video).

3. Connect the direct test lead to the drive output jack.
4. Connect the direct test lead to the video input of the RF modulator.
5. Observe the picture on playback monitor.
6. Adjust drive level control for best picture.

Note: If the picture appears distorted, use the opposite polarity drive signal.

If you obtain a picture when injecting a signal, the modulator is good. If you do not obtain a picture, check for B+ and control voltage to the modulator. If these voltages are both present, the RF modulator is defective.

To check the RF modulator (audio portion):

1. Apply power to the VCR.
2. Set the VC93 as follows:
   ~ Modulation to "color bars" or "external."
   ~ Drive signal to "audio."
   ~ Output signal monitor/dvm to "drive signal."
   ~ Adjust the drive level for "0.400" on the output signal level LCD readout.
3. Connect the direct test lead to the drive output jack.
4. Connect the direct test lead to the audio input of the RF modulator.
5. Listen to the playback monitor for audio.
6. Adjust the drive level control for best audio.

If you obtain audio when injecting a signal, the modulator is good. If do not obtain sound, check for B+ and control voltage to the modulator. If these voltages are both present, the RF modulator is defective.

Super-VHS, super-Beta, and hi-8 testing

Super-VHS, super-Beta, and hi-8 formats extend the FM deviation and frequency range of the FM luminance signal to enhance the picture quality. These VCRs will play videotapes recorded with the standard format signal and the enhanced signals. The following conditions must be present for these machines to go into their special enhanced mode:

Super-VHS VCRs

A special super-VHS tape is required to record or play back in the super-VHS mode. This tape has a small hole that's detected by super-VHS machines. Super-VHS machines also contain a circuit that monitors the FM signal and detects the higher frequency super-VHS signal. If both conditions are present, the machine will switch to the super-VHS mode. (Note the block diagram in Fig. 11-17.)

To check super-VHS circuits:

1. Insert a super-VHS tape in the machine or defeat the super-VHS tape detection circuit.
2. Press the "play" button.
3. Set the VC93 VCR format switch to "super-VHS."
4. Set the VC93 playback signal switch to "lum."
5. Inject the signal before the super-VHS detector takeoff point.
6. Check that the machine's "super-VHS" indication comes on.

Note: The super-VHS takeoff point is usually located right after the A/B headswitcher.

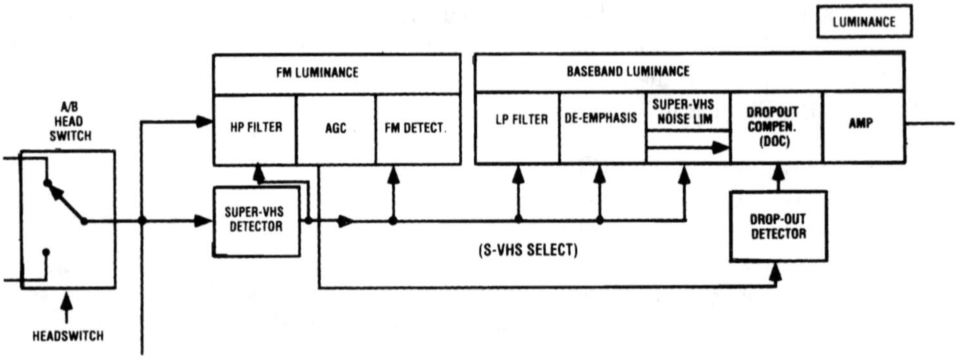

11-17 A Super-VHS VCR checks for the presence of the Super-VHS signal before switching to the Super-VHS mode.

Super-Beta VCRs

Super-Beta machines don't use a special tape. The FM detector automatically detects the presence of the Super-Beta signal.

To check super-Beta circuits:

1. Insert tape into the machine.
2. Press the "play" button.
3. Set the VC93 VCR format switch to "super-Beta."
4. Set the VC93 playback signal switch to "lum."
5. Inject the signal before the FM detector.
6. Check that the machine's "super-Beta" indication comes on.

If you don't see a super-Beta indication, move your injection point closer to the video heads.

Hi-8 VCRs

Hi-8 VCRs use a special tape. The hi-8 cassette contains a "detect hole" that tells the machine if the proper tape is being used. In the playback mode, a hi-8 signal detector verifies that a hi-8 signal was recorded on the tape. To check hi 8 CIRCUITS:

1. Insert a hi-8 tape in the machine or defeat the hi-8 tape cassette detection circuit.
2. Press the "play" button.
3. Set the VC93 VCR format switch to "hi-8."
4. Set the VC93 playback signal switch to "lum."
5. Inject the signal before the FM detector.
6. Check that the machine's "hi 8" indication comes on

Troubleshooting the VHS-C format

The only main difference between VHS and VHS-C machines is the configuration of the video heads. The VHS-C format was developed to make the mechanical section physically smaller to use in camcorders. To get this size reduction, the video head cylinder was reduced in size. To make the information recorded on the tape compatible with standard VHS machines, VHS-C uses four video heads instead of two.

Use the same procedures to troubleshoot and signal inject into VHS-C machines as you use for standard VHS machines. The only procedure that's different is injecting a signal between the rotary transformer and the headswitcher. Because VHS-C machines have four outputs instead of two, you'll need to jumper the corresponding video heads into pairs so you can inject the signal into all four heads simultaneously. Figure 11-18 shows the connections for injecting at the output of rotary transformers in VHS-C machines.

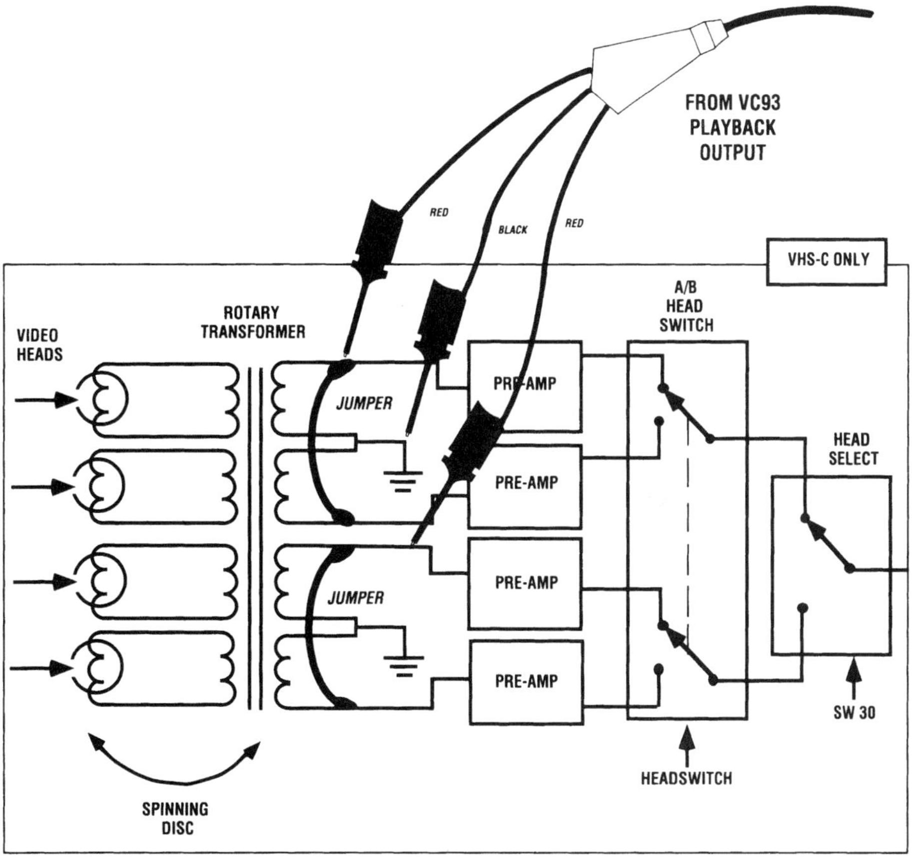

11-18 VHS-C VCRs use four heads and a combination of switches to select the correct video heads. Jumper the corresponding video heads into pairs to inject a signal.

Chroma troubleshooting

Let's now look at the signals required for proper color operation. The chroma portion of the NTSC composite video signal is frequency converted and processed before being recorded onto the tape. This minimizes crosstalk between tracks and compensates for errors introduced in playback. The playback circuits convert the signal back to NTSC standards. The chroma playback circuits need four key signals to return the color to its original form. Each of these signals is available from the VC93.

Play back a test tape that has a color video signal recorded on it. The most likely color symptoms are no color or poor color. Either problem occurs when any of the four key signals is missing or incorrect. Refer to these key blocks in Fig. 11-19.

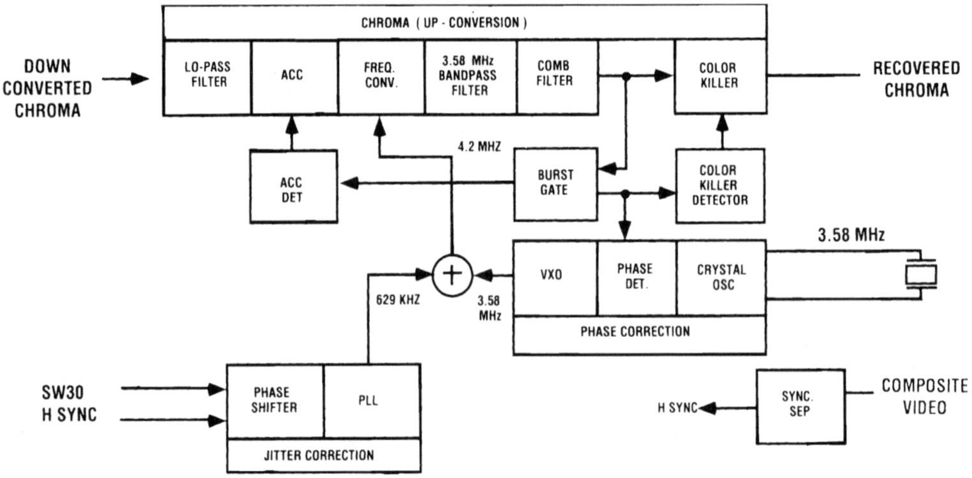

11-19 The chroma conversion section requires four key signals for proper operation.

Down-converted chroma

Down-converted chroma is the converted chroma signal picked up from the tape. It's separated from the luminance information after the A/B headswitcher using either a low-pass or comb filter.

3.58-MHz reference signal

The 3.58-MHz reference is used to reestablish the correct NTSC standard video signal.

30 Hz

Several formats change the phase of the recorded color information to minimize crosstalk and interference between tracks. The vertical sync pulse provides the recorded phase reference. During playback, a 30-Hz signal is the reference to cancel the phase shifting.

Composite video (with jitter)

Formats that use phase shifting to minimize crosstalk also change the color phase for each horizontal line. Horizontal sync pulses, taken from the luminance circuits, reference the timing of the phase shifting during playback. The horizontal sync pulses are either separated before the chroma circuits, or they're separated from composite video by the chroma circuits.

Checking overall chroma operation

The first step in troubleshooting a chroma problem is to be sure that the luminance circuits are working properly. If they are, supply a known-good reference signal from the VC93 playback output to the input to the chroma circuitry. Because the color circuits also require composite video from the luminance circuits, you also must inject the playback signal before the point where the luminance and chroma signals separate. To make an overall chroma operation check:

1. Insert a blank tape in the VCR and press "play." (Refer to the test setup in Fig. 11-20.)

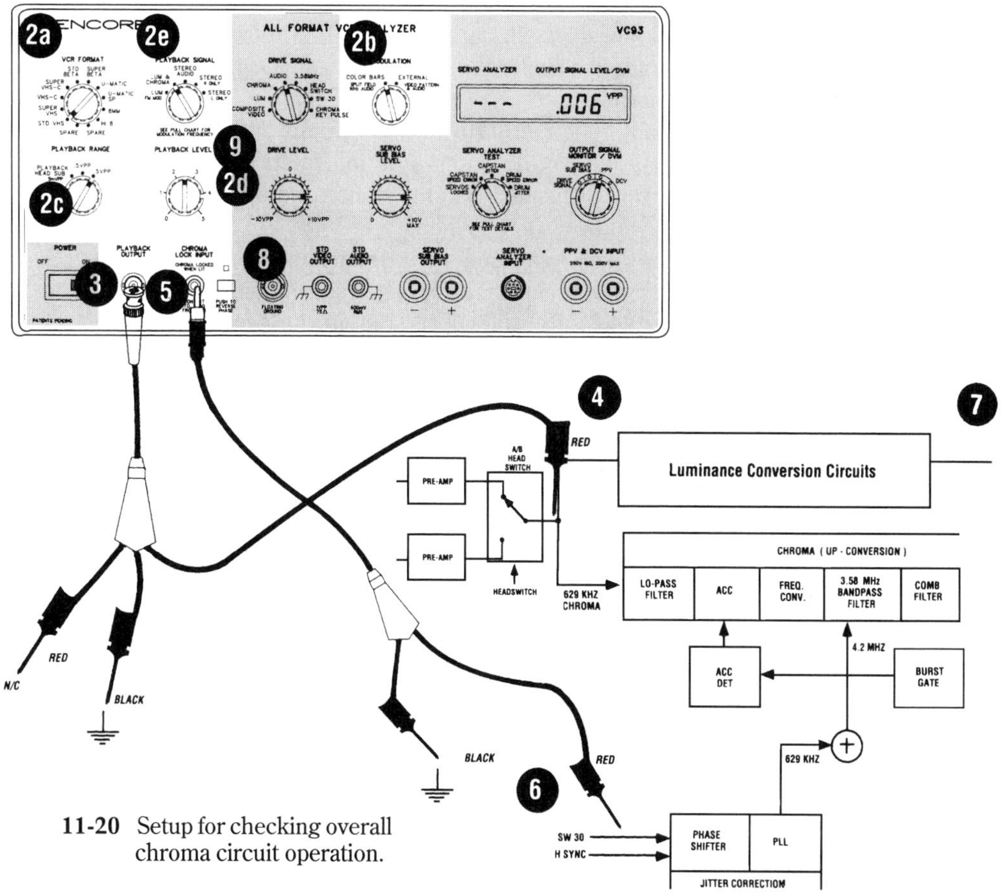

11-20 Setup for checking overall chroma circuit operation.

2. Set the VC93 as follows:
   - ~ Format to format being serviced.
   - ~ Modulation to "color bars" or "external."
   - ~ Playback range to "5VPP."
   - ~ Playback level to midrange.
   - ~ Playback signal to "lum & chroma."
3. Connect the head substitution test lead to the playback output jack.
4. Connect the head substitution test lead to the output of the A/B headswitcher.
5. Connect the chroma lock test lead to the chroma lock input jack.
6. Connect the chroma lock test lead to the SW30 test point in the VCR. Note: The chroma lock indicator will light up if you're connected to the correct test point.
7. Observe the monitor for a color picture.
8. If you don't see color, press the chroma lock phase button to select the other SW30 phase.
9. Adjust the playback level control for the best color.

If you see color, all the playback color circuits are working. If you can't obtain color when playing a test tape that contains a color pattern, the chroma signal is getting lost before the VC93 injection point.

If the heads are operating normally , then either the color circuits are defective or one of the chroma signals is missing or incorrect. For information on troubleshooting "no color" symptoms, refer to the section on "Troubleshooting chroma conversion problems" in chapter 4.

Note: Some VCRs separate the chroma and luminance signals inside the enclosure containing the head preamps. These VCRs have separate luminance and chroma signal output pins. Connect one of the "red" head substitution test leads to the luminance pin and the other "red" lead to the chroma pin. (Refer to block diagram in Fig. 11-21).

Identifying head-related chroma problems

A wrong or missing color problem can be caused by a frequency response problem in a stage before the chroma and luminance processing circuits. To test the stages before the chroma conversion circuits:

1. Set up the VC93 as in the "Checking overall chroma operation" test in chapter 4.
2. Set the playback range control to "playback head sub."
3. Move the head substitution test lead to the rotary transformer output. (Note the test hookup in Fig. 11-22).
4. Observe the playback monitor for color.
5. Adjust the playback level control for the best color.

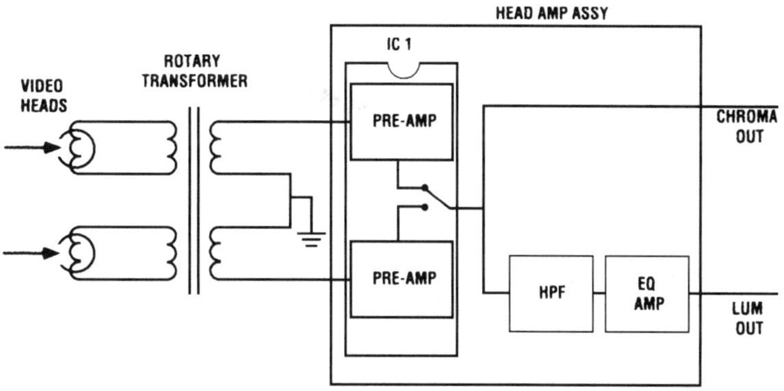

11-21 Some VCRs separate the luminance and chroma signals in the head preamp circuits. Connect one of the head substitution leads to the chroma output and the other to the luminance output.

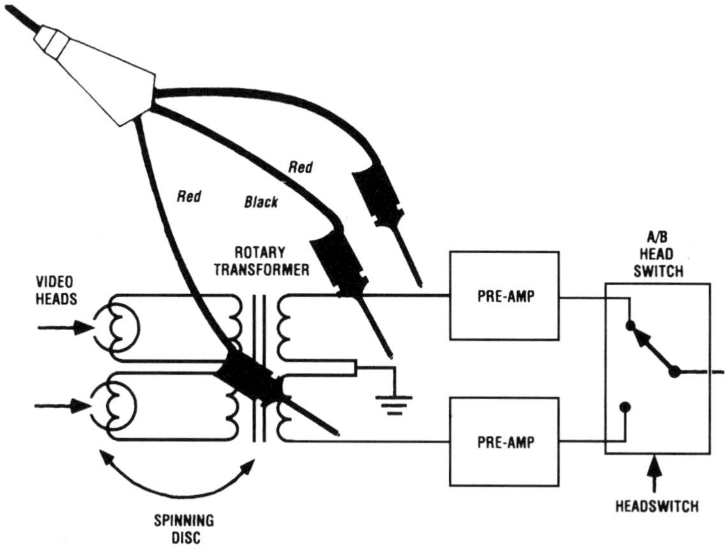

11-22 If you cannot obtain color from a VCR tape, but the VC93 proves that the color circuits are working, move the test leads to head preamp input to determine if the preamps are operating.

If you see color, either the video heads or the rotary transformer are defective. A likely cause is a worn video head. If you don't obtain color, the problem is between your injection point for the "overall chroma operation" test. Use the VC93 to step through the remaining stages to isolate the defective stage.

Note: Remember to increase the level of the playback signal after the video preamps. Turn the playback range switch to "0.5 VPP" or "5 VPP" and adjust the playback level for the best picture on the monitor.

Troubleshooting chroma conversion problems

If the "overall chroma operation" test doesn't produce color, you need to troubleshoot the chroma conversion circuits. You should now check the following items:

1. Power supply voltage to the color conversion IC.
2. Grounds to the color conversion IC.
3. The record/playback selector signal.

If these check out, you need to troubleshoot the color conversion section. First, supply a down-converted chroma signal from the VC93 to the VCR chroma conversion circuits. Next, do the following tests until you locate the problem: Substitute the 3.58-MHz color oscillator; substitute the chroma key pulse; substitute the SW30-Hz pulse.

Substituting the 3.58-MHz color oscillator

A missing or off-frequency 3.58-MHz reference oscillator will result in no or incorrect playback color. Use the VC93 3.58-MHz drive signal to substitute for the reference signal.

Note: Some VCRs use the color 3.58-MHz signal to control the servos. A missing or wrong 3.58-MHz reference will not allow the VCR to function. Injecting the 3.58-MHz signal into the chroma section will return the machine to operation. To substitute for the 3.58-MHz reference:

1. Connect the VC93 to the chroma circuits as described in the section entitled "Checking overall chroma operations" in chapter 4. (See Fig. 11-23).
2. Set the drive signal switch to "3.58 MHz."
3. Connect the direct test lead to the drive output jack.
4. Connect the direct test lead to the 3.58-MHz oscillator.
5. Adjust the drive level control while watching the monitor for color.
6. If you don't see color, press the chroma lock phase button to select the other SW30 phase.

If color returns, the 3.58-MHz signal is either missing or defective. If color doesn't return, remove the 3.58-MHz drive signal and go to the "Substituting the chroma key pulse signal" section.

Substituting the chroma key pulse signal

VHS, Beta, and 8mm VCRs use a horizontal sync pulse to operate the chroma phase shift circuits. (Note the block diagram shown in Fig. 11-24). No playback color or poor playback color occurs if the phase shift circuits aren't functioning. Some VCRs feed the composite video from the luminance circuits to the color IC

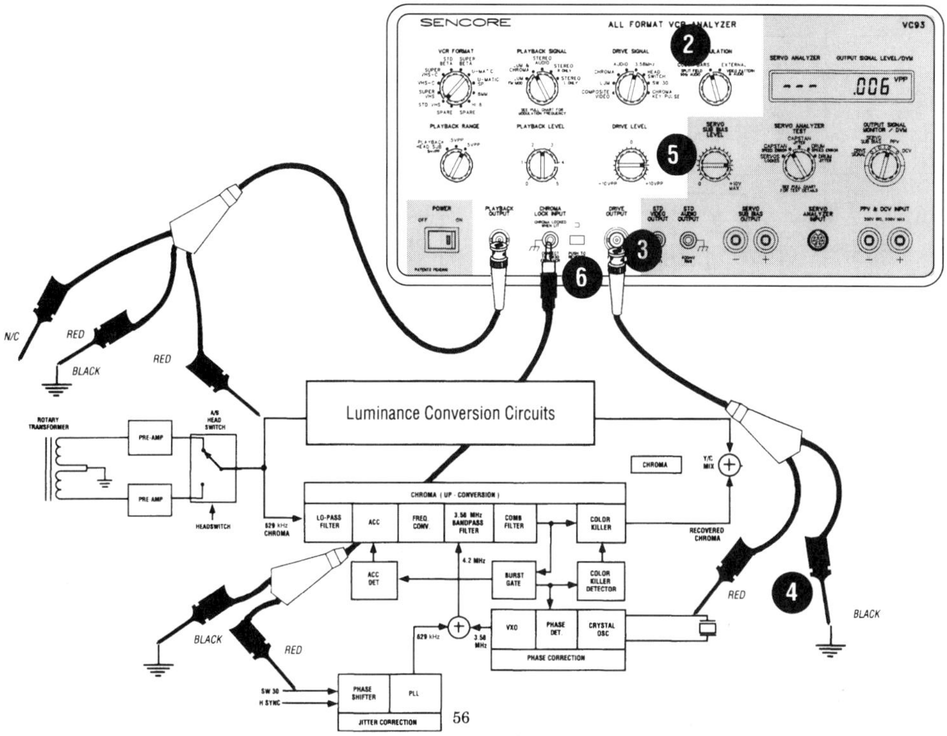

11-23 Setup for injecting the 3.58-MHz signal.

and extract the horizontal sync pulses inside the IC. Other VCRs have the sync separator outside the chroma IC. To substitute horizontal sync (sync separator external to color circuits):

1. Connect the VC93 to the chroma circuits as found in the section on "Checking overall chroma operation" in chapter 4. (Note the test setup shown in Fig. 11-25.)
2. Set the drive signal switch to "chroma key pulse."
3. Connect the direct test lead to the horizontal sync input on the chroma IC.
4. Adjust the drive level control"clockwise" ("+" polarity) while watching the playback monitor for color.
5. If color doesn't return, rotate the drive level control "counterclockwise" ("–" polarity).
6. If you don't see color, press the chroma lock phase button to select the other SW30 phase and repeat steps 4 and 5.

If color returns, troubleshoot the cause of the bad or missing sync pulse. If color doesn't return, remove your injection signal and go to the section entitled "Substi-

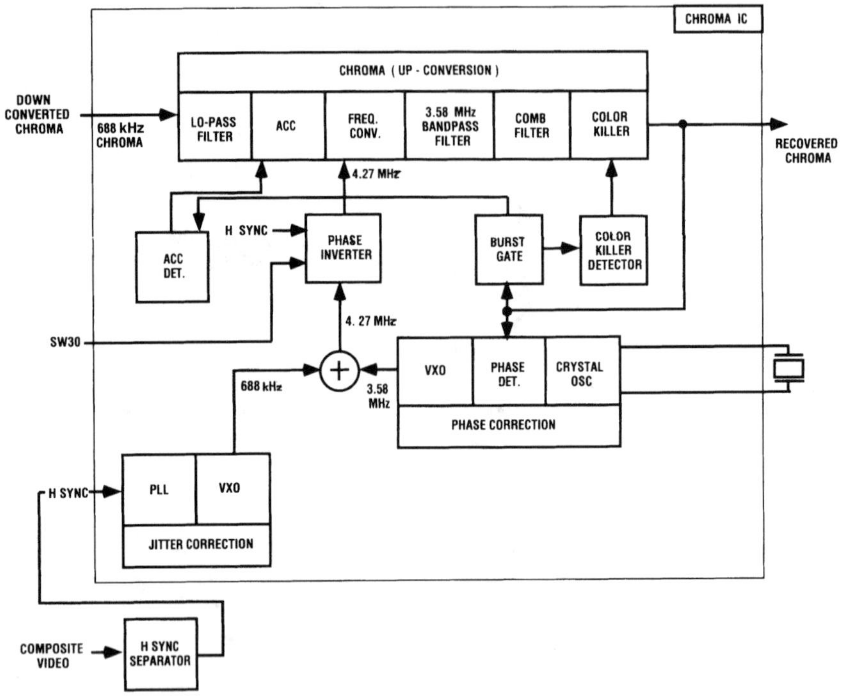

11-24 Some VCRs use a sync separator to contain the horizontal sync pulse for the chroma circuits.

tuting the 30-Hz phase switching signal" in chapter 4. To substitute composite video (sync separator internal to color IC):

1. Connect the VC93 to the chroma circuits as described in the section entitled "Checking overall chroma operation" in chapter 4. (Refer to the test setup in Fig. 11-26.)
2. Set the drive signal switch to "composite video."
3. Connect the direct test lead to the composite video input on the chroma IC.
4. Adjust the drive level control"clockwise" ("+" polarity) while watching the playback monitor for color.
5. If color doesn't return, rotate the drive level control "counterclockwise" ("–" polarity).
6. If you don't see color, press the chroma lock phase button to select the other SW30 phase and repeat steps 4 and 5.

If color returns, troubleshoot the cause of the bad or missing composite video signal. If color doesn't return, remove your injection signal and go to the section entitled "Substituting the 30-Hz phase switching signal" in chapter 4. To substitute composite video (sync separator internal to color IC):

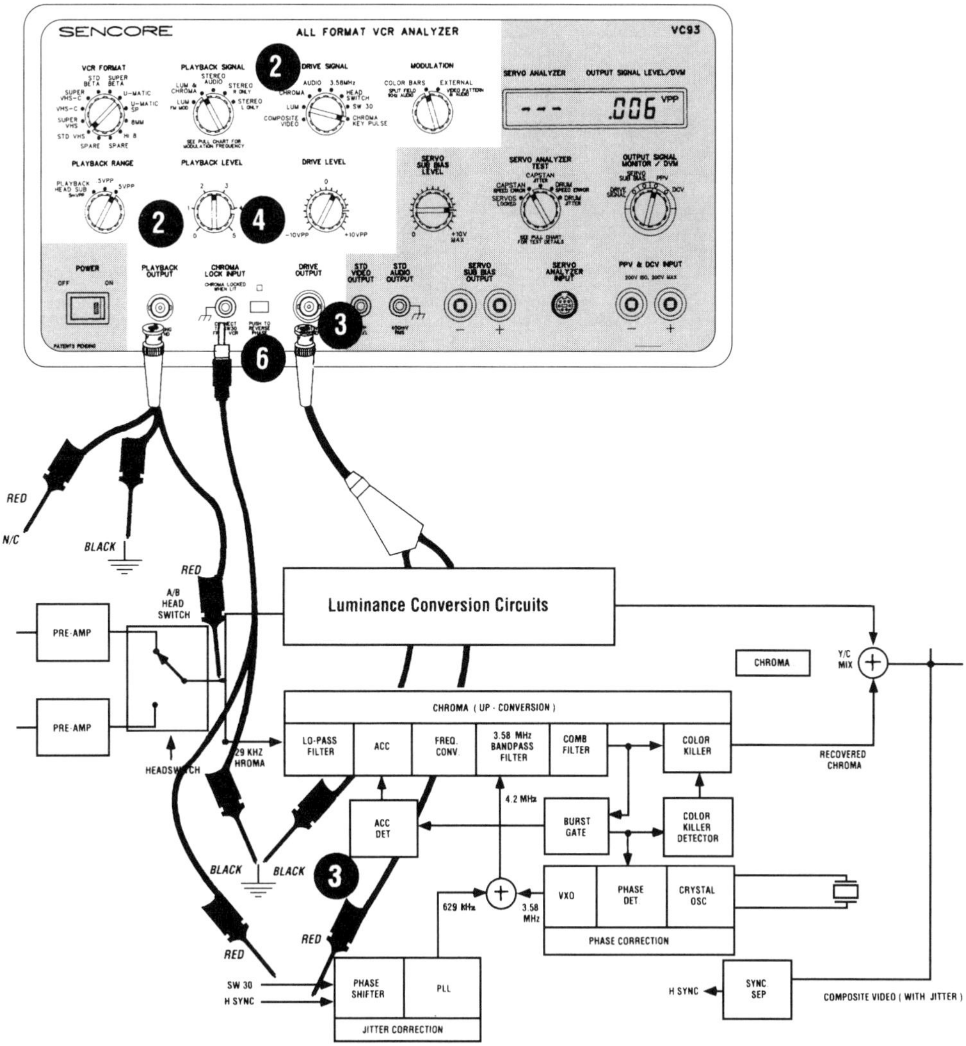

11-25 Setup for injecting horizontal sync pulses into the chroma circuits.

1. Connect the VC93 to the chroma circuits as described in the section entitled "Checking overall chroma operation" in chapter 4. (Refer to the test setup in Fig 11-26.)
2. Set the drive signal switch to "composite video."
3. Connect the direct test lead to the composite video input on the chroma IC.
4. Adjust the drive level control "clockwise" ("+" polarity) while watching the playback moitor for color.

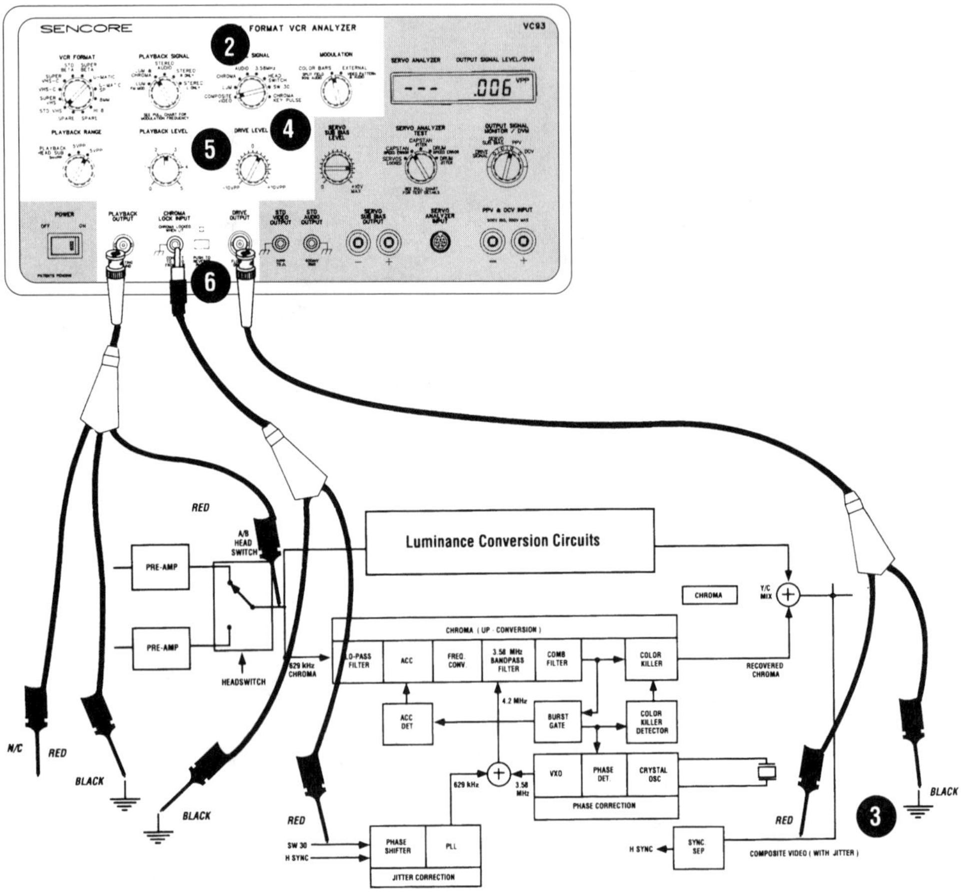

11-26 Setup for injecting composite video into the chroma circuits.

5. If color doesn't return, rotate the drive level control "counterclockwise" ("–" polarity.)
6. If you don't see color, press the chroma lock phase button to select the other SW30 phase and repeat steps 4 and 5.

If color returns, troubleshoot the cause of the bad or missing composite video signal. If color doesn't return, remove your injection signal and go to the section entitled "Substituting the 30-Hz phase switching signal."

Substituting the 30-Hz phase switching signal

VHS, Beta, and 8mm formats also use a 30-Hz signal to reference the phase rotation circuits. If this signal is missing, you'll get no or a wrong playback color. To substitute the 30-Hz phase switching signal:

1. Connect the VC93 to the chroma circuits as shown in Fig. 11-27.
2. Set the drive signal switch to "SW30."

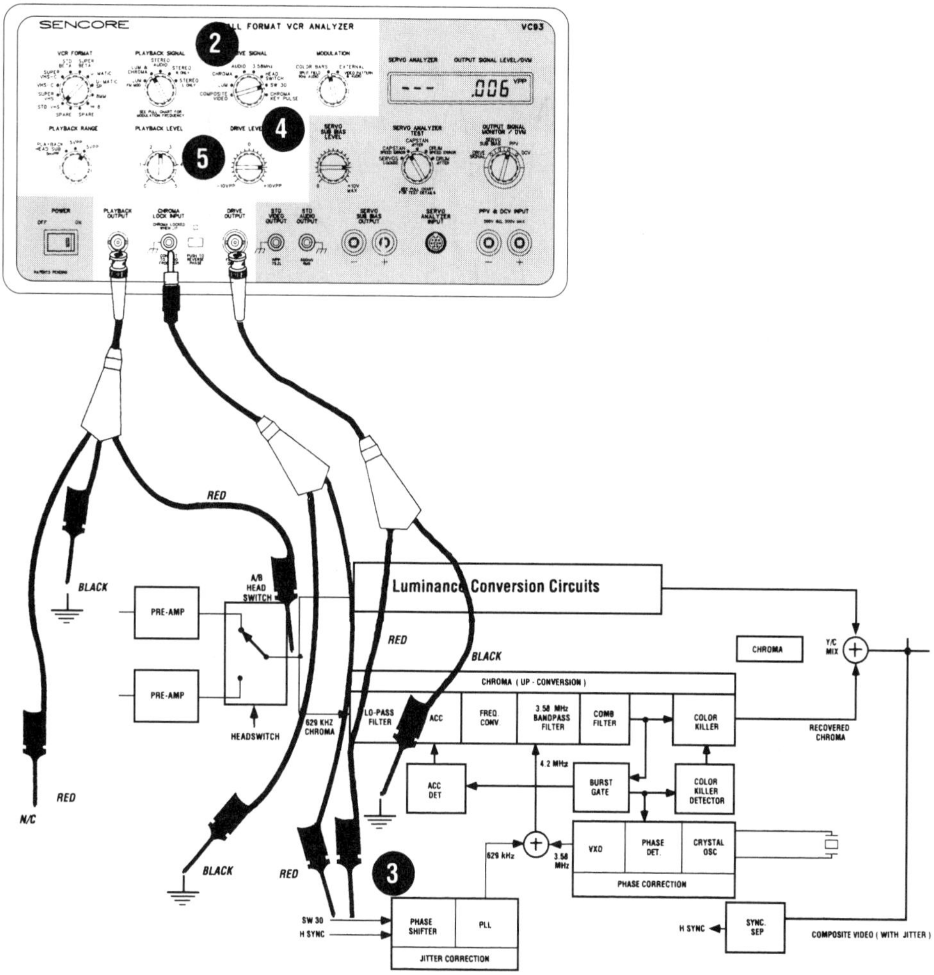

11-27 Setup for injecting the SW30 signal into the color circuits.

3. Connect the direct test lead to the SW30 input on the chroma IC.
4. Adjust the drive level control"clockwise" ("+" polarity) while watching the playback monitor for color.
5. If no color appears, rotate the drive level control "counterclockwise" ("–" polarity).

If color returns, troubleshoot the cause of the bad or missing composite video signal. If color doesn't return, remove your injection signal and go to the section entitled "Substituting the 30-Hz phase switching signal."

Note: The SW30 signal is used throughout the VCR. Substituting for the SW30 signal might cause the VCR to shut down. This is normal. Simply remove your injection signal and press the "play" button on the VCR. If color returns, find the reason for the missing SW30 signal.

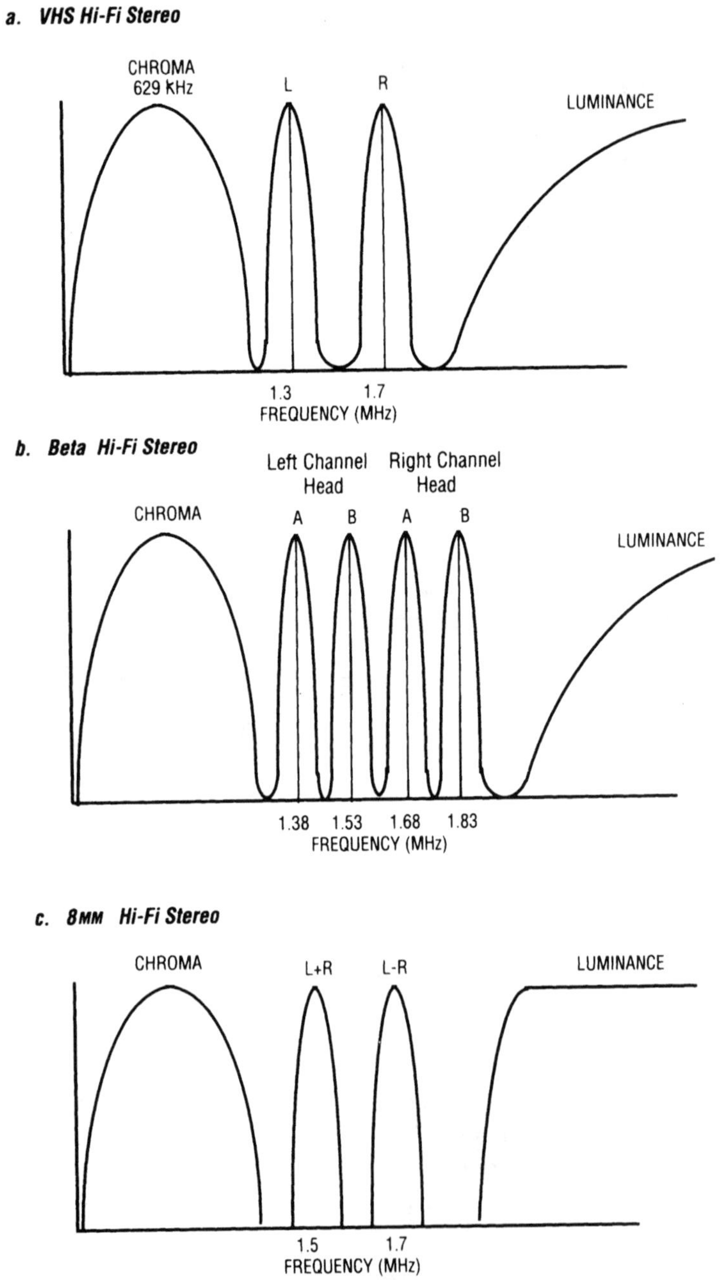

11-28 Parts A, B, and C show various VCR audio signal formats.

Audio troubleshooting in VCRs

VCR and camcorder audio is recorded in several ways, so you must select the correct audio test signal. Audio was originally recorded only in a linear track. Linear audio is

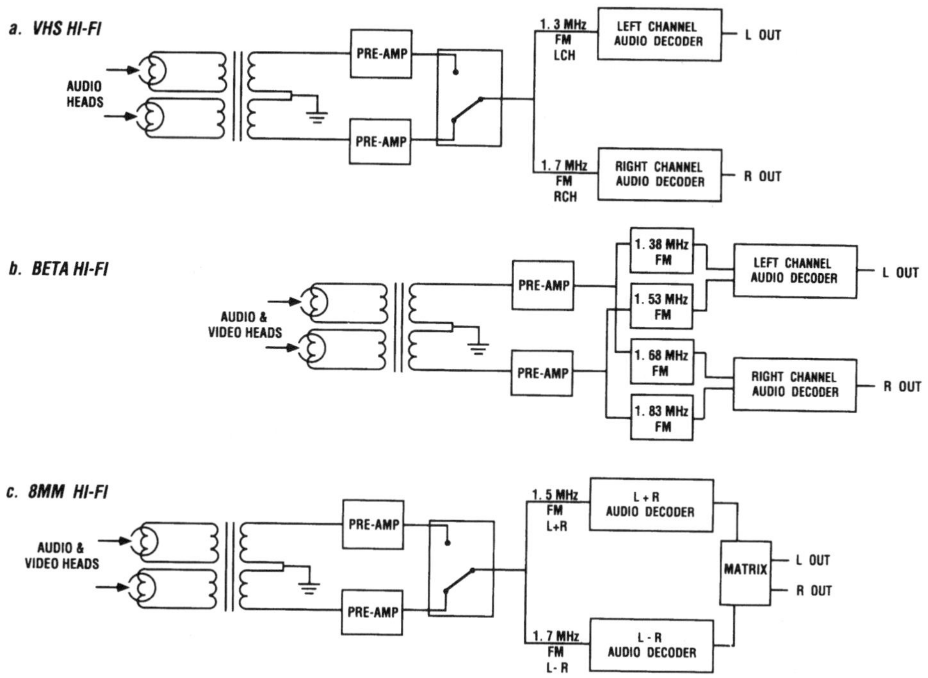

11-29 Parts A, B, and C show block diagrams for VHS, Beta, and 8mm stereo audio systems.

limited in frequency response and has wow and flutter. Stereo hi-fi recording uses spinning heads and FM carriers, much like the video signal, to improve audio fidelity.

These are three different FM hi-fi stereo VCR systems: VHS hi-fi stereo, Beta hi-fi stereo, and 8mm hi-fi stereo. Eight-millimeter mono is recorded using a similar FM system. The frequency waveforms for different audio systems are shown in Figs. 11-28A, B, and C. Figures 11-29A, B, and C give the block diagram of the different audio recording schemes.

The VC92 playback signals duplicate the FM audio signals coming from the spinning heads. Use them to substitute in any hi-fi stereo circuit from the rotary heads to the FM detector. Use the VC93 drive signals to inject into any audio stage after the stereo hi-fi FM detectors and anywhere that baseband (15-Hz to 20-kHz) audio is present.

Note: Some 8mm VCRs use pulse code modulation (PCM) hi-fi stereo. This system uses digitally coded information. Troubleshoot these systems by using an oscilloscope to check for proper data flow.

Identifying defective hi-fi stereo audio heads

All three hi-fi stereo systems (VHS, Beta, and 8mm hi-fi stereo) use rotating heads to pick up the FM modulated hi-fi stereo signal from the tape. In Beta and 8mm, the same head that picks up the video signal also picks up the audio signal. Thus, any head-related problem that affects the audio also affects the video. Before suspecting a head problem in these types of VCRs, check to be sure that the video is correct.

The VHS hi-fi stereo signal is picked up off the tape using a separate set of audio heads and rotary transformer windings. In this format a problem with the audio heads or audio rotary transformer windings will not affect the video.

Note: The VHS hi-fi stereo audio headswitch signal is derived from the video headswitch signal. Thus, a video headswitch signal problem will affect the audio headswitch signal. The audio headswitching point is more crucial than the video headswitching. Therefore, headswitching problems in VHS hi-fi stereo machines will affect the audio before the video signal.

Troubleshoot problems associated with the rotating audio heads the same way you would troubleshoot video head problems. The only difference is that you use the "stereo audio" playback signal and listen for improvements in the audio.

Checking audio rotary transformer windings

In the Beta and 8mm formats, the same rotary transformer windings transfer both the audio and the video signals. A defective rotary transformer will affect both the playback audio and video. The VHS format uses separate audio rotary transformer windings, meaning a bad rotary transformer will affect either the playback audio or video, but not likely both.

Check the audio rotary transformer using the same procedures used to check out video rotary transformers. The only difference is that you use the "stereo audio" playback signal.

Checking audio head preamps

Beta and 8mm format machines use the same head preamps to amplify the audio and the video. A defective head preamp will affect both the video and the audio playback signal. VHS format machines use separate audio and video preamps.

Troubleshoot audio head preamp problems using the same procedures you use to check out the video head preamps. The only difference is that you use the "stereo audio" playback signal and listen for improvements in the audio. The various audio pick-up heads and preamp circuits are shown in Figs. 11-30A, B, and C).

Injection after the A/B headswitcher

Beta hi-fi stereo VCRs separate the audio signal from the luminance signal before the A/B headswitcher. The audio signal undergoes several steps of processing before reaching the headswitcher. 8mm and VHS format VCRs, however, combine the FM audio signals using an A/B headswitcher before processing the signal. In either case, you can inject the VC93 "stereo audio" playback signal anywhere in the audio signal path before the audio FM detector.

Misadjustment of the audio headswitching can cause the symptom of no hi-fi stereo in VHS format machines. The VC93 "stereo audio" playback signal is a continuous carrier signal and is not affected by incorrect headswitching. If you obtain good

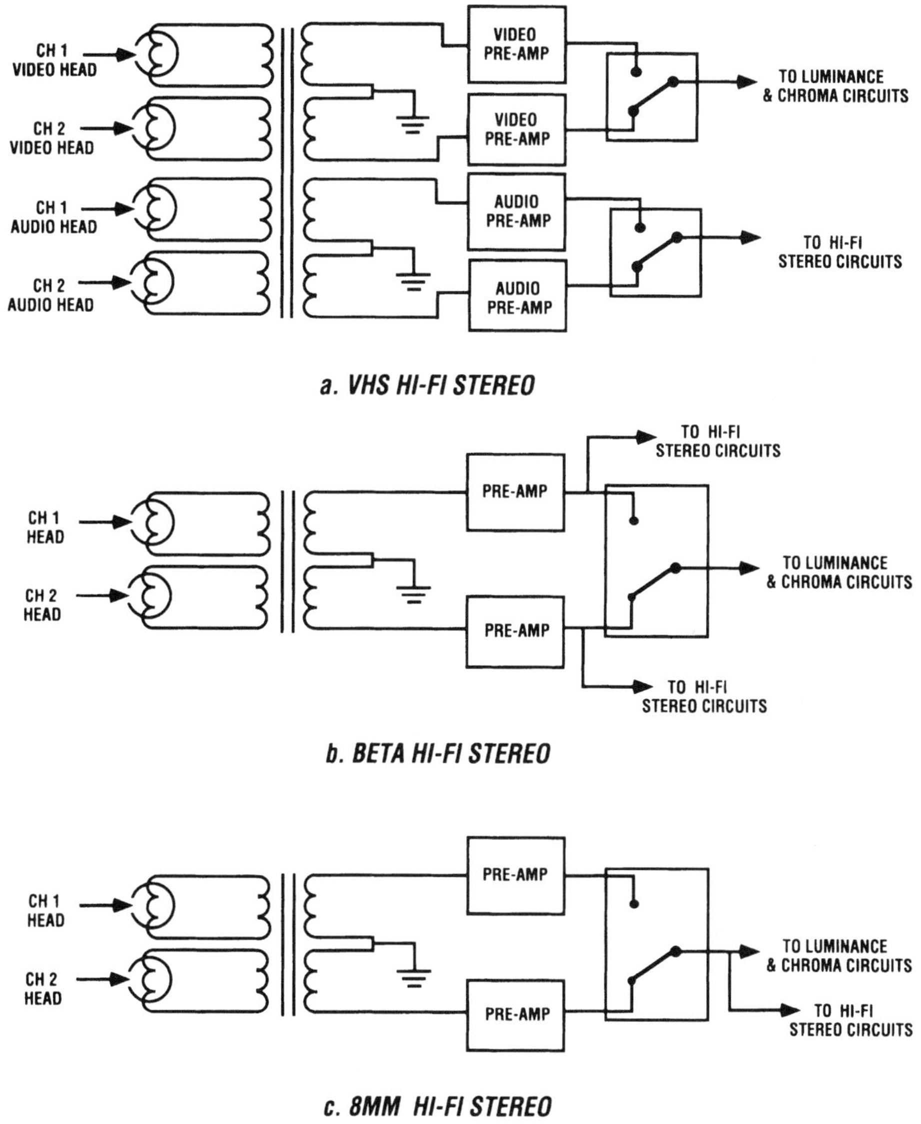

11-30 VHS hi-fi systems use separate rotary heads for the hi-fi audio signals. Hi-fi Beta and 8mm systems use the same rotary heads for video and audio.

audio when injecting the playback signal, check the headswitch adjustment before troubleshooting the rest of the hi-fi audio signals.

Note: Troubleshoot the FM portion of the hi-fi stereo circuits in 8mm and hi-8 machines by selecting the "hi-8" VCR format. This provides both FM carriers needed for hi-fi troubleshooting. If the 8mm or hi-8 VCR you're servicing has mono FM only, select the 8mm VCR format to troubleshoot the mono FM circuits.

Baseband audio troubleshooting

All VCR formats, except 8mm, record and play back linear audio. Hi-fi VCRs record linear audio as a backup to the hi-fi signal. Use the VC93 audio drive signal to inject into any audio stage from the linear audio heads to the audio output jacks. Also, use the audio drive signal to troubleshoot the hi-fi stereo circuits after the FM detector. To inject into the linear audio circuits (all formats):

1. Place a work tape in the VCR and press "play." (Refer to Fig. 11-31 for this test setup.
2. Set the VC93 as follows:
   - Modulation to "color bars" or "external."
   - Drive signal to "audio."
   - Drive level to "0."
3. Connect the direct test lead to the drive output jack.

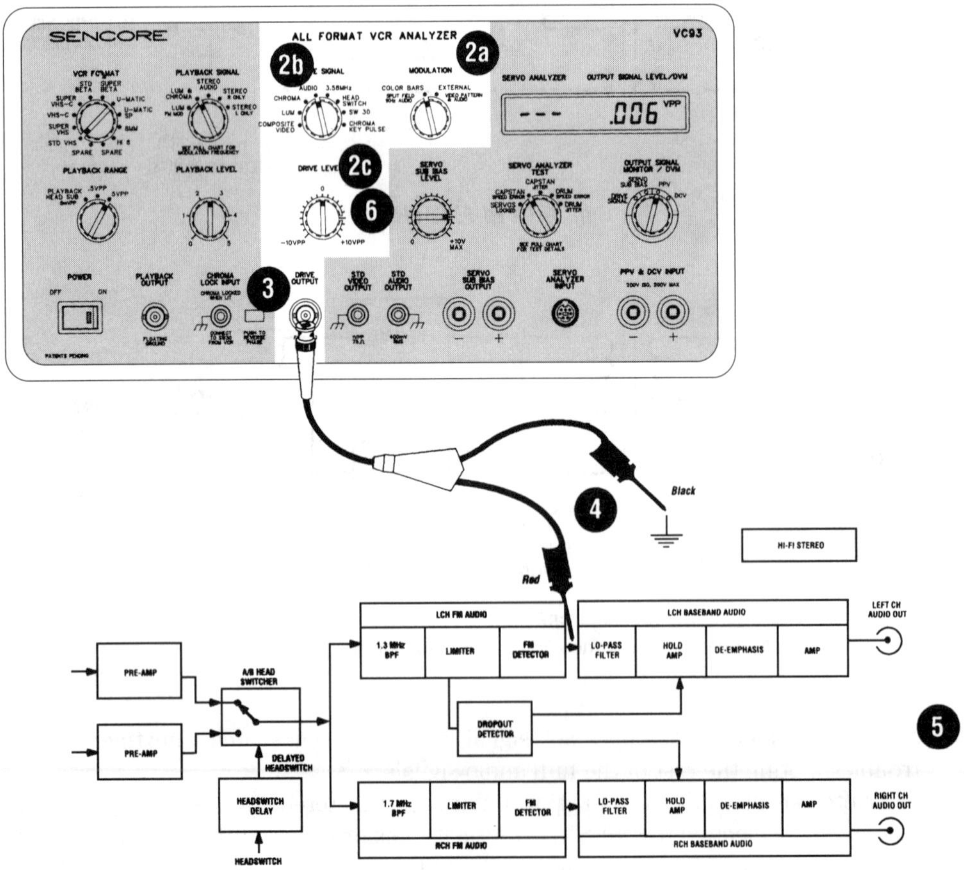

11-31 Setup for injecting into the linear audio circuits, and the hi-fi stereo circuits after the FM detectors.

4. Connect the direct test lead to the injection point.
5. Monitor the playback audio.
6. Adjust the drive level for the best sound.

If you obtain audio, the circuits between the injection point and the output are good. If you don't obtain audio, there's a problem between your injection point and the output. Locate the defective stage by moving your injection point closer to the audio head until you no longer hear an output. You're now injecting before the defective stage.

Note: Due to the high gain of the audio amplifiers, the audio will distort if you inject directly into the linear audio head. If you need to inject directly into the audio heads, inject into the output of the first amplifier, or build a voltage divider to reduce the amplitude of the signal .

Servo troubleshooting

Servo problems can be more difficult to identify and troubleshoot than other types of problems. Many servo problems produce symptoms of poor audio and/or poor video that can mislead you into troubleshooting circuits that have no defects.

The VC93 provides five servo analyzer tests. Do the tests in sequence, starting with the servos locked test. Use the results of each test to zero in on the defect. Here's a review of each servo test:

Servos locked test Determines if the servo capstan phase loop and drum phase loop are locked to the reference signal.

Capstan speed test Determines if the capstan servo is operating at the correct speed. It identifies speed-select circuit problems and other speed-related problems.

Capstan jitter test Measures how constant the capstan movement is. It helps identify capstan phase problems and mechanical problems, including motor bearings and bad idlers.

Drum speed test Determines if the drum is operating at the correct speed. It helps identify a drum that's operating too fast or too slow.

Drum jitter test Measures how constant the drum rotation is. It helps identify drum-related problems, including bad motor bearings and bad drum phase loop.

The next part of this chapter explains how to use the results of the servo tests for troubleshooting. A summary of the servo tests and possible causes of "bad" test results is provided in Fig. 11-32.

Understanding the servos locked test

The servos locked test compares the change in phase of the CTL pulse to the SW30 pulse when using the "troubleshooting test lead," or the phase of the audio and vertical sync pulse when using the "performance test lead" and "performance test tape." VCRs lock the capstan and drum phase circuits to a common REF 30 source. Usually the REF 30 source is internal to the servo IC and can't be viewed or measured. Because the capstan and drum phase loops are locked to the same REF 30, they also must be locked to each other. The VC93 uses this fact to check servo lock. Note: The

SERVOS LOCKED	CAPSTAN SPEED ERROR	CAPSTAN JITTER	DRUM SPEED ERROR	DRUM JITTER	MOST LIKELY DEFECT
GOOD	GOOD	GOOD	GOOD	GOOD	NO SERVO DEFECTS*
GOOD	GOOD	GOOD	GOOD	BAD	DRUM MECHANICAL
GOOD	GOOD	GOOD	BAD	N/A	REFERENCE FREQUENCY
GOOD	GOOD	BAD	GOOD	GOOD	CAPSTAN MECHANICAL
GOOD	BAD	N/A	GOOD	GOOD	REFERENCE FREQUENCY
GOOD	BAD	N/A	BAD	N/A	REFERENCE FREQUENCY
GOOD	BAD	N/A	GOOD	BAD	REFERENCE FREQUENCY
BAD	GOOD	GOOD	GOOD	GOOD	CAPSTAN PHASE LOOP or DRUM PHASE LOOP
BAD	BAD	N/A	GOOD	GOOD	CAPSTAN SPEED LOOP or CAPSTAN MECHANICAL
BAD	GOOD	BAD	GOOD	GOOD	CAPSTAN PHASE LOOP or CAPSTAN MECHANICAL
BAD	GOOD	GOOD	BAD	N/A	DRUM SPEED LOOP or DRUM MECHANICAL
BAD	GOOD	GOOD	GOOD	BAD	DRUM PHASE LOOP or DRUM MECHANICAL
BAD	BAD	N/A	BAD	N/A	REFERENCE FREQUENCY
BAD	BAD	N/A	GOOD	BAD	REFERENCE FREQUENCY

*NOTE: A noise bar that occurs periodically at a rate of one minute or greater could be a capstan or drum phase problem.

11-32 Use the results of the five servo tests to determine the most likely cause of a servo problem.

SW 30 pulse is used for this test, since it's derived from the PG pulse and is more universal to all VCRs.

Results of the servos locked test are displayed as a percentage reading that indicates how close the two servos are locked. This reading varies between brands and models. After testing several machines, however, you'll develop a feel for what reading is typical for the machine you're servicing. Generally, older VCRs, which use analog servo circuits, show a higher percentage reading, indicating a looser lock. Newer, digital machines typically have a much smaller percentage reading, indicating a more precise lock.

The "good" and "bad" servos locked readings are based on extensive research that shows that bad capstan or phase servo circuits produce errors greater than 2 percent. No "good/bad" reading is displayed if the servos are varying widely. If you see a "bad" or no indication, suspect either a bad capstan or drum phase circuit. A "good" indication means that the phase circuits are operating properly.

Regardless of the servos locked test results, do all of the servo analyzer tests before you begin to troubleshoot the problem. The results of the other tests might further isolate the problem to the capstan or drum phase loop.

Notes on the servos-locked test

Note 1. Some newer machines have very tight capstan drum speed loops that operate marginally even with a bad phase loop. The servos locked test has a minimum resolution and will not identify a problem where the reference signals change phase slower than one cycle per minute. If you observe a noise bar that comes and goes slower than once a minute, suspect a servo phase-loop problem.

Note 2. Some machines use a combined PG/FG drum signal. The servos in these machines can lock incorrectly if the drum PG signal is missing, locking instead to the FG signal. Since the FG signal is a higher frequency, the servos can lock to one of several transitions. Depending on which transition it locks to, the drum will lock, but noise or a noise bar occurs in the picture. The VC93 servos locked test will indicate that the servos are locked. This is a correct diagnosis, as the servo phase loops are locked. (Refer to the waveform drawing in Fig. 11-33.)

The best way to prove this problem is to stop and start the VCR several times while observing the position of the noise bar. Because the servos can lock on any FG transition, the noise bar will change position each time the machine is placed into "play." If you observe this symptom, check the combined FG/PG pulse for a missing PG pulse.

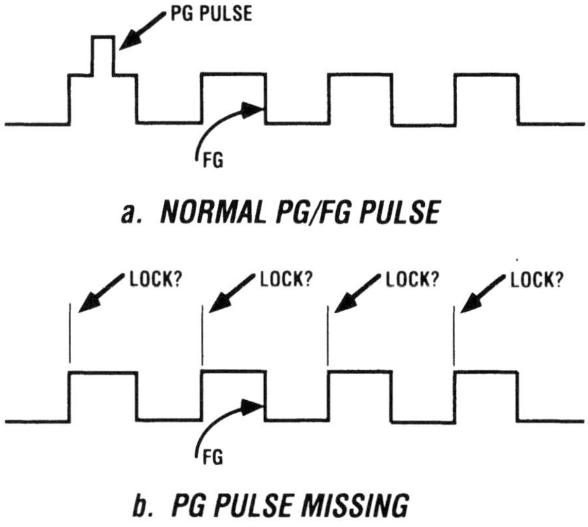

11-33 Some VCRs use a combined FG/PG pulse (A). If the PG pulse is missing, the VCR will lock to any cycle of the FG pulse.

Understanding the capstan speed error test

The capstan speed error test checks how fast the tape is being pulled through the machine. The VC93 analyzes the frequency of the CLT pulse when using the "troubleshooting test lead" (or the playback audio frequency when using the "performance test lead" and "performance test tape") and compares it to an internal 29.97-Hz reference. This test checks the actual tape speed, not just the capstan motor speed. This dynamic test locates mechanical problems such as tape drag and slippage, in addition to capstan circuit problems.

The capstan speed error percentage readings tell how far off the actual tape speed is from the desired speed. The percentage reading is very helpful in identifying the source of a speed error problem. (See Fig. 11-34.)

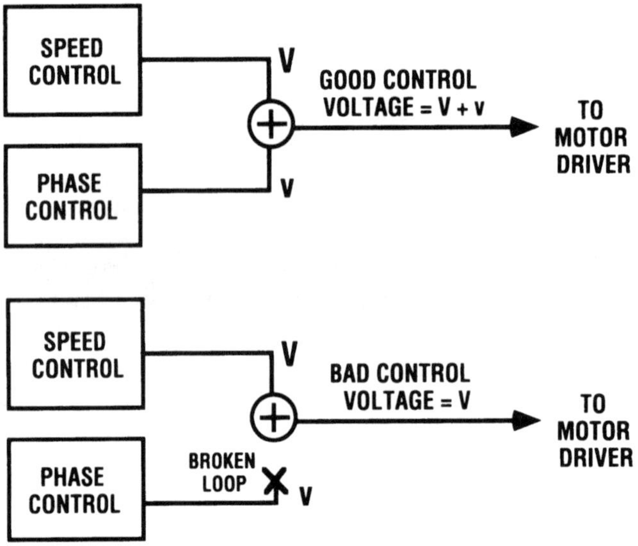

11-34 The phase control loop adds a small amount of correction voltage to the speed control voltage. A defective phase loop will cause the speed to be off slightly.

Readings that are off more than 10 percent suggest a problem in the capstan speed loop. Readings that are within 10 percent are caused by either a defective capstan speed loop or a defective capstan phase loop. (A defective phase loop will produce a wrong correction voltage. This voltage adds to the speed loop correction voltage to control the capstan motor. Because the capstan motor control voltage is off slightly, the motor speed is also off.)

To determine which loop is defective, play back a tape that contains segments of different tape speeds. If the speed circuits are working properly, you'll obtain low capstan speed error percentages at all speeds. However, if the speed circuits aren't working properly, you'll obtain large percentage errors at some tape speeds. Note: The slowest tape speed will likely have a larger error than faster tape speeds, even in a correctly functioning machine.

If the capstan speed error percentages are only slightly greater than 2 percent for all tape speeds, the problem is likely to be a mechanical problem such as excessive oxide buildup on the capstan, a hard capstan idler, or a loose capstan motor belt. If these check out okay, the problem is likely to be a defective capstan phase loop.

The "good/bad" readings are based on extensive research that shows that capstan speed errors greater than ± 0.5 percent usually can't be corrected by the phase-loop circuitry. No reading will be displayed if the tapes speed is varying widely. If you see a "bad" indicator, or no indication at all, use the percent readings to help determine what the capstan problem is. A technician is performing the capstan speed test with the VC93 in Fig. 11-35.

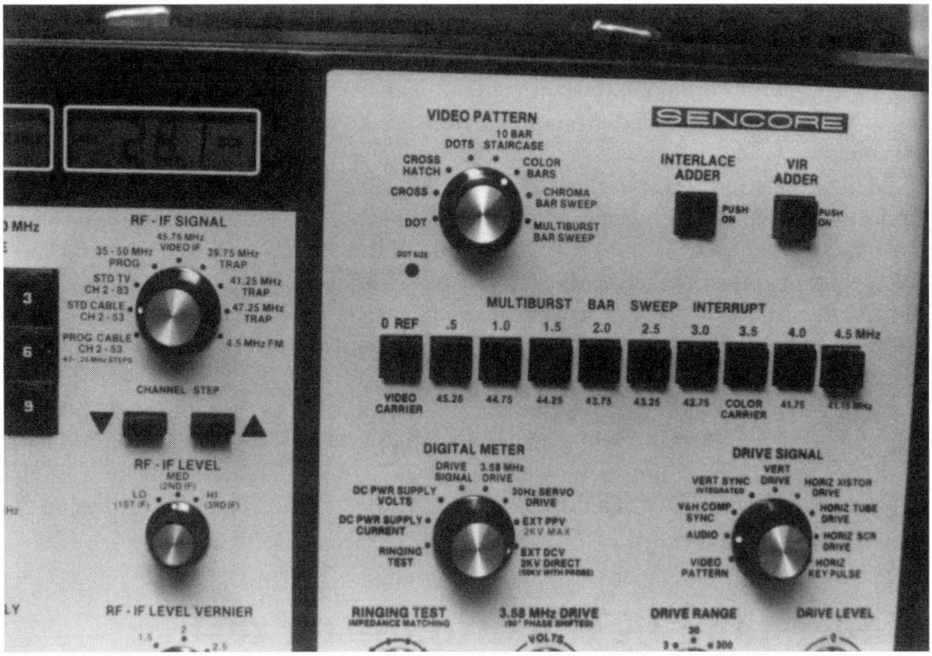

11-35 A technician is using the Sencore VC93 to check out a VCR loop control system.

Understanding the capstan jitter test

The capstan jitter test analyzes for minute speed variations in the CLT pulse when using the "troubleshooting test lead," or in the playback audio when using the "performance test lead" and "performance test tape." This test actually checks tape speed variations and not just capstan motor speed variations. Thus, it checks for mechanical problems such as tape drag and slippage, as well as for failures in the capstan control circuits.

The capstan jitter percentage readings indicate the amount of tape speed variation. All machines have some speed variation due to tape stretch, the tightness of the capstan loops, and other mechanical problems. Excessive speed variation causes the linear audio to vary excessively in pitch. The "good/bad" indications are based on research that shows that capstan speed variations greater than 0.5 percent give unacceptable audio performance and can affect the video performance.

Likely causes of "bad" capstan jitter are: a defective capstan phase loop, a slipping capstan belt or idler, or a bent or dirty capstan shaft. Also suspect a possible bad capstan motor. A bad motor bearing or motor winding can cause the capstan to "catch" as it turns. The lack of a "good/bad" reading indicates that one of the potential defects is causing severe capstan jitter.

Understanding the drum speed error test

The drum speed error test analyzes the frequency of the SW30 pulse when using the "troubleshooting test leads" or the playback video vertical sync pulse when using the servo performance test lead and the "performance test tap." It compares the signal to an internal 29.97-Hz reference. The SW30 signal is universal between most VCRs and is derived from the drum PG signal that's used by the drum servo circuits.

The drum speed error percentage reading displays how far the drum speed is off from the desired speed. Wrong drum speed will cause the horizontal sync pulses in the playback video to occur at the wrong time. If the drum speed is close to correct, the symptom appears as a misadjusted horizontal hold control on the playback monitor. A large drum speed error will cause the playback video to be so bad that no conclusion can be made by observing the playback monitor. Newer VCRs will mute the playback video when the drum speed is far off.

The good/bad readings are based on the amount of frequency offset that can be tolerated by most TV receivers. A speed error more than 0.10 percent will produce a "bad" reading. If the drum speed is varying widely, no "good/bad" indication is displayed. The most likely reason for a "bad" drum speed error reading is a missing FG pulse from the drum motor.

Understanding the drum jitter test

The drum jitter test analyzes for minute speed variations in the SW30 signal or the playback video vertical sync pulse when you use the servo performance test lead. The percentage reading indicates how much the speed of the revolving drum varies. Excessive drum speed variation causes the picture to appear to "breath" in and out.

The "good/bad" indication is based on research showing that drum speed variations greater than 0.10 percent give unacceptable picture quality. No "good/bad" reading is displayed if the speed is varying widely. This indicates a problem in the drum servo. A "bad" indication also means that the drum servo is bad. The most likely reason is a problem in the drum phase circuit or a mechanical defect such as a bad motor bearing, a drum that's out of balance, or a spot of dirt or oxide on the drum.

Making up a work tape

For the highest-quality work tape, always use a direct connection to the video and audio inputs of the VCR instead of using the antenna input. Always record your work tape using a high-quality tape and the highest quality VCR available. A four-head or five-head VCR is better than a two-head VCR because it uses the ideal head width for each tape speed. For fast troubleshooting, record three separate work tapes. Record the first tape with the performance and troubleshooting tests running at the SP speed. Record a second tape at the SLP speed and the third tape at the LP speed. This will speed troubleshooting of speed-related VCR problems.

The following sequence uses the VA62A Video Analyzer as the source for the audio and video signals. Connect the VA62A to the VC93 as shown in Fig. 11-36. Now perform the following steps:

1. Set the VA62A controls:
   - ~ Video pattern switch to multiburst bar sweep.
   - ~ All multiburst bar sweep interrupt buttons to the on position.
   - ~ Audio to 1 kHz.
2. Set VC93 modulation switch to external.
3. Connect the STD video output jack to the video in jack on the VCR.
4. Connect the STD audio output jack to the audio in jack on the VCR.
5. Place a blank tape in the VCR.
6. Set the VCR speed to fast speed.
7. Press the record button on the VCR.

Record the speed select test sequence on the tape in the order shown in Fig. 11-37, changing the tape speed on the VCR every 10 seconds. Switch the VCR to the SP mode and record the performance test sequence shown in Fig. 11-37, changing the video pattern and the audio on the VA26A every 20 seconds. Finally, record the troubleshooting test sequence, changing the video pattern and audio signal on the VA62A every two minutes.

You should make a separate tape following that sequence, but place the VCR in the medium speed for the performance test sequence and the troubleshooting test sequence. Do the same with a third tape, but record the last two sequences at the slowest tape speed.

Note: If you don't have a VA62A, then use the color bars pattern and a 1-kHz tone contained in the VC93. Record the speed select test sequence using the color bars pattern and switching the speed on the VCR every 10 seconds. Record the rest of the tape by switching the speed every 20 seconds. The key VCR format characteristics are shown in Fig. 11-38. (Portions of this chapter courtesy of Sencore.)

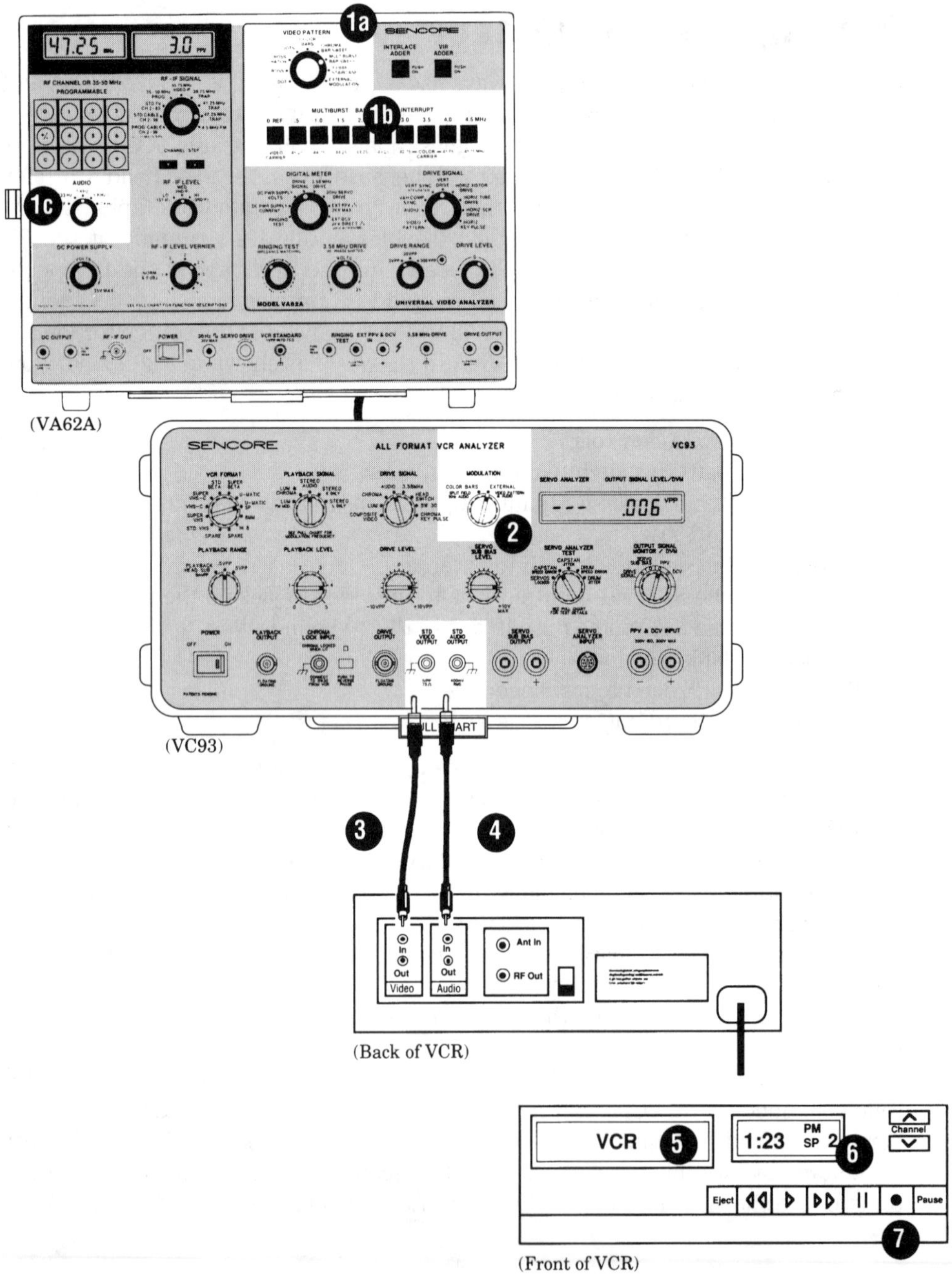

11-36 Equipment setup to make a work tape.

TYPICAL WORK TAPE

PURPOSE	LENGTH	VIDEO PATTERN	AUDIO
Speed Select Test Sequence	10 seconds	Multiburst Bar Sweep (Fast Speed-SP)	1kHz
	10 seconds	Multiburst Bar Sweep (Medium Speed-LP)	1kHz
	10 seconds	Multiburst Bar Sweep (Slowest Sweep-SLP)	1kHz
Performance Test Sequence **(single speed)**	20 seconds	Color Bars	333Hz
	20 seconds	Chroma Bar Sweep	1kHz
	20 seconds	Multiburst Bar Sweep	5kHz
	20 seconds	10-Bar Staircase	7kHz
Troubleshooting Test Sequence **(single speed)**	2 minutes	Color Bars	333Hz
	2 minutes	Chroma Bar Sweep	1kHz
	2 minutes	Multiburst Bar Sweep	5kHz
	2 minutes	10-Bar Staircase	7kHz

11-37 Information that should be found on a typical work tape.

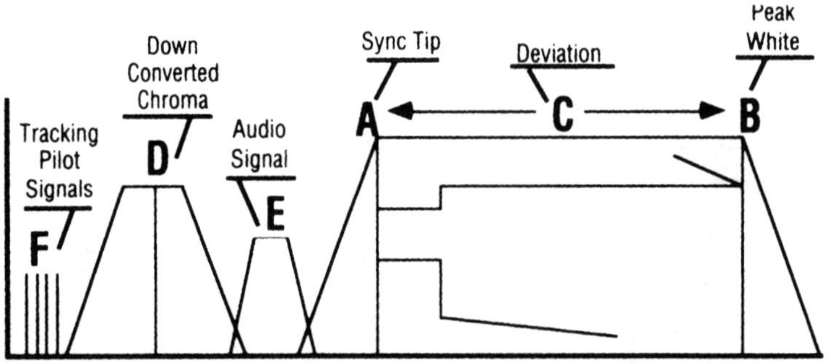

FORMAT	A (MHz)	B (MHz)	C (MHz)	D (kHz)	E (MHz)	F (Hz)
VHS	3.4	4.4	1	629	1.3 ; 1.7	N/A
SUPER VHS	5.4	7	1.6	629	1.3 ; 1.7	N/A
VHS-C	3.4	4.4	1	629	1.3 ; 1.7	N/A
SUPER VHS-C	5.4	7	1.6	629	1.3 ; 1.7	N/A
BETA	3.5	4.8	1.3	688	1.38;1.53 1.68;1.83	N/A
SUPER BETA	4.4	5.6	1.2	688	1.38;1.53 1.68;1.83	N/A
U-MATIC	3.8	5.4	1.6	688	N/A	N/A
U-MATIC SP	5.0	6.6	1.6	688	N/A	N/A
8mm	4.2	5.4	1.2	743	1.5	102,544 118,951
HI-8	5.7	7.7	2	743	1.5 ; 1.7	165,210 148,689

11-38 Key VCR format characteristics.

12
Sencore VG91 universal video generator and TVA92 video analyzer

THIS CHAPTER INTRODUCES SENCORE'S NEW TECH CHOICE VIDEO ANALYZER Systems. These new test instruments will perform all of the tests that the VA62A Video Analyzer will do—plus much more. In the next year or two, I hope to complete a service manual on these new instruments, after I've had more troubleshooting experience with them and can check out various other techniques with some of the more advanced video and VCR electronic equipment now coming down the pike.

VG91 Universal Video Generator

Video was once considered solely in terms of the home TV receiver. Television has grown to become a major and indispensable means of entertainment. However, today's video has expanded far beyond the living room. Today, NTSC video, in the form of television receivers, video monitors, and VCRs, is used extensively in educational and industrial applications, as well as in the home entertainment center.

Consumers place great demands on today's video display systems. Projection and large-screen direct-view color TV with features such as improved definition and stereo sound are mainstays in today's high-performance video. Higher-quality video and audio signals are delivered by broadcasters and cable systems, and the viewer has an ever-increasing number of channels to choose. Video recorders bring far-off places and hi-tech training to the classroom or meeting room, and camcorders that offer near broadcast quality in a palm-sized package allow the consumer to capture a piece of history on videotape. All of these revolutionary changes offer new opportunities and place growing demands on the technicians who are responsible for servicing and maintaining today's high-tech NTSC TV/video products.

The Sencore VG92 Universal Video Generator (see Fig. 12-1) is an all-channel, all-pattern, all-purpose NTSC video/MTS audio generator. It provides all the off-air TV and cable channels needed to verify customer complaints, recreate trouble symptoms, and isolate receiver problems. Each channel is variable in signal level and can be modulated with video patterns and mono or MTS audio. Video IF signals, standard S-video (YC), composite video, and baseband audio signals are provided to isolate problems to any video component. All of the test signals needed to thor-

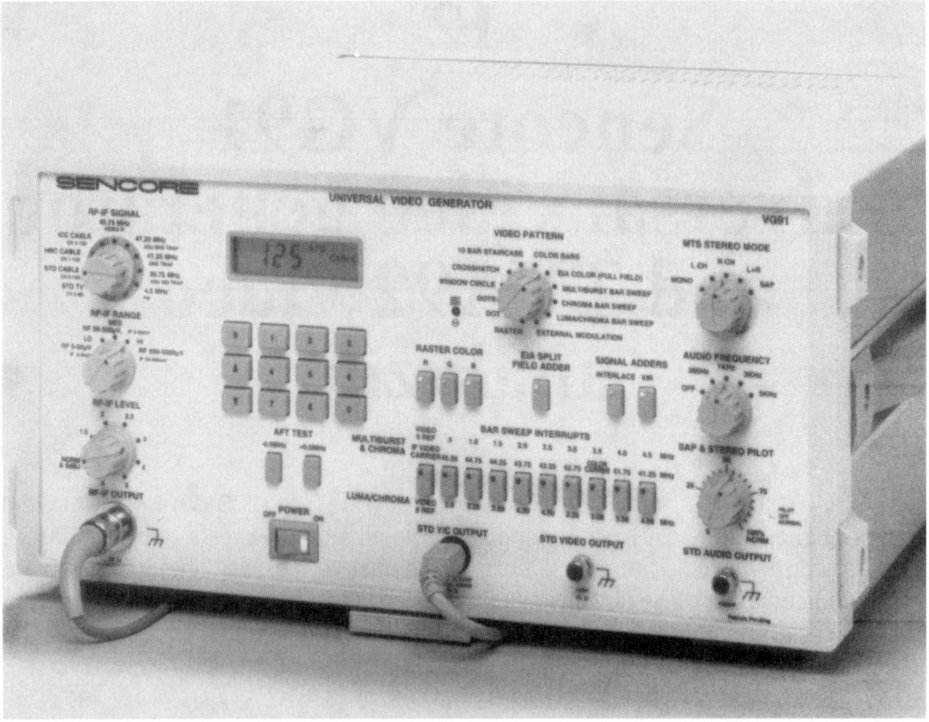

Sencore

12-1 Photo of the Sencore VG91 generator.

oughly and accurately test, align, and troubleshoot NTSC TV-video systems are provided in one complete, easy-to-use instrument.

The VG92 serves as the signal for the Sencore VC93 All-Format VCR Analyzer and the TV92 TV Video Analyzer. To resist obsolescence, the VG92's signals are common to FCC and NTSC specifications. As video systems change, the VG91 can be updated with other accessories or companion analyzers to meet the servicing needs of new video systems.

VG91 features

The VG91 produces all of the allocated VHF and UHF off-air TV channels and 125 standard, HRC, or ICC cable-designated channels. The TV-RF channels are digitally synthesized and match all off-air and cable-channel-designated frequencies to test "cable-ready" tuners. An RF attenuator provides accurate signal levels to duplicate weak or strong signals.

A crystal-accurate 45.75-MHz IF generator provides a modulated IF carrier to substitute for tuners and isolate defects in video-IF circuits. Exclusive trap signals dynamically check adjacent channel rejection and provide fast and accurate trap alignment. A modulated 4.5-MHz audio IF signal is provided for audio troubleshooting. An AFT test provides a quick performance check of the receiver's AFT circuits to simplify troubleshooting and alignment.

The VG91's NTSC video pattern generator provides standard patterns for performance testing and alignment. An exclusive, patent-pending luma/chroma bar sweep pattern analyzes circuits with extended bandwidths such as comb filters and wideband I color circuits.

An MTS stereo audio generator provides signals to verify the performance of MTS circuits. The MTS signal is available on all RF channels, as well as the 45.75-MHz IF and 4.5-MHz audio IF signals. Y/C (S-Video), composite video, and baseband audio outputs provide signals for testing video and audio circuits through standard input jacks.

The VG91 teams up with specialized Sencore video analyzers such as the VC93 "All-Format VCR Analyzer" or TVA92 "TV Video Analyzer." The team specializes in analyzing and pinpointing defects in specific products such as VCRs or TVs. The VG91 provides the necessary synchronizing signals and video patterns for these and other future Sencore video analyzers.

Controls

Some of the controls and their location on the front panel are as follows (also see Fig. 12-2):

1. Video pattern switch.
   - ~ Raster.
   - ~ Dot.
   - ~ Dots.
   - ~ Window circle. This new pattern consists of a circle centered within a box and cross lines for centering and linearity adjustments.
   - ~ Crosshatch.
   - ~ 10-bar staircase.
   - ~ Color bars.
   - ~ EIA color full field.
   - ~ Multiburst bar sweep.
   - ~ Chroma bar sweep.
   - ~ Luma/chroma bar sweep.
   - ~ External modulation.
2. MTS stereo mode switch.
   - ~ Mono.
   - ~ L channel.
   - ~ R channel.
   - ~ L + R.
   - ~ SAP.
3. Audio frequency switch.
4. SAP & stereo pilot.
   - ~ Control adjusts MTS stereo pilot level.
   - ~ Pilot off norm indicator.
5. Std audio output jack.
6. Signal adders buttons.
   - ~ Interlace

    ~ VIR
7. Bar sweep interrupts buttons.
    ~ This row of numbers indicates the video frequency of the bar sweep pattern.
    ~ IF frequency of the chroma bar sweep.
    ~ This row of numbers indicates the video frequency of the video pattern.
8. Std video output jack.
9. EIA split field adder.
10. Raster color.
11. Std Y/C output jack.
12. Pull chart.
13. Power on/off switch.
14. AFT test buttons.
15. RF channel keyboard.
16. RF-IF output jack.
17. RF-IF level control.
18. RF-IF range switch.
19. RF-IF signal switch.
    ~ Std TV.
    ~ Std cable.
    ~ HRC cable.
    ~ ICC cable.
    ~ 45.75-MHz video IF.
    ~ 47.25-MHz ADF SND trap.
    ~ 41.25-MHz SND trap.
    ~ 39.75-MHz ADJ VID trap.
    ~ 4.5-MHz FM.
20. RF-IF digital readout.
21. Dot size control.

Instrument operation

VG91 accessories

The VG91 supplies dc voltages, video, audio, and sync signals to companion units or any accessories. The RF-IF generator portion of the VG91 provides the signals used in the stages before the video or audio detectors in a receiver that tunes television signals.

RF generator

The RF generator section has 4 groups of channels that duplicate the frequency assignments of over-the-air broadcasts and cable TV signals. Refer to the front panel layout of the RF-IF generator portion in Fig. 12-3. All channels are provided through the RF-IF output jack and can be modulated with a video carrier as selected by the video pattern switch and an audio carrier as selected by the MTS stereo mode and audio frequency controls.

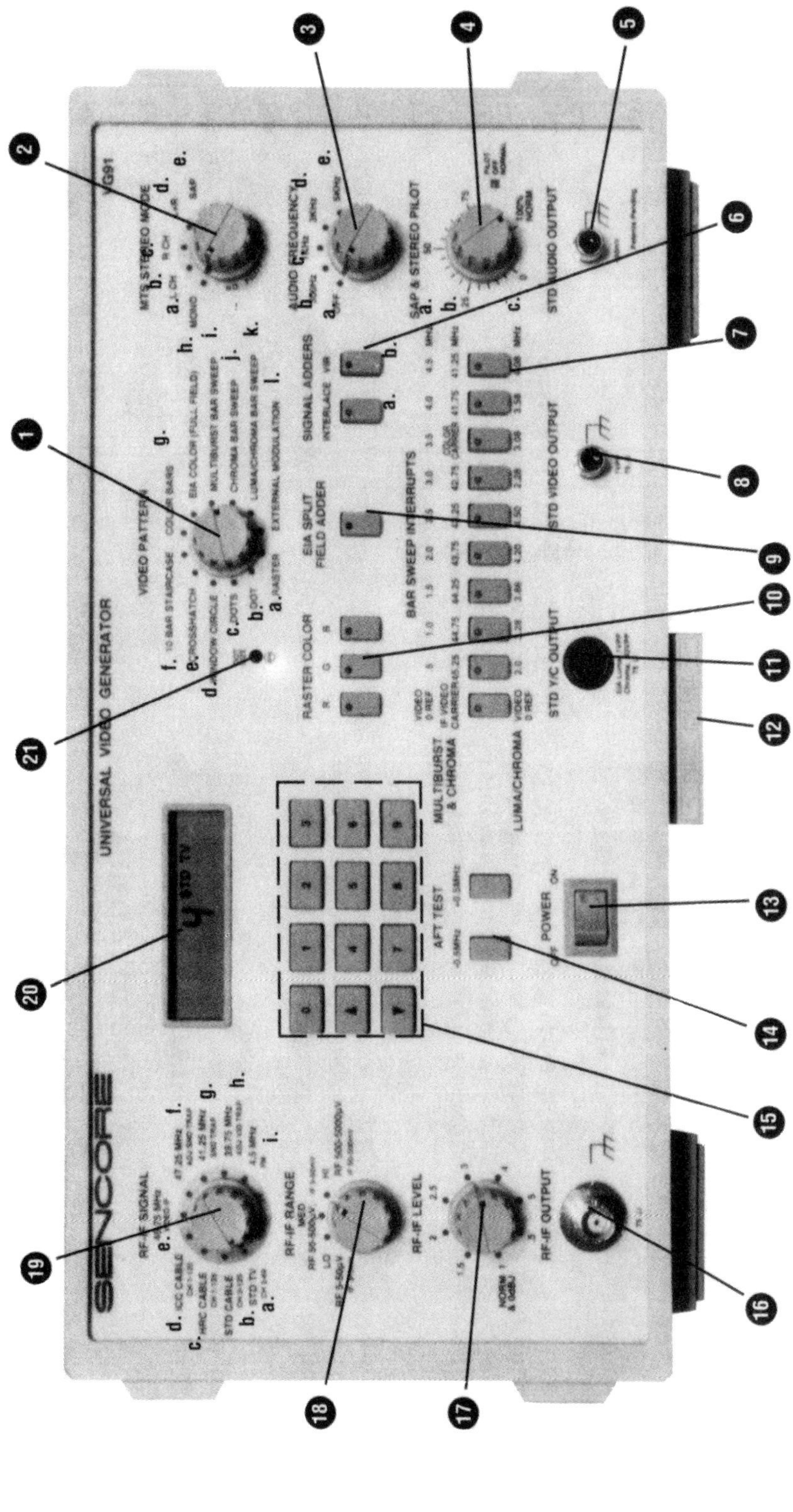

12-2 Front panel control location of the Sencore VG91 generator. Sencore

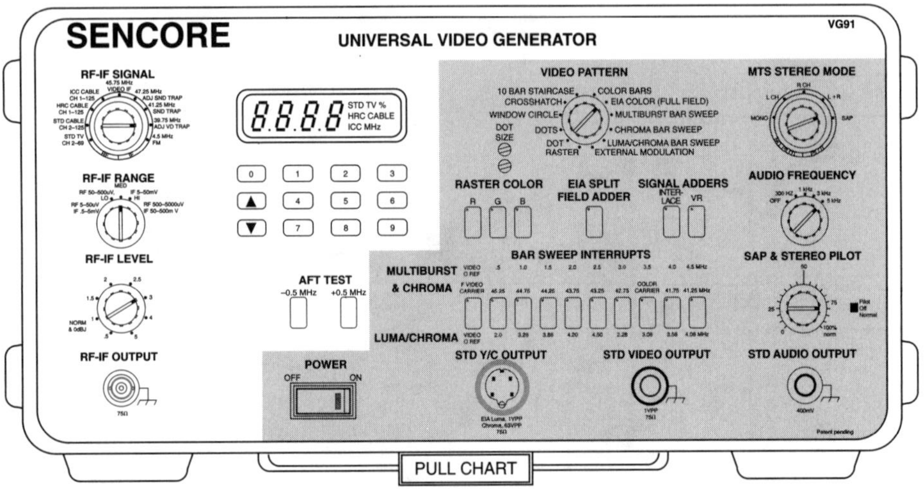

12-3 The RF-IF generator portion provides the signals used to test, troubleshoot, and align the stages before the video and audio detectors. Sencore

IF generator

The VG91 IF generator provides modulated signals to troubleshoot or align the video IF and sound IF stages of a television-signal receiver. The IF generator has three main sections:

1. 45.75-MHz video-IF.
2. Video IF setting signals.
3. 4.5-MHz sound IF generator.

Video patterns

The video pattern generator section of the VG91 produces standard test patterns and several unique analyzing patterns. Each pattern provides specific information concerning the operation, performance, and alignment of video equipment. The video pattern switch determines the video pattern that modulates the RF and IF video carriers and feeds the Y/C and video output jacks. Refer to the front panel location of the video pattern controls in Fig. 12-4.

Raster

The raster pattern provides a raster display. The color of the blank raster can be changed from black to grey or any primary or secondary color using the raster color switch.

Dot pattern

The dot pattern provides a standard single dot that's centered within the raster. This pattern is used to set the center, static convergence of a color receiver or monitor.

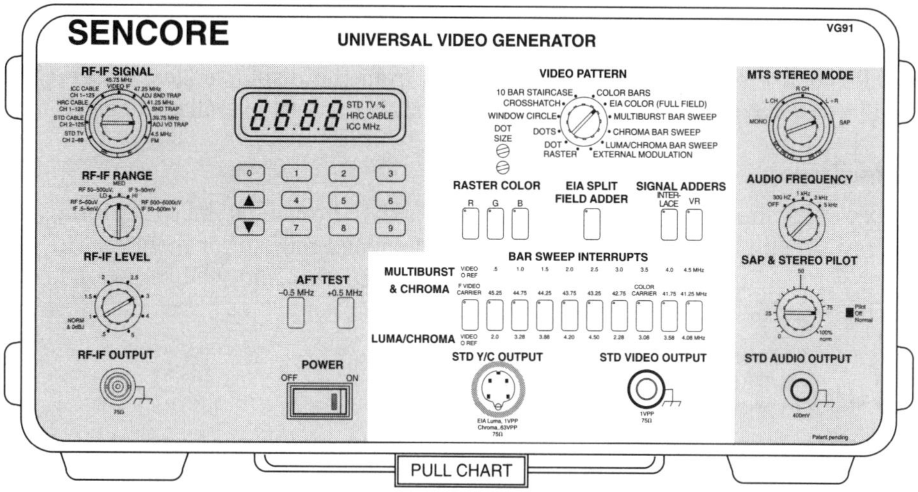

12-4 The video patterns modulate the RF and IF video carriers and feed the Y/C and standard video output jacks. Sencore

Dots pattern

The dots pattern provides the standard dynamic convergence pattern that's recommended by most video manufacturers.

Window circle

The window circle pattern consists of several patterns combined into one useful pattern: a cross, a box, and a circle all centered within the raster. This pattern can be used for centering width, height, linearity, pincushion, and other deflection-related adjustments and tests. The size of the vertical lines is adjustable with the dot size control, and the background color can be set using the raster color switches. The window circle pattern is shown in Fig. 12-5.

The cross marks the center of the screen. Use it to check and adjust picture centering. The top, middle, and lower horizontal lines provide a convenient reference

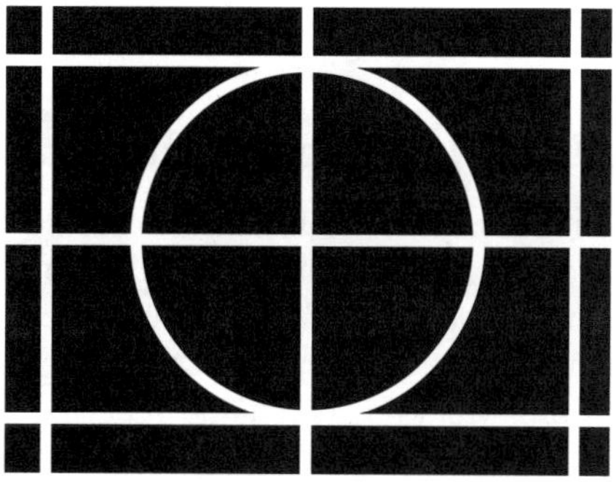

12-5 The window circle video pattern. Sencore

for rotating the deflection yoke. Use the box as a reference for making deflection adjustments such as vertical height, linearity controls, horizontal width, and pincushion. Position the lines about five percent in from the display edges. Observe the circle when making these adjustments to maintain the best overall symmetry.

Crosshatch

The crosshatch pattern produces 21 vertical and 15 horizontal lines that form symmetrical squares on the screen of a properly adjusted receiver or monitor. The size of the vertical lines is adjustable with the dot size control, and the background color can be set using the raster color switches.

The MTS/audio generator

The MTS audio generator section of the VG91 provides both standard audio and MTS stereo signals for testing and troubleshooting monaural and MTS stereo television-signal receivers. The audio generator has three main functions:

1. It provides modulation for the RF and IF signals with the selected audio frequency and composite stereo mode.
2. It amplitude modulates the IF trap carriers to provide improved resolution when performing IF trap alignment.
3. It provides an audio frequency signal to apply standard audio input connectors.

The front panel in Fig. 12-6 shows the MTS stereo mode, audio frequency, SAP & stereo pilot, and std audio output controls. The MTS/audio generator provides modulation for the RF, IF, and trap signals as well as an audio out signal.

Computer-automated bus operation

The VG91 universal video generator can be used for computer-controlled, automated testing. A simple computer program can be developed to automatically set up any of the front panel controls (except the AFT test buttons and power switch). The VG91 can be used with either the IEEE-488 bus and the Sencore IB72 accessory or the RS232 bus and the Sencore IB78 accessory. Both accessories connect to the interface accessory jack located on the rear of the VG91.

Using the VG91 as a listener

The main use of the VG91 in an automated test system is as a listener. As such, the settings of the front panel controls can be automatically set with a computer to predetermined conditions. This can be especially useful as an aid in performance testing, as the computer can prompt a nontechnical person through the test.

The first valid bus command received by the VG91 "locks out" the front panel controls, and the VG91 responds only to bus commands. All of the front panel con-

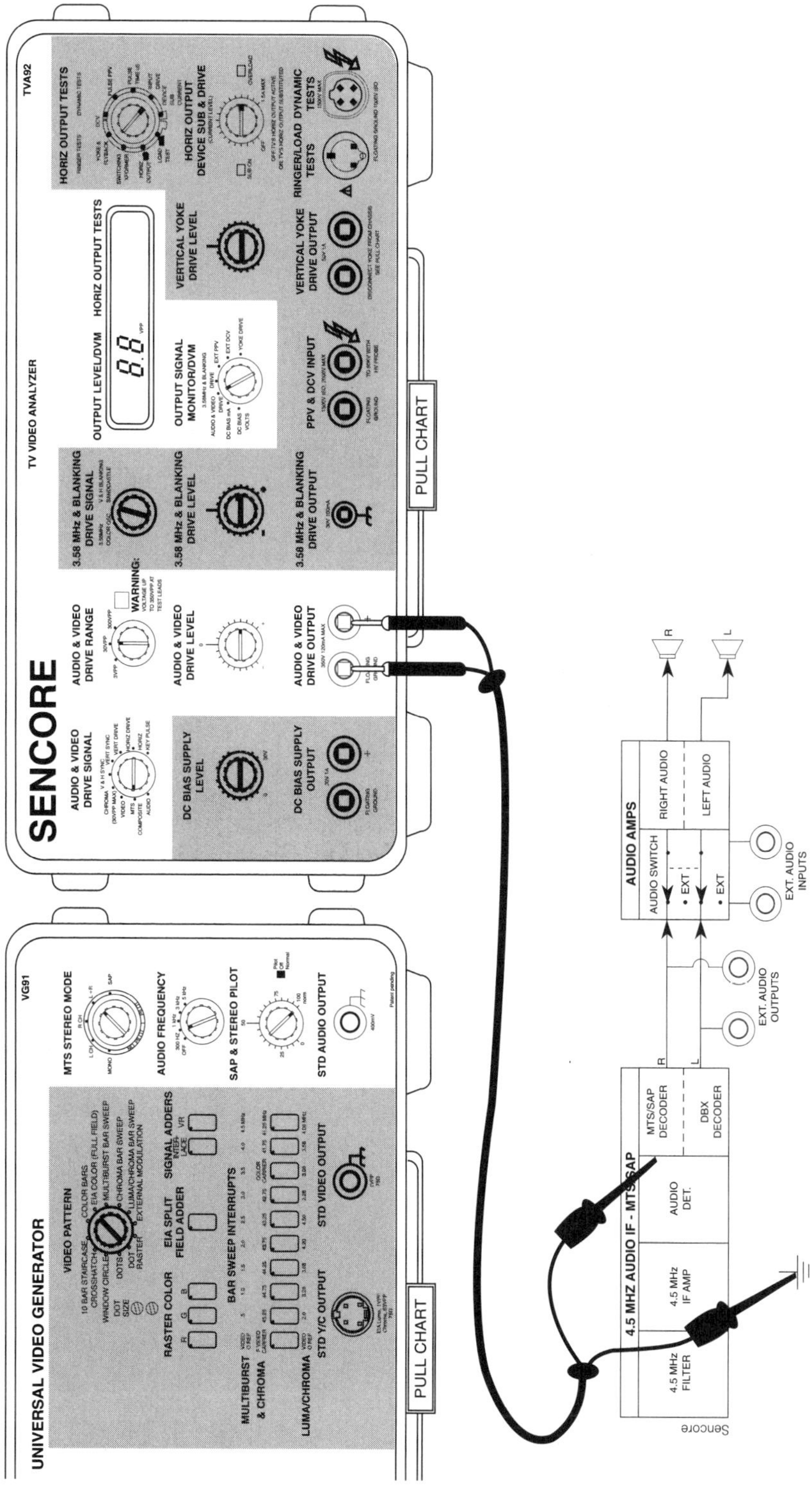

12-6 Setup for injecting into the audio circuits after the audio detector.

trols remain locked out until a "CPO" (control panel on) command is received from the controller. The VG91 functions the same under bus control as it does during manual control, and the digital meter continues to display the RF-TV channel or selected IF signal.

Commands

The VG91 uses a series of commands to control the front panel setup. The entire command must be sent in order for the VG91 to properly respond.

Talker mode

In the talker mode, the VG91 continually returns data to the computer regarding the setting of the front panel controls. The front panel controls can be read all at once, or individually. Sending an individual function command instructs the VG91 to return the setting of only that control. The "CPS ALL" command instructs the VG91 to return the settings of all front panel controls. If you send the CPS ALL command, your controller must be able to receive 13 data fields, one after another, with a carriage return and line feed between them. The data returned by the VG91 is in a standard data format, and the data string consists of three parts of digital information.

Error codes

The VG91 will return an error code when it receives an invalid command. The error code can be triggered by an erroneous channel, level, or bus command.

Analyzer applications

Let's now look at some troubleshooting applications using the VG91 instrument.

TV/video functional blocks

A basic video system has two sections: video and audio. Ahead of those sections, a TV-RF system adds three stages: video IF, tuner, and audio IF. A functional block diagram for a TV/video system is shown in Fig. 12-7. The TV tuner stage converts the RF signal to an IF frequency; the video-IF stage amplifies and filters the IF signal; the video stage processes the luminance and color information; the audio IF recovers the RF audio; and the audio stages amplify the detected sound signal. A TV/video product might contain only a few of these operational sections or all five. Identifying these sections and the common functional blocks within each greatly simplifies testing and troubleshooting of TV/video products.

Performance testing

The first step in troubleshooting any TV-RF or video system is to confirm symptoms or customer complaints by performance testing. To ensure total operation of the equipment, performance testing also should be the final step after you repair the

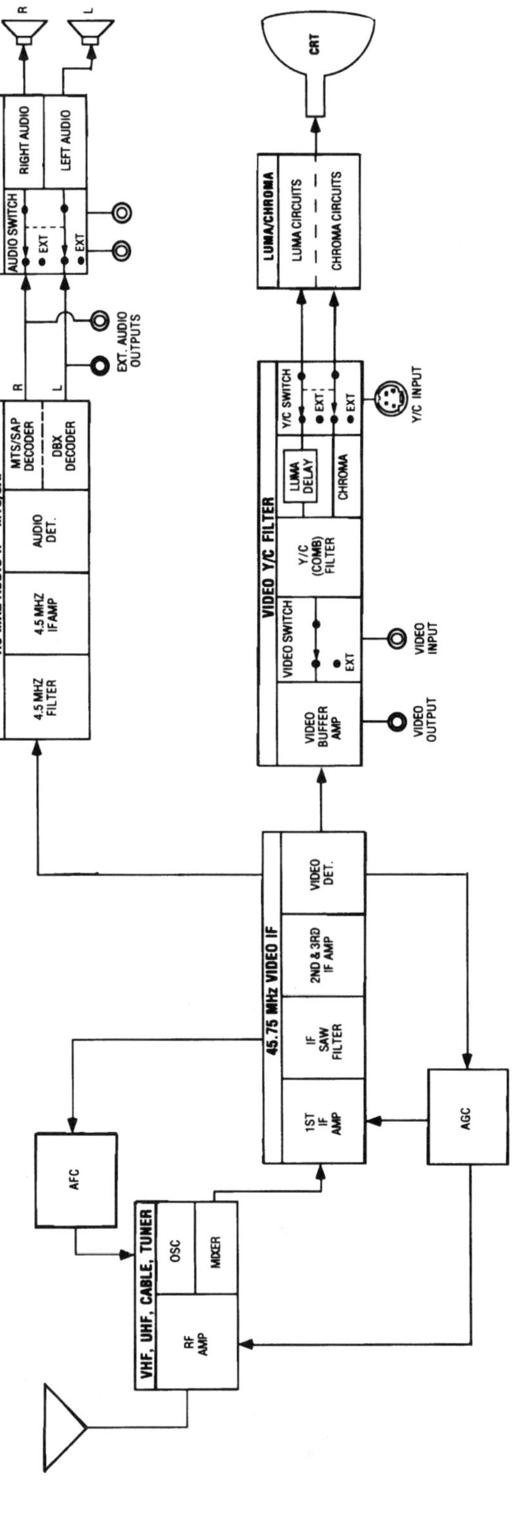

12-7 Typical TV-video system block diagram. Sencore

chassis. The VG91 provides modulated reference signals so you can observe the operation at the outputs using the system's CRT, an external monitor, or an oscilloscope.

To reliably test the performance of a system, you need to apply a known-good signal at a known level. Over-the-air and cable signals vary in signal level, video content, and signal quality. These variations make it extremely difficult to properly evaluate the system's performance. The special patterns, adjustable RF-IF levels, AFT test frequencies, and standard outputs supplied by the VG91 allow you to evaluate the operation of each functional block that's common to TV-video systems. The VG91 performance tests fall into two general categories:

1. Testing receiver tuner and IF circuits.
2. Testing video and audio processing circuits.

Tuner systems

There are two basic types of tuners used in TV-RF systems: cable-ready and non-cable-ready. When testing non-cable-ready tuners (tuners that only receive off-air VHF/UHF channels) you only need to use the std TV mode on the VG91. Most new digital tuners, however, are "all-channel cable-ready" and can select all off-the-air or cable signals. (See Fig. 12-8.) This is done with a user switch or programmable setting commonly called "normal TV" or "cable." These tuners require more extensive testing to ensure correct operation.

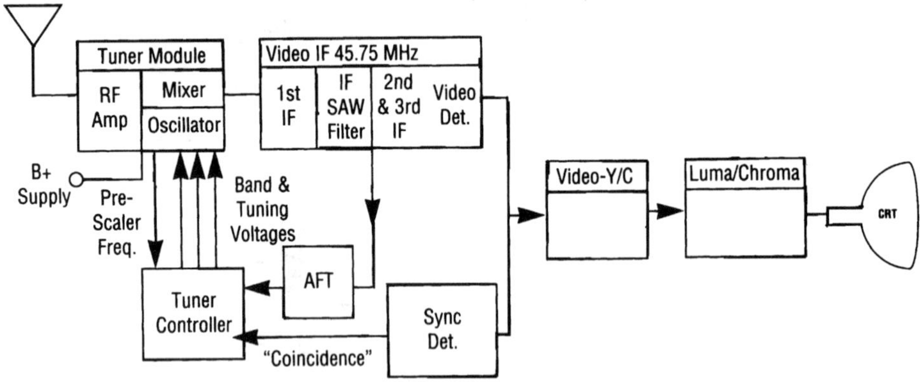

12-8 Block diagram of a typical "all-channel" digital tuning system. Sencore

Cable TV systems can shift some or all of the channel carriers as much as ±2 MHz from the standard broadcast frequencies. Conventional tuners require the customer to adjust the fine tuning to match the offset carrier. Digital tuners use extra control circuits to do this automatically. As a result, digital tuners operate differently in the "cable" mode than in the "normal TV" mode.

In the "normal TV" mode, digital tuners go to the center frequency of the TV channel. The AFT voltage fine tunes the tuner to center the IF signal. In the "cable" mode, digital tuners auto-search for a TV-RF carrier up to ±2 MHz from the standard center frequency of the channel. When a TV signal is found, the tuner locks to the channel and begins normal AFT action. If no signal is found after several searches, the tuner returns to the center channel frequency.

Testing IF response

The IF bandpass requirements differ between TV-RF receivers. A large-screen television typically has a bandpass of 3.5 to 4.0 MHz to display good picture detail. Small-screen televisions usually have about 3 MHz of bandwidth, while black-and-white TV receivers have only a 2- to 2.5-MHz bandwidth. (The narrow bandwidth in a black-and-white TV is necessary to prevent the color signals from interfering with the luminance signals.)

If bandwidth is determined by the alignment of LC-tuned circuits or by the fixed response of a SAW filter, LC-tuned circuits easily drift or change alignment due to thermal or mechanical stresses. SAW filters can also become defective and drift.

The VG91's multiburst bar sweep video patterns quickly test the complete response of video-IF stages. These patterns provide various video frequencies that allow you to test the IF response by visually observing the output of the video detector. The basic setup of the VG91 for injecting a signal into the IF stages is shown in Fig. 12-9.

Using either a monitor connected to the receiver's video out jack (the CRT in a television) or an oscilloscope connected to the video out jack or video detector output, you can determine if the receiver is passing a video frequency. Each frequency bar of the multiburst bar sweep pattern produces a series of black-and-white vertical stripes. As the video frequency increases, the stripes become closer together. The last band of visible stripes shows the highest video frequency passed by the receiver.

Synchronous detector testing

Most of today's TV-RF receivers use a synchronous video detector. Synchronous detectors offer many advantages over the conventional diode detector. Several major advantages are as follows:

1. This type of detector is less prone to pickup of unwanted signals.
2. It generates very few beat products.
3. It's linear and gives good contrast range.
4. It produces no output when a carrier is not present, which results in a black raster.

However, the synchronous detector must be properly aligned to fully use these advantages. If it's misaligned, it can severely degrade the video.

The synchronous detector requires an IF carrier for operation and therefore can't be aligned or tested with a sweep and marker generator. The VG91 provides an IF carrier that can be modulated with a linear 10-bar staircase pattern to quickly and

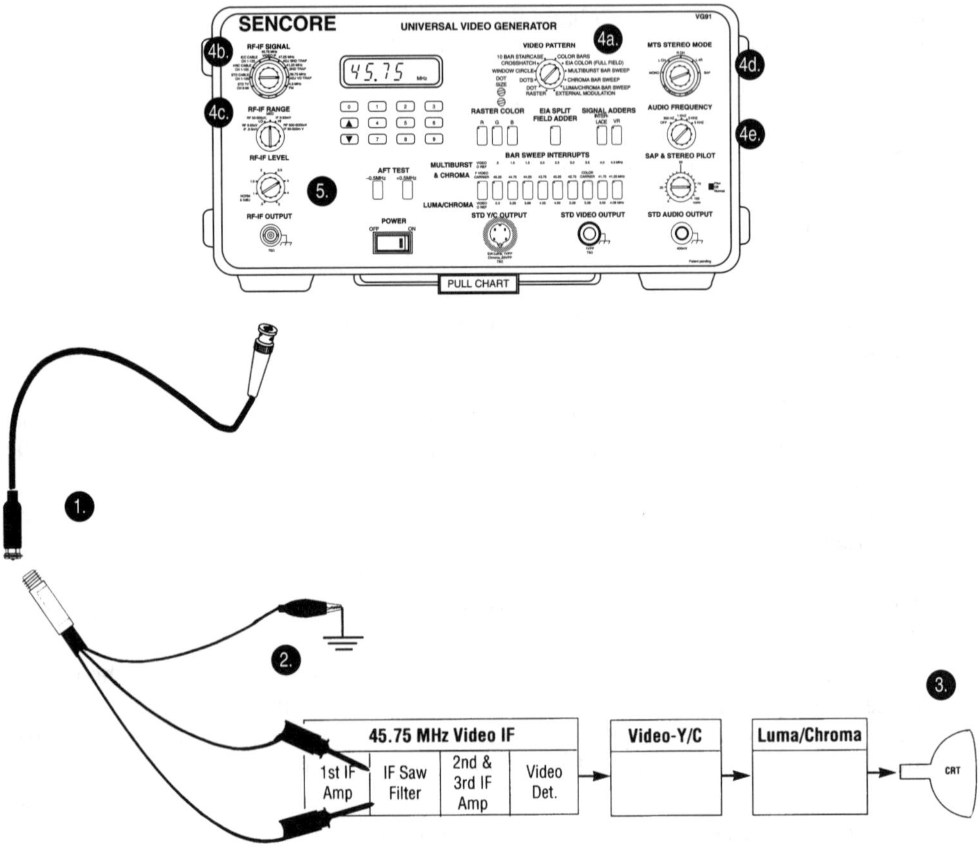

12-9 Basic setup for injecting into the IF stages. Sencore

dynamically test and align a synchronous video detector. To align the synchronous detector coil, apply the VG91's 45.75-MHz video IF signal into the IF stages and observe the output. You can use a video monitor to judge the performance, but an oscilloscope allows the best adjustment.

Video stages

Most TV-video products, including VCRs, monitors, and television receivers, process the luminance and chroma portion of the demodulated video signal. Some video products, such as monitors, only have video stages and rely on some other component to tune and demodulate the RF signal. The VG91's video patterns provide special reference signals to analyze and test the performance of any NTSC video stage. Some patterns are specifically designed to test the performance of the luminance stages, while others test the chroma stages.

To provide the best test of the video stages, you should apply the VG91's signal as close as possible to the video circuits. (Figure 12-10 shows the basic setup for testing video stages.) Otherwise, defects in earlier stages might cause degraded performance. When possible, apply the signal to the unit's Y/C input. If the unit doesn't have a Y/C connector, use the video line in jack. Only use the RF input if the chassis doesn't have a Y/C or video in connector.

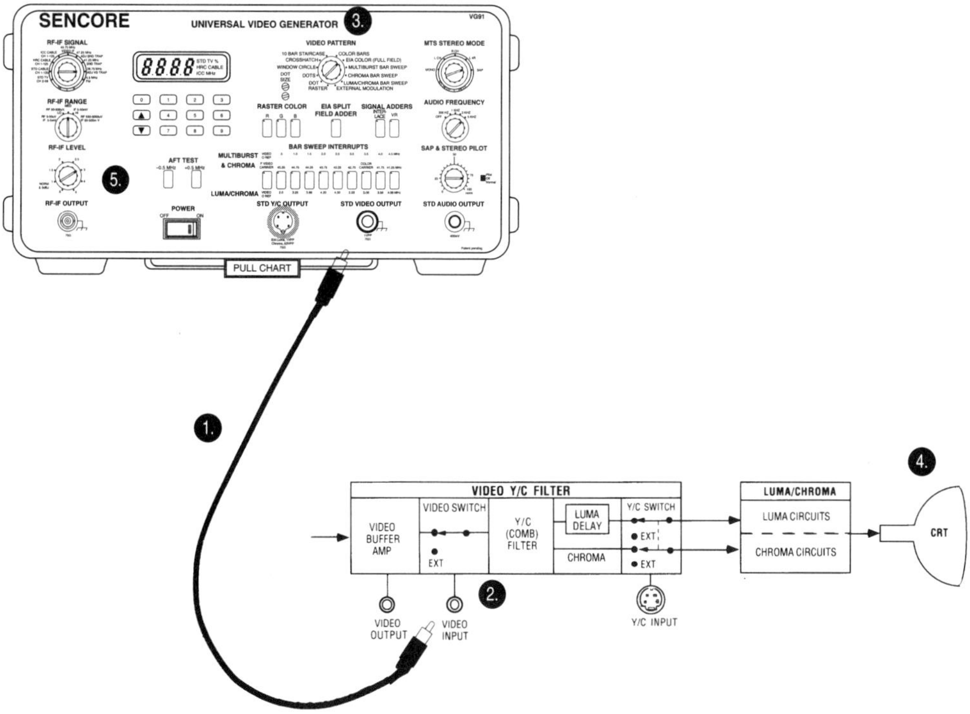

12-10 Basic setup for testing the video stages. Sencore

Testing luminance frequency response

Use the multiburst bar sweep pattern to test the frequency response of the video luminance stages. This pattern checks the video luminance circuits over their entire frequency range to locate any stage that might be restricting the video resolution.

Some service literature references "lines" of resolution. A *line of resolution* is a measure of video frequency response and is the smallest horizontal detail that can be reproduced. Lines of resolution can be determined from the multiburst bar sweep pattern by noting the highest frequency bar that still shows individual lines. Find the frequency of this bar (marked in MHz by the multiburst bar sweep Interrupt button) and multiply by 80 to convert to the approximate lines of resolution. A 2-MHz bar, for example, equals 160 lines of resolution. The multiburst bar sweep lines of resolution pattern is shown in Fig. 12-11.

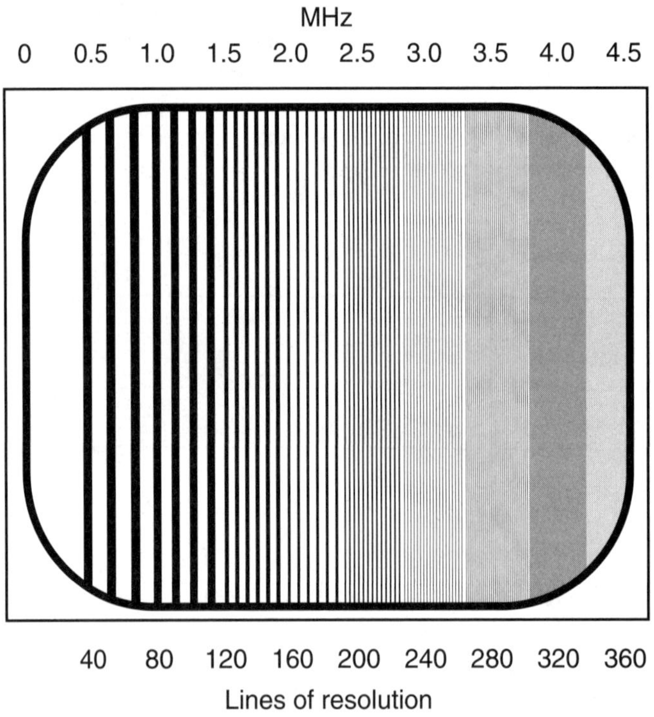

12-11 Use the multiburst bar sweep pattern to determine "lines of resolution." Sencore

Audio troubleshooting tips

Generally, when the problem is only with sound reproduction in a TV receiver, the defect will usually be found in the audio section. Of course, there's always an exception to this statement. The most common exception would be buzz or hum in the sound. If both sound and picture raster are affected, you should suspect the fault to be in a common stage. Also, if the set is a remote model, the sound problem could be in the remote control system.

The following symptoms usually indicate trouble in the sound section where the picture has not been affected: no sound, distorted sound, weak sound, motorboating, microphonics, and buzz or hum in the sound.

Sometimes trouble in the sound section will also affect other sections of the receiver. For example: interference in the picture due to radiation from the sound section, sound bars in the picture, or no picture and no sound caused by a loss of a common B+ supply feed.

Troubles in the sound section are usually caused by faulty components, loose connections, and/or poor solder joints. Improper alignment of the sound IF amplifiers and detector can cause a complete loss of sound, weak or distorted sound, or hum or buzz in the sound reproduction. When all other checks have failed to turn up or pinpoint the problems just covered, an alignment response check should be performed on the sound IF amplifiers and FM detector stages.

Some "sound" troubleshooting tips

Use a logical troubleshooting procedure to locate a sound problem, just as you would for any other section of the TV receiver circuits. This includes preliminary checks and test-instrument checking if the defect is not located in a reasonable time with these preliminary checks.

You can generally assume (but not always correctly) that there's a fault in an audio section circuit if only a sound problem is indicated. A visual inspection of the chassis should reveal any obvious defects, such as tubes not lit, burned parts, loose connections, or hot-to-the touch ICs. Vary the volume control and note any effect. For a tube-type set, check or replace all tubes in the sound section. For a solid-state set, check all transistors and replace any plug-in chips. If the trouble has not been found at this point, connect the VG91 generator and inject the audio test signals. The type and order of these checks will depend on the type of audio symptoms.

IC sound systems

In some modern-day TV receivers you'll find a complete sound system from the IF amplifier to the audio amplifier located on one IC. Because these chips contain so many different circuits, many technicians use component substitution as a way of finding audio trouble. The Analyzer helps you eliminate that costly step of soldered-in IC substitution because it has all the signals you need to inject from the sound IF all the way to and including the speaker. The Analyzer provides a 4.5-MHz modulated IF signal and four audio tone signals for the IC injection points.

A good place to start is at the speaker. Inject the 1000-Hz tone and listen for sound coming from the speaker. Then just work your way to the input of the audio amplifier, listening for the 1000-Hz tone from the speaker. In this manner, you can be sure the audio portion is now operating correctly and is not the problem. The 4.5-MHz modulated IF signal is then injected into the input of the IF amplifier and to both sides of the ceramic filter to determine where the signal is being lost. Thus, the Analyzer lets you go through the complete sound system to isolate the problem in a short time as compared to installing another IC or other component.

Audio performance tests

All television receivers and most VCRs can receive the audio portion of the television RF signal. Additionally, many receivers have MTS decoders and can receive stereo and secondary audio program (SAP) audio signals. Problems in the tuner, IF stages, sound detector or MTS decoder can affect the quality of the recovered audio signal. A full performance test of a stereo TV RF receiver includes the following tests:

1. Mono operation.
2. Stereo operation.
3. Stereo separation.
4. Stereo signal-to-noise test.
5. SAP operation.

6. SAP signal-to-noise test.
7. SAP threshold test.

Over-the-air signals can't be used to test the performance of MTS receivers because they're constantly varying in separation and signal strength. The VG91 universal video generator provides a controlled signal to accurately and reliably test MTS stereo receivers. The performance tests should be made by applying a signal into the tuner. This will check the entire receiver's audio circuits under actual operating conditions. Any RF channel can be used to check the operation of the audio stages. As with the video performance test, you should do a quick operational test of mono and stereo operation before and after a receiver repair. The VG91 basic setup and connections for doing audio performance tests is shown in Fig. 12-12.

Stereo separation test

This test checks how the receiver can reproduce a signal in one channel while rejecting the signal in the other channel. The modulation is set to produce a signal in just one channel. Any signal that appears in the other channel is crosstalk and indicates poor separation. The best separation occurs when the difference between the two audio output channels is the greatest. Many factors cause poor stereo separation:

1. Improper IF alignment.
2. A defective or misaligned stereo decoder.
3. A defective matrix.
4. A defective or misaligned dBx decoder.
5. A problem in the power supply.

Stereo separation should be tested at two different audio frequencies because the dBx noise reduction circuits process low frequencies differently than high frequencies. Thus, problems can develop that affect the low-frequency separation, and vice versa. The VG91 provides several audio frequencies to check both low-frequency and high-frequency separation.

Stereo frequency response

All TV RF receivers use a 75-μsec deemphasis network to reduce the output of the high audio frequencies. This deemphasis network reverses the preemphasis higher-audio frequency boost that was added at the transmitter. In mono receivers, the deemphasis network is after the FM detector output and, in stereo receivers, it's in the L+R signal path.

Introduction to MTS theory

Commercial TV broadcasting in the United States is based on NTSC standards. In the NTSC system, each broadcast channel has a 6-MHz bandwidth. The video signal is broadcast as an AM signal that's 1.25 MHz above the lower channel boundary. The audio (aural) signal is an FM signal with a carrier located 4.5 MHz above the visual

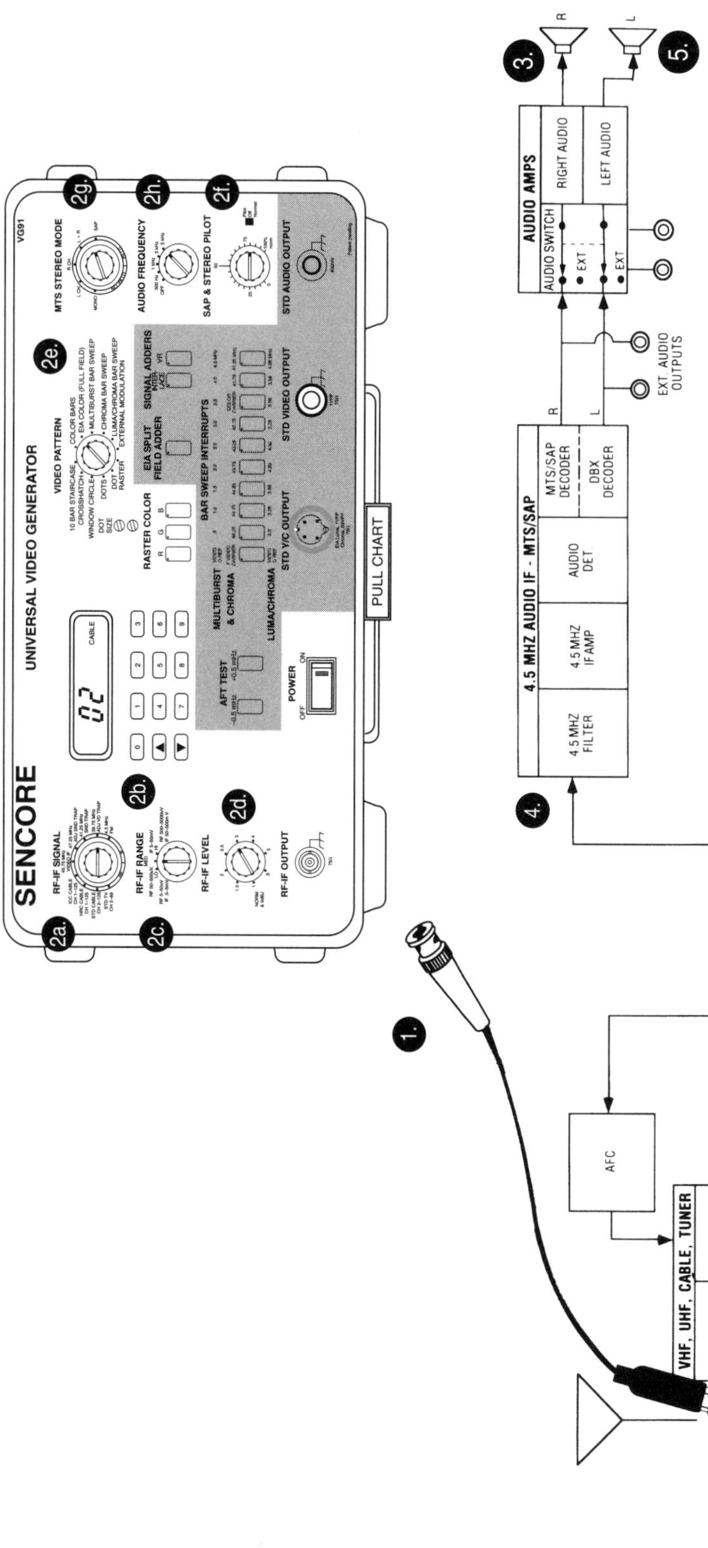

12-12 Basic setup for audio performance tests. Sencore

carrier. The monaural audio can produce a maximum instantaneous carrier deviation of 25 kHz and have a frequency response from 50 Hz to 15 kHz. The audio also uses a 75- µsec preemphasis.

With the advent of FM stereo radio and the increased popularity of musical programming on television, the need for a standard stereo signal format has developed. One requirement of the new system was compatibility of the stereo signal with existing mono systems. The FCC adopted a stereo TV signal format that's compatible with existing mono receivers and uses an advanced noise reduction system, developed by dBx, to solve some of the problems associated with stereo FM radio.

The MTS signal

The signal format adopted for stereo television includes more than the stereo audio signal. It includes the mono channel, a stereo subchannel, a second audio channel, and a professional channel. Thus, the correct term for this system is Multichannel Television Sound (MTS). Figure 12-13 shows the makeup of the composite audio signal. This signal consists of the following.

Mono channel The mono (L+R) channel is identical to the monaural signal used in the original NTSC format.

Stereo subchannel The stereo (L+R) subchannel (sometimes called the difference channel) is similar to the stereo (L–R) subchannel used in FM stereo. In the MTS system, however, the (L–R) signal is processed by a noise reduction system before it's broadcast. To obtain the most efficient transfer of information, this channel is broadcast as an amplitude-modulated, suppressed-carrier signal. The noise reduction system masks noise that's inherent in the FM signal.

Pilot signal The pilot signal at 15,734 Hz is used to tell the receiver that a stereo signal is present. It's also the reference for demodulating the (L–R) information.

Secondary audio program This channel allows other audio, such as a second language, to be broadcast along with the stereo signal.

Professional channel This channel is included to transfer telemetry, such as information from remote television crews, back to the main production studios. This channel is not used to transmit broadcast information.

Generating the MTS signal

In order to better understand how stereo TV works, let's look at how the entire MTS signal is generated. Then we can see how it converts back to the original audio information at the stereo TV receiver.

The main channel audio (L+R signal) is the sum of the left and right audio signals and occupies the same bandwidth as a monaural signal. All of the audio information is present in the main channel audio. This permits nonstereo receivers to obtain the full audio information. The main channel audio frequency modulates the aural carrier, and as in the original NTSC system, it also undergoes a 75-µsec preemphasis.

The stereo (L–R) subchannel, or the difference channel, is the difference between the left and the right audio information. The difference signal first goes

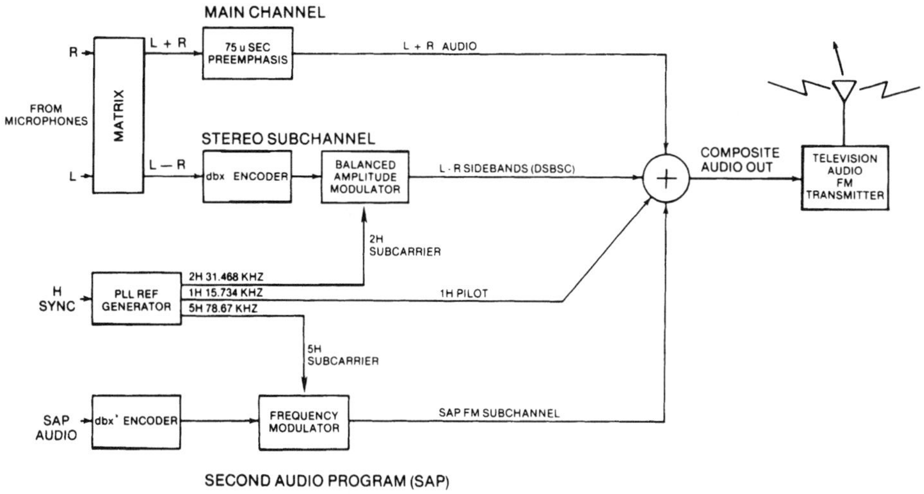

12-13 The makeup of the TMTS signal at the transmitter. Sencore

through a noise reduction circuit. It's then processed by a balanced modulator using a 2-H (31.468 kHz) subcarrier to produce a double-sideband suppressed-carrier (DSBSC) signal. Because the audio information is contained in the sidebands, it's not necessary to transmit the carrier. Eliminating the carrier allows all the transmitter power to be used to transmit the information. At the receiver, the carrier must be reinserted to demodulate the (L–R) information.

In order to properly demodulate the stereo (L–R) subchannel, a reference signal is needed that's phase locked to the original carrier. This reference signal is provided by the pilot signal. The pilot is a continuous sine-wave signal at a frequency of 1 H (15.734 kHz). The pilot does two things: it tells the receiver that a stereo signal is being broadcast, and it provides a phase reference to demodulate the stereo (L–R) subchannel. The 1-H pilot signal, the 2-H (L–R) carrier, and the 5-H (SAP) carrier are all phase locked together by a PLL reference generator at the transmitter.

The secondary audio program channel (SAP) carries another audio program such as a second language. The SAP audio is sent through a noise reduction system that's identical to the one used for the stereo (L–R) subchannel. The signal is then frequency modulated using a 5-H subcarrier.

Demodulating the composite audio signal

In the receiver, the modulated RF is converted to a 45.75-MHz IF signal by the tuner. Two different IF systems are used. In one system, the 45.75-MHz IF signal is sent through a single IF section. The aural signal is then converted into a 4.5-MHz signal. The other system is called split IF because a separate 45.75-MHz IF section is used for the aural signal, as shown in Fig. 12-14.

Once the IF signal has been converted to 4.5 MHz, it's amplified and demodulated like it is in a standard TV sound system. The output of the 4.5 MHz sound de-

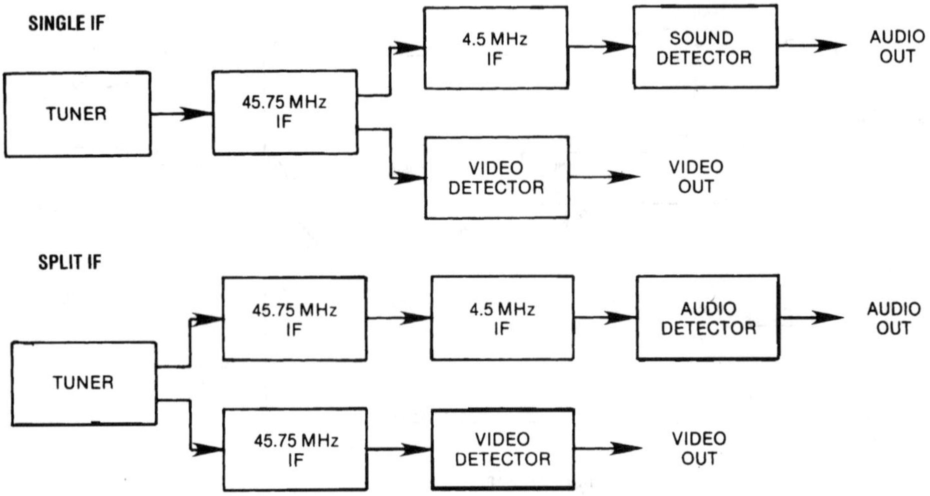

12-14 The IF section of a typical MTS receiver. Some MTS receivers use a "split IF," where the audio IF is passed by a separate 45.75-MHz IF. Sencore

modulator contains the audio baseband signal, also called the *composite audio signal*. This signal includes the mono (L+R) channel, the stereo (L–R) subchannel, the stereo pilot, and the SAP channel.

The composite audio signal is separated into separate signal paths for processing, as shown in Fig. 12-15. One signal path is through a 1-H low-pass filter. This filter passes only the 50-Hz to 15-kHz signal of the mono (L+R) channel. The mono signal is deemphasized and phase compensated before going on to the matrix. The deemphasis compensates for the preemphasis done at the transmitter. The phase compensation corrects the phase delay produced when the stereo (L–R) subchannel passes through the noise reduction circuits.

The composite audio signal also passes through a 2-H bandpass filter. This filter passes the information contained in the L–R subchannel. The L–R subchannel signal is then sent to a balanced demodulator. An oscillator, phase locked to the stereo pilot, demodulates the L–R information. After demodulation, the L–R signal is sent through a noise reduction circuit for processing and then to the matrix, where it's combined with the L+R signal to obtain the original left channel and right channel audio.

A third path for the composite audio signal is through a 5-H bandpass filter. This filter passes the secondary audio program (SAP) information. The SAP signal is demodulated and sent on to the noise reduction circuits, stereo matrix, amplifiers, and speakers.

The TVA92 TV video analyzer

Today's television/video circuits include many advancements such as comb filters, multichannel TV sound (MTS), integrated flybacks, startup/shutdown circuits, LSI

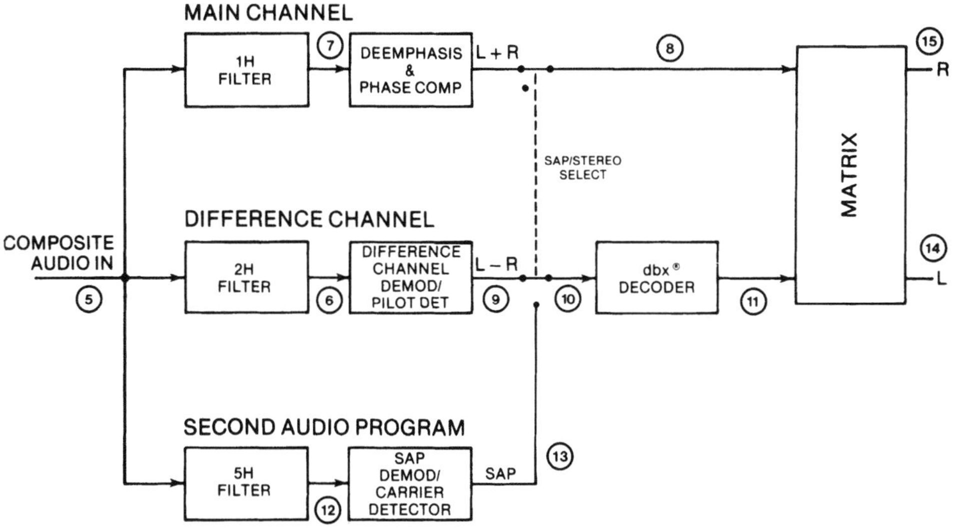

12-15 The MTS receiver processes the main channel, stereo channel, and SAP channel signals separately. Sencore

ICs, and switch-mode power supplies. These advancements provide customers with better products, but these products also present unique challenges to video service technicians. The TVA92 video analyzer can help to meet these challenges. (See Fig. 12-16.)

The TVA92 is designed to be used with the VG91 universal video generator. The VG91 serves as the heart of a multifaceted video analyzing system and provides the

Sencore

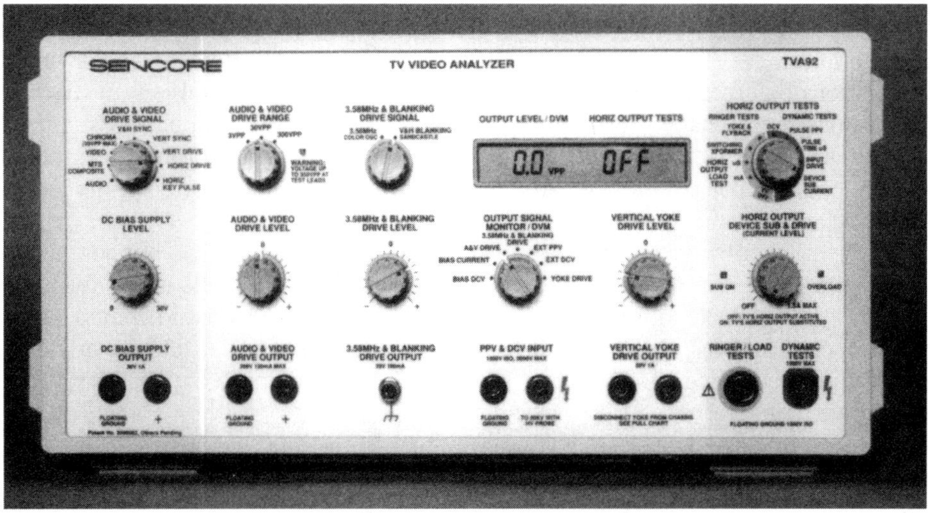

12-16 Photo of the TVA92 Video Analyzer.

signals needed by other analyzers. The TVA92 specially adapts this video analyzing system for isolating faulty stages in the sync, video, chroma, deflection, and audio sections of an NTSC television receiver or video monitor, from the detectors to the CRT or speaker. The TVA92's special substitution signals and analyzing tests greatly cut troubleshooting time and increase troubleshooting efficiency. Together, the VG91 and TVA92 fully equip the video service technician to troubleshoot and test the performance of modern television receivers and NTSC video monitors.

TVA92 features

The TVA92 consists of three general sections:

1. The audio, video, and chroma drive signals.
2. The signal monitor/dvm.
3. The horizontal output tests.

The drive signals provide a known-good signal that can be injected into any audio, video, or color stage after the video or audio detectors. The drive signals are low impedance and swamp out the existing signal so that you seldom need to disconnect components or disable the signals already in the circuit. The drive level is adjustable and can be monitored with an internal digital meter so you can match the substitute signal to the level required by the circuit.

A separate drive output provides the 3.58-MHz color oscillator and V&H blanking (sandcastle signal) drive signals so they can be used simultaneously with the video drives. This allows you to quickly isolate luminance, color, or blanking problems. The vertical yoke drive signal places the yoke under normal operating conditions. This provides fast, conclusive visual results on the raster of the vertical yoke's operation.

Special horizontal output tests dynamically analyze the horizontal output and deflection stages. These are provided to test and locate abnormal loading or timing conditions before power is applied to the chassis. The horizontal output dynamic tests analyze the chassis horizontal output transistor, B+, flyback pulse PPV and duration, and drive signal for a fast diagnosis of startup/shutdown problems, intermittents, or improper parameters. A patented ringer test finds shorted turns among flybacks, yokes, and switching transformers.

The horizontal output device sub and drive substitutes for the horizontal output device to power up the horizontal output and flyback circuits under full operating potentials and currents. The output current is monitored and controlled for safe operation. This unique TVA92 feature allows you to substitute for any horizontal output device and isolate problems before you damage expensive replacement components.

An autoranged digital meter allows you to monitor all of the drive signals to prevent overdriving. The meter also allows you to measure external dc and peak-to-peak voltages up to 2 kV. The front panel control locations for the TVA92 are shown in Fig. 12-17.

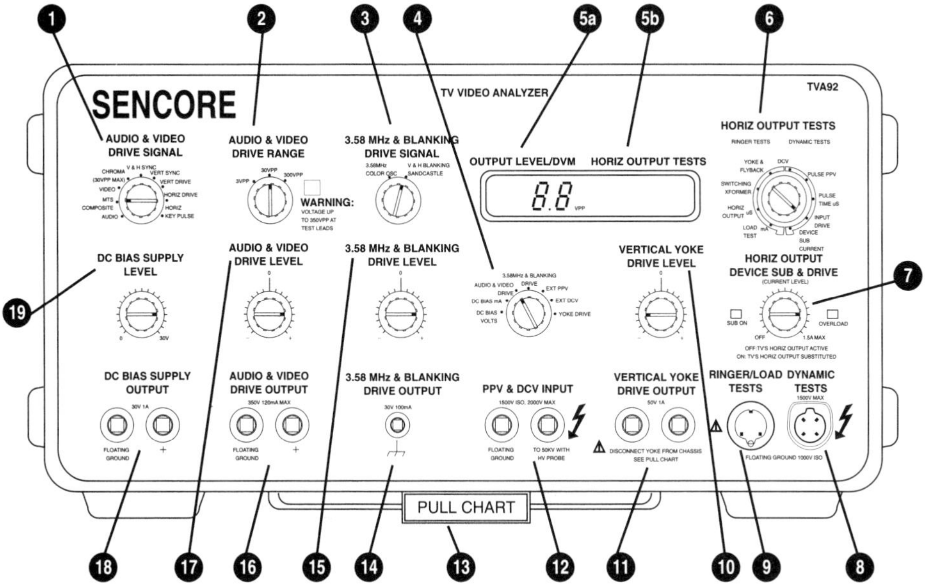

12-17 TVA92 front panel control location and features. Sencore

Operation of the VG91 with the VA62A Video Analyzer

The TVA92 TV Video Analyzer must be used with either the VG91 Universal Video Generator or with the VA62A Universal Video Analyzer. The TVA92 is not a stand-alone unit; it requires input signals and voltages through the synchronizing interconnect jack. The TVA92's drive signals and voltages are supplied by the VG91 or VA62A. This allows you to use the RF and IF signals from the VG91 or VA62A together with the drive signals from the TVA92. The TVA92 is turned on and off with the power switch on the VG91 or VA62A.

The operation of the TVA92 is identical using the VG91 and VA62A, with the exception that the VA62A doesn't provide the MTS composite audio signal. Follow the instructions in the service manual for substituting the VA62A for all references to the VG91.

Connecting the TVA92 to the VG91

The TVA92 is connected to the VG91 Universal Video Generator using the supplied 39G266 synchronizing interconnect cable. This cable connects between the TVA92's synchronizing signal input jack and the VG91's synchronizing signal output jack, or the accessory output jack on the VA62A.

The TVA92 provides all the signals that are necessary to completely troubleshoot all of the audio, video, and chroma stages between the audio and video detector outputs and the CRT or speaker. The signals closely match the waveshape and amplitude of the signals normally found in the circuit to return near-normal operation when injected into a working stage located after the defect. By injecting known-good signals, you're able to isolate defects to a single functional block.

Understanding the horizontal output device sub and drive

The TVA92's horizontal device sub and drive provides a substitute horizontal output transistor that enables you to power up and test the horizontal output, high-voltage, and scan-derived circuits at full operating potentials to confirm their operation before you invest time and money in a new replacement output transistor. Let's now look at how the horizontal output device sub and drive works and how to use it. (The control setup is shown in Fig. 12-18.)

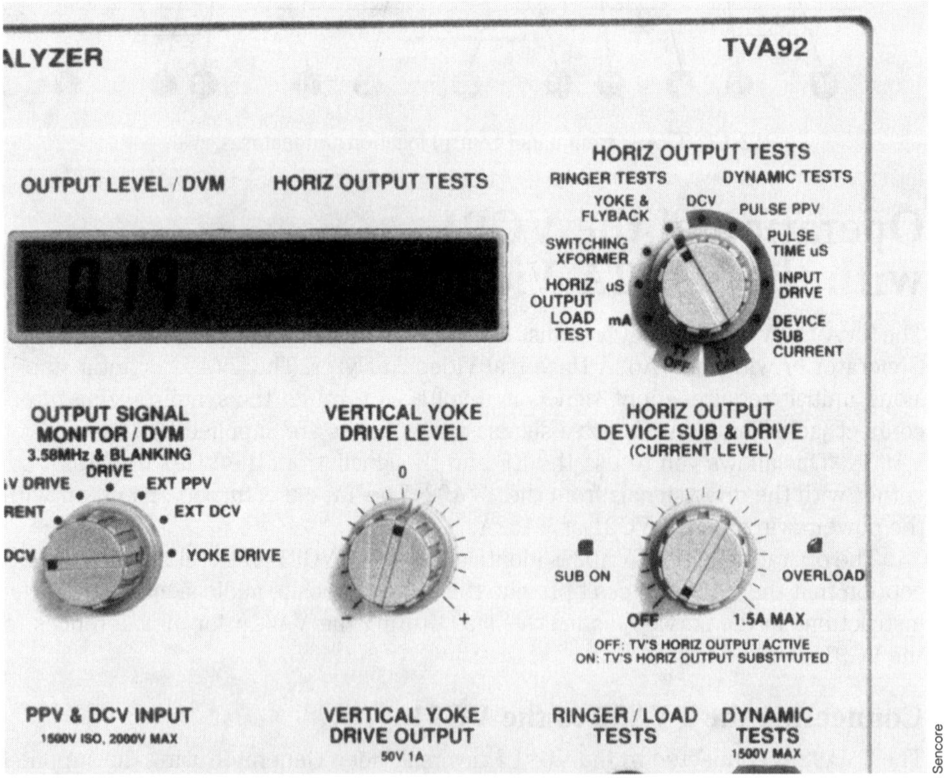

12-18 Settings for yoke and flyback checks.

Using the horizontal output sub and drive

Use the horizontal output device sub and drive when the chassis horizontal output transistor (HOT) is defective. You'll want to use the output sub before you order a replacement transistor or calculate a repair estimate because it allows you to test the entire output and high-voltage circuits at full operating potentials, as well as test the entire chassis by viewing the picture on the set's CRT screen.

The horizontal output sub and drive is very useful for isolating difficult horizontal problems such as insufficient drive current or low transistor gain (beta). These problems cause reduced high voltage and horizontal deflection, which are symptoms of a horizontal output stage problem.

Flyback pulses are used by the scan supplies to provide the dc voltage to power the horizontal oscillator and driver stages. Without flyback pulses, there will be no dc to run the oscillator, and without the oscillator there will be no flyback pulses. The horizontal output device sub and drive powers the output stage to troubleshoot these problems.

Finally, use the horizontal output device sub and drive to isolate problems that affect the frequency or waveshape of the drive signal. Drive problems alter the switching action of the HOT, so it can't match up the resonant timing of the output stage. This causes transistor failure, flyback squealing, deflection foldover, and other unusual symptoms.

Understanding horizontal output device operation

To better understand how the HOT (or similar output device) can be substituted with the TVA92, let's briefly review the function of the horizontal output transistor. (The operation of the HOT is illustrated in Fig. 12-19.)

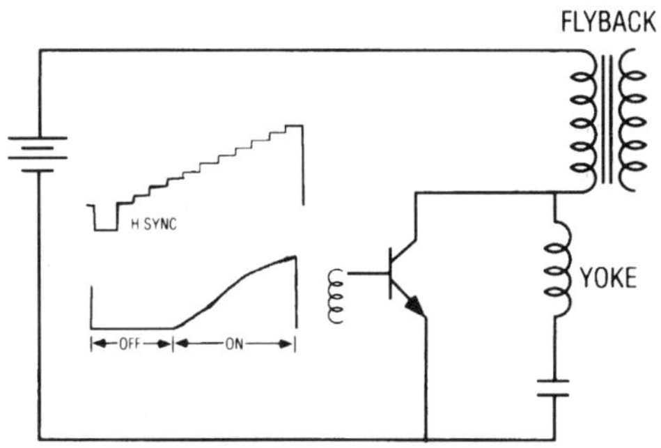

12-19 The HOT turns on 30–35 μsec before horizontal sync, providing a conduction path for flyback and yoke currents. Sencore

The HOT acts as an on/off switch for the current in the primary of the flyback transformer. The transistor is switched on and off at a 15,734-Hz rate by a current drive signal applied to its base. It's turned on approximately 30 to 35 μsec before horizontal sync.

When turned on, the transistor provides a path for current to flow from the B+ power supply into the flyback and yoke. Because they're inductive components, the current increases at a linear rate, but when the transistor is switched off, the current suddenly drops. The sudden change in current produces a large amplitude flyback pulse at the collector of the HOT.

The HOT remains off during the retrace period while the resonant action of the output circuit reverses the current flow. The HOT stays switched off for a brief time following retrace while the damper diode conducts.

The TVA92's horizontal output device sub and drive provides a substitute horizontal output transistor, as illustrated in Fig. 12-20. This internal subbing transistor works just like the HOT in the chassis as it switches on and off to complete the current path for flyback and yoke.

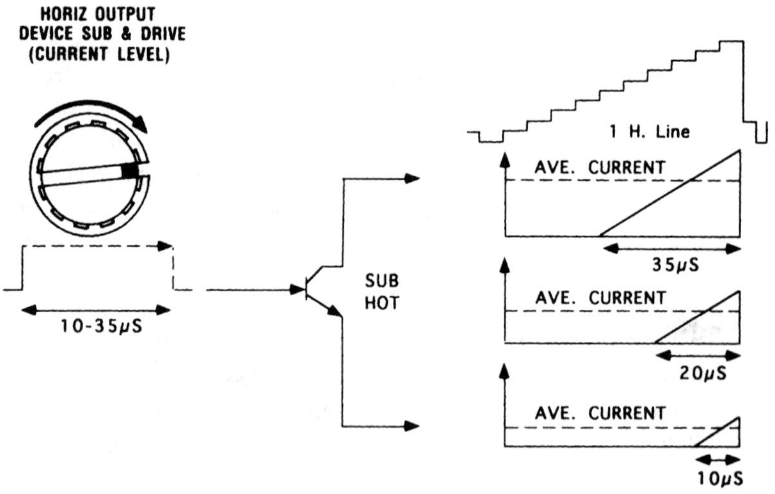

12-20 The TVA92 provides a substitute horizontal output transistor switched by a variable horizontal drive. Sencore

The sub transistor is switched by a drive signal generated inside the TVA92. This drive signal is controlled by the horizontal output sub and drive control. In the "off" detent position, the subbing transistor remains out of the circuit and the chassis is free to operate without influence from the TVA92.

Turning the horizontal output device sub and drive control clockwise connects the TVA92's subbing transistor to the chassis and increases the pulse width of the drive signal, as shown in Fig. 12-21. This makes the sub transistor conduct longer and increases the amount of current that flows through the subbing transistor and through the flyback primary circuit. This current is displayed in the horizontal output test's readout. By varying the conduction time, you can slowly increase the current level (power) to the output stage while monitoring for problems.

Note that the current flowing through the subbing transistor is a result of the chassis B+ supply, horizontal output stage, and flyback circuit. The TVA92 doesn't supply current to the chassis. Different chassis will produce different current levels at the same setting of the horizontal output device sub and drive control. The current reading is the average current in the circuit. The peak current in the circuit might be as high as 3 or 4 amps.

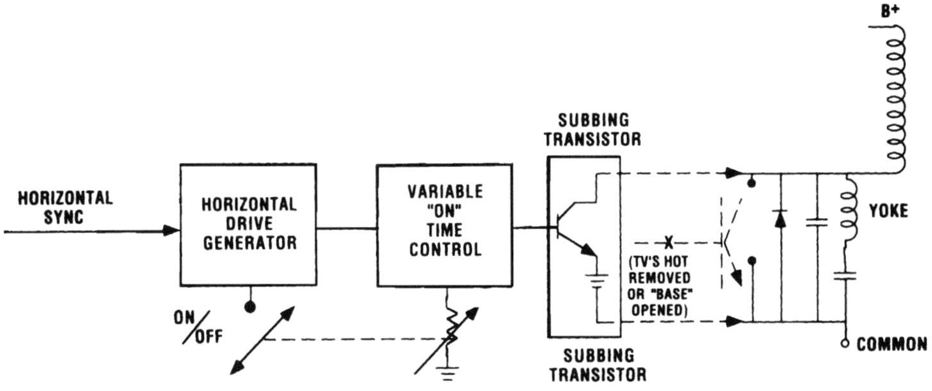

12-21 Increasing the conduction time of the subbing transistor allows higher current to pass through the subbing transistor. Sencore

The current allowed to pass through the subbing transistor is limited to 1.5 amps. This level allows you to adequately test most horizontal output stages but reduces the possibility of damaging either the chassis or the sub transistor.

The switching of the subbing transistor is synchronized to the video signal that's supplied to the chassis by the TVA92 and VG91. Thus, the chassis horizontal output stage is in sync with the video. A properly working chassis will produce near normal high voltage and the CRT will show near normal video, color and deflection when you're subbing with the horizontal output device sub and drive. Because deflection timing varies slightly between chassis, the video on the CRT might be shifted on some chassis.

You might use the "DCV," "PPV" and "Us" Dynamic Tests to monitor the horizontal output circuit while you're using the horizontal output device sub and drive. These measurements will closely agree with results when a replacement output transistor is installed in a working chassis.

Preparation for sub transistor

Improper loading, timing, or B+ voltage can produce excessive voltages or currents that might damage TV or TVA92 circuits when using the horizontal output device sub and drive. Before you substitute with the output device sub and drive, always do the following.

1. Disable the chassis horizontal output transistor.
2. Check for any abnormal loading or timing problems.
3. Test the B+ power supply voltage.

You can disable the chassis HOT by either removing it from the chassis or by opening the circuit path to the base. If the chassis HOT is shorted, it must be removed. A shorted transistor is indicated by a reading of approximately "250 mA" in the horizontal output load test, or by a B+ voltage of only a few volts. (The test setup for horizontal output transistor testing is shown in Fig. 12-22.)

Note: By switching the "base" test lead connection to ground when the device sub and drive is used, the TVA92 provides some protection against a chassis HOT that's not disabled. Don't rely on this to protect the chassis or TVA92 circuits. The transistor can't be fully disabled in this manner, and damage will occur if the test leads become detached while subbing. Always disable the chassis horizontal output transistor.

Check for abnormal loading or timing problems when using the horizontal load tests. If you detect any problems, correct them first. Also test the B+ power supply voltage with the "DCV" dynamic test and correct any regulation problems.

Connecting to the chassis

The horizontal output device sub and drive is connected to the chassis HOT using the dynamic tests lead. The connections for the device sub and drive are the same as for the other TVA92 dynamic tests.

BLUE/"B"	Base of output transistor (or gate of FET).
RED/"C"	Collector of output transistor (or drain of FET).
BLACK/"E"	Emitter of output transistor (or source of FET).

If the chassis HOT was removed, connect to the circuit board points or components, which are the same as the elements of the transistor.

Understanding the device sub current readings

The "device sub current" position of the horizontal output tests switch allows you to monitor the average current flowing through the subbing transistors. Remember that this is the average current flowing in the yoke and flyback primary. Component failures in the horizontal output stage or flyback secondary, such as a flyback shorted turn, leaky filter capacitor, or an excessive load on a flyback secondary, will cause an increased power demand on the output stage. These failures result in increased conduction current through the sub transistor. Use the current readout to detect unusually high current levels.

The normal sub current varies between chassis and depends on the CRT size, the number of scan-derived circuits, and the efficiency of the output stage. Typical current in a small black-and-white chassis ranges from 300 to 500 mA, while current in a full-size color chassis typically ranges from 500 to 1000 mA. Older full-size color TVs with many flyback scan-derived powered circuits might have conduction currents as high as 1.5 amps.

The normal current flowing through the HOT is a rising sawtooth. When the transistor is conducting, the current rises to a peak of several amps. However, current flows through the transistor for only approximately one-half of one horizontal line. Therefore, a current meter in series with the HOT will measure the average conduction current. This current is only a fraction of the peak current flowing through

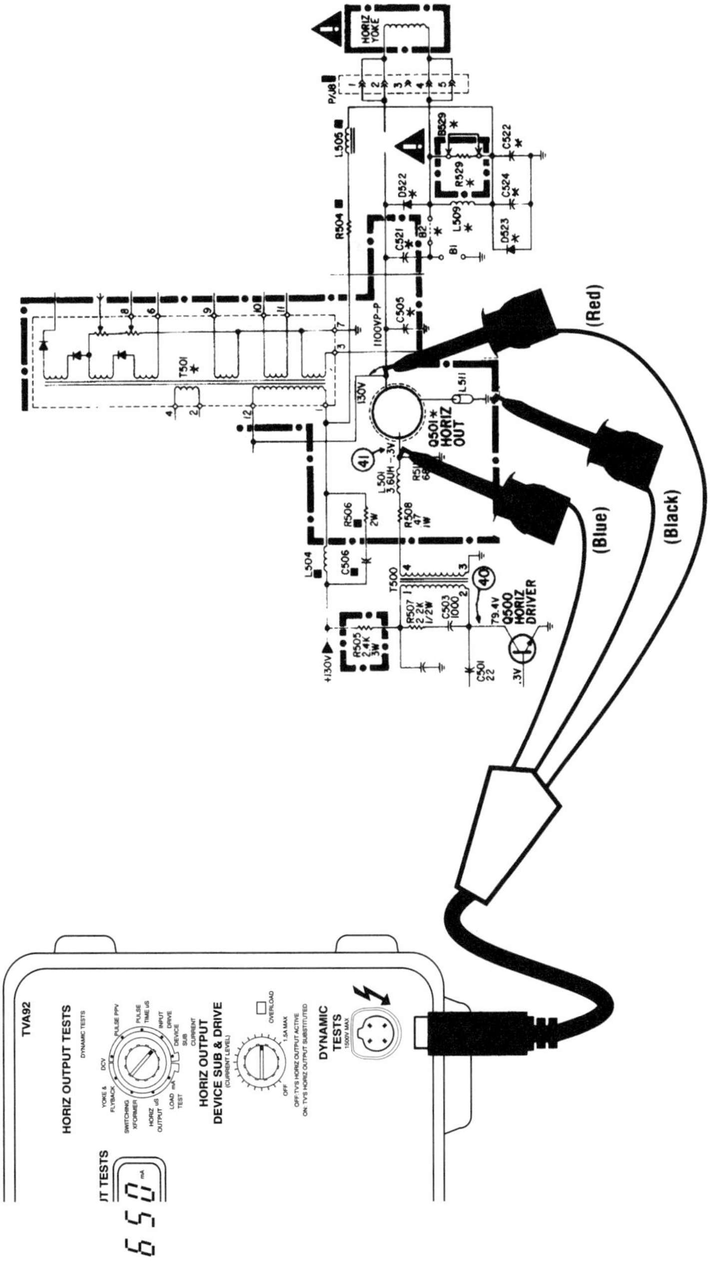

12-22 Connect the dynamic test leads to the chassis horizontal output transistor or equivalent circuit points. Sencore duction current. This current is only a fraction of the peak current flowing through the transistor.

The average conduction current of the subbing transistor is not the same as the current supplied to the output stage by the B+ supply. The B+ power supply only provides a fraction of the total current flowing through the output transistor's collector. This is because the yoke and flyback are inductive components that induce current back into the circuit as the current oscillates in the tuned output stage. The B+ power supply merely replenishes the circuit with additional current each time the transistor conducts.

Sub current comparison

The collector current measured by the TVA92 provides the most accurate indication of the current (power) in the output stage. It also provides the best means (better than emitter current) for monitoring for normal or abnormal conditions while subbing. Emitter current (as is typically indicated on schematics) will flow through the sub transistor whenever drive is applied, whether or not current is flowing through the output stage.

Many schematics show the normal emitter current through the horizontal output transistor. You can compare the TVA92's current readout to the service information to determine if the output stage is demanding extra current, or if it's operating normally. However, a small compensation is required to compare the two values.

To compare the TVA92's reading to a schematic, you need to subtract the base current from the emitter current that's shown on the schematic. For example, the schematic in Fig. 12-23 shows a current of 1.2 amps. This chassis will read about 1 amp when tested with the TVA92 horizontal drive test (1.2 amps to 200 mA). Note: If the chassis horizontal output transistor contains a damper diode, the TVA92's device sub current readout might show up to twice the emitter current value indicated by the schematic.

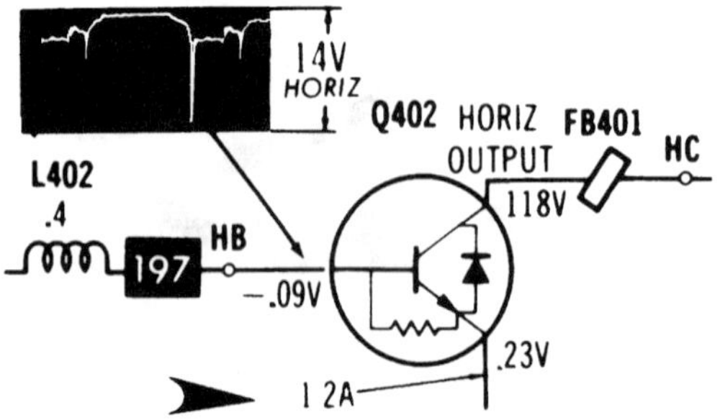

2-23 Some service schematics record the horizontal output transistor's emitter current level.
Sencore

The current measured by device sub current readout will vary with the operating conditions of the chassis. For example, the high-voltage section delivers current for the CRT guns, and the gun current varies with the dark and light picture objects. When comparing the TVA92's device sub current readout to the schematic, select the color bars video pattern and set the chassis brightness and contrast controls to midrange.

Setting the horizontal output device sub and drive level

Slowly increase the horizontal output device sub and drive control while monitoring the device sub current readout. Continue increasing the level until normal high voltage and deflection is reached. You can confirm proper operation by monitoring the CRT for proper raster and deflection. Monitor the center of the raster, and adjust the horizontal output device sub and drive control slightly beyond the level where normal video occurs.

Watch the current readout for unusual conditions. If you see the current climb over 1 amp, or if the overload light comes on as you increase the drive control, there's a severe problem in the output stage. Turn the horizontal sub and drive to "off" and use the horizontal output load test to isolate the problem. Note: Most CRT filaments are powered from the horizontal output stage. Allow a few seconds for the filaments to heat.

Most functioning output stages will operate normally with the horizontal output device sub and drive control set to about midrange. Of course, variations from this are normal, depending on the normal current demand of the output stage.

The sub current readout also indicates if the horizontal output stage has some resistive losses that are making it inefficient or are loading it down. The flyback and yoke are typically high Q components with low resistive losses, and the secondary circuits draw a normal level of power that's easily replenished by the horizontal output stage. Therefore, increasing the horizontal output device sub and drive control beyond the level needed for full deflection and high voltage shouldn't cause the sub current reading to significantly increase. However, if the horizontal output stage does have high resistive losses or abnormal secondary loading, the current will continue to increase as the horizontal output device and drive control is increased. This indicates a leaky output component or abnormal loading. Some possible causes include a leaky output transistor, retrace cap, damper diode, and scan-derived load or rectifier.

Using the horizontal output device sub and drive

To substitute for the horizontal out transistor:

1. Remove ac power from the chassis.
2. Use the horizontal output load test to identify abnormal loading or timing problems and correct them.
3. Disable the chassis horizontal output transistor by removing it or by disconnecting the base drive signal.
4. Connect the dynamic test leads to the dynamic tests jack and to the horizontal output transistor or equivalent circuit test points.
5. Set the horizontal output device sub and drive control to "off."
6. Apply ac power to the chassis.
7. Use the "DCV" dynamic test to confirm proper B+ power supply voltage and correct if defective.

8. Set the horizontal output tests switch to "device sub current."
9. Increase the horizontal output device sub and drive control setting.
   - Monitor the device sub current readout for excessive current. Stop subbing if the current exceeds 1.5 amp or if the overload light comes on.
   - Monitor the CRT and adjust the level to just above the point where foldover disappears.
10. Read the digital readout to measure the subbing-current level.

If normal high voltage is produced with a good video on the CRT, you can be sure that the horizontal output, high voltage, RF-IF, video, color, vertical, and CRT circuits are functioning properly. If the CRT doesn't produce a raster, measure the high voltage and use the dynamic tests to analyze the horizontal output stage. If the high voltage is normal, the chassis has problems in a stage other than the horizontal output or high voltage stages. You might replace the chassis HOT, or continue to use the horizontal output device sub and drive with the other TVA92 drive signals and tests to further isolate the problem.

The horizontal output device sub and drive doesn't confirm that the horizontal oscillator and driver stages are functioning properly. Use the input drive dynamic test to confirm base drive. If the input drive test confirms that a drive signal is present, all of the stages are functional. If the input drive test shows that the drive is missing, troubleshoot the horizontal oscillator, driver, or safety shutdown circuits.

Troubleshooting with the horizontal output load test

Problems that cause excessive loading on the B+ power supply or flyback, or that affect the crucial timing of the horizontal output stage, often strain or damage output components when the ac power is applied to the defective chassis. Timing problems can also cause the safety shutdown circuits to shut down the chassis. In either case, the "dead" horizontal output stage leaves the technician with no way to easily pinpoint the problem.

The TVA92's horizontal output load test is specifically designed to troubleshoot these difficult horizontal output stage loading and timing problems. This tip covers how to use the horizontal output load test to troubleshoot B+, horizontal output, and flyback secondary circuits with the chassis power turned off.

When to use the horizontal output load test

The most effective time to use the horizontal output load test is when you're troubleshooting start-up or shutdown symptoms. These symptoms can be caused by the switched-mode power supply (SMPS), B+ power supply, horizontal output circuit, or safety shutdown circuits. However, because the chassis is dead, the actual cause is difficult to determine.

The horizontal output load test allows you to quickly check the B+ load current and the resonant action horizontal output and flyback circuits without ac power applied to the circuit. If the horizontal output load test "mA" and "US" readouts indicate "good," you know that the B+ power supply is not severely loaded and the operation and resonant timing of the horizontal output circuits is okay. You can then use the dynamic tests to further isolate the problem with the chassis ac power applied.

If either of the horizontal output load test readouts indicate "bad," then you know there's a problem somewhere in the output stage. You can use the output load test to help further isolate the defect.

Understanding the mA readout

The horizontal output load test applies 15 volts to the chassis B+ test point while it simulates the normal switching action of the horizontal output stage. The mA readout indicates the total current that's being delivered by the 15-volt supply to the output stage.

The 15-volt supply is approximately 1⁄10 of the normal B+ voltage that's found in most full-size color television receivers. Therefore, the mA reading can be used to determine the approximate current load on the chassis B+ supply. The circuits operate as they would if the full B+ voltage were applied, but they draw approximately 1⁄10 of the current. To approximate what the actual load current would be at full voltage, simply multiply the mA reading by 10.

In most receivers, the actual B+ current load will be slightly higher than this, due to CRT beam current. A resistance check with an ohmmeter at the chassis B+ test point, on the other hand, will only show dc shorts or leakage resistance. Readings greater than 80 mA indicate an excessive current load on the chassis B+ power supply that's approaching 1 amp or more.

Note: Some older TV sets use a normal B+ supply voltage that's less than 100 volts. These receivers can show more than 80 mA when tested with the output load test. When testing or troubleshooting these receivers, use the dc bias supply to provide a voltage that's 1⁄10 of the chassis B+ supply.

Isolating dc loading problems

There are three common types of short and leakage problems that cause excessive B+ supply current.

1. A low resistance path for direct current (dc short to ground).
2. A dc leakage (higher resistance short).
3. An ac short or leakage path.

See the horizontal output load test "mA" reading as an indicator of the severity of the loading, and monitor the current as you troubleshoot the problem.

The B+ current supplied by the horizontal output load test to the chassis is limited to 250 mA. Therefore, a current reading of approximately 250 mA indicates a low-resistance dc short from the B+ power supply to ground. Before you assume a short, however, confirm that you've connected the test leads to the proper B+ and ground test points in the chassis.

The most likely cause of a dc short on the B+ supply is a shorted horizontal output transistor. Disconnect the "B+" (orange) ringer/load test lead from the chassis and remove the horizontal output transistor. Then reconnect the "B+" ringer/load test lead and note the mA reading. If the mA readout is 10 mA or less, you've confirmed that the horizontal output transistor is shorted. If opening the horizontal output transistor doesn't remove the short, continue opening the possible dc short-circuit paths shown in Fig. 12-24 until you've isolated the defective component.

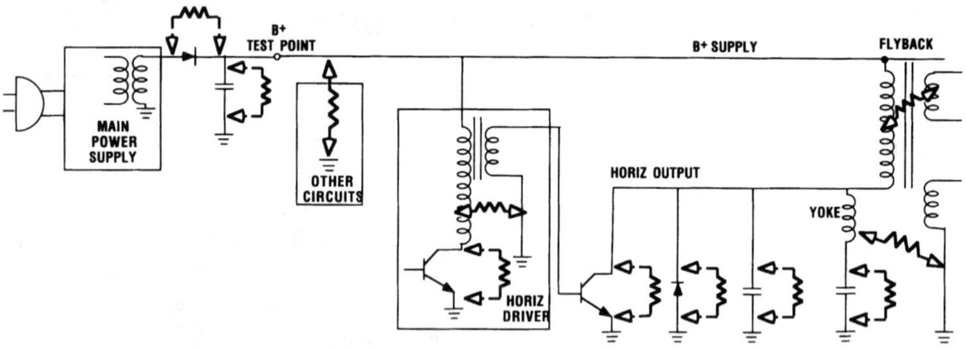

12-24 Possible dc short or leakage paths that can load down the B+ power supply. Sencore

Note: The horizontal output load test can be performed with the horizontal output transistor and damper in or out of the circuit. These components, if good, will not change the mA or μsec readings.

The loading problems, other than a low-resistance short, produce current readings ranging from 80 mA to 200 mA. The first step in isolating these loading problems is to determine if the added current is caused by a dc load or by an ac load. Do so by disconnecting the "collector" or yellow ringer/load test lead from the horizontal output stage, as shown in Fig. 12-25. Removing the collector lead stops the switching action of the output stage and removes all of the ac currents in the flyback and yoke. All power transfer to the flyback secondary circuits also stops.

Check the mA readout with the collector lead disconnected. The current now is only the dc current to circuits that are powered by the B+ supply. This includes the horizontal output stage and perhaps the horizontal driver stage and oscillator. These stages typically draw less than 10 mA during the output load test. If the current is much higher than this, suspect a dc short or leakage path on the B+ power supply. If the current drops to normal, the excessive load is caused by ac leakage.

There are many possible dc leakage paths, as shown in Fig. 12-24. A dc short or leakage path can be caused by a leaky filter capacitor or diode in the B+ supply itself, or by shorts or leakage paths in the horizontal output stage or other stages connected to the B+ power supply. To isolate a dc short or leakage path, open the B+ supply circuit path and hook the B+ (orange) test lead directly to the B+ side of the

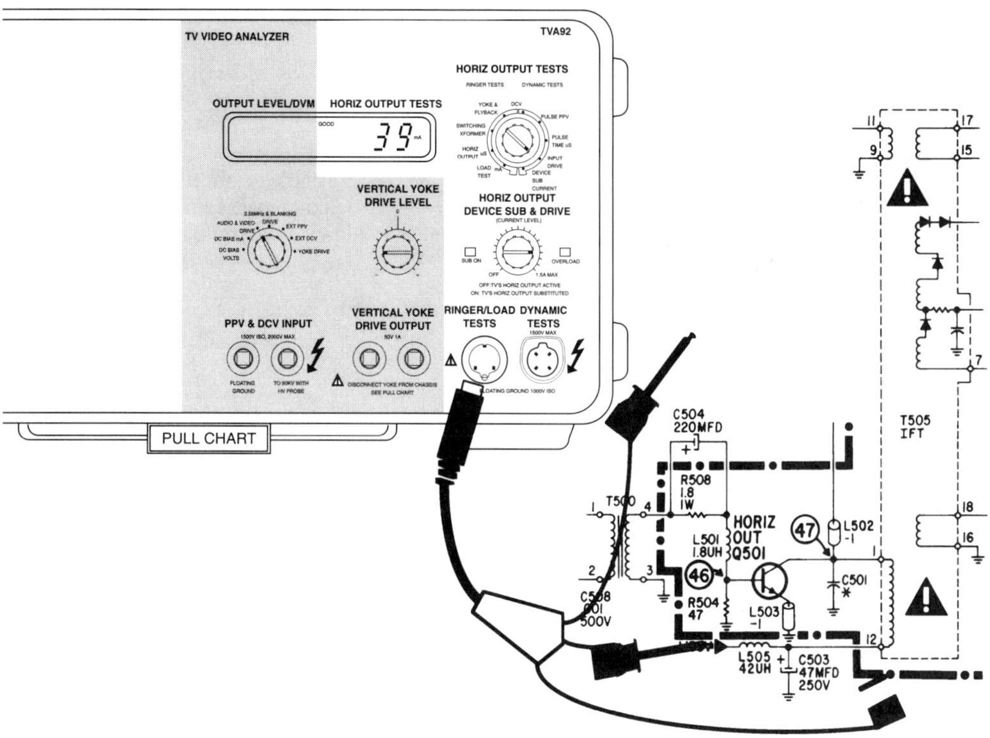

12-25 To determine if the loading is caused by dc or ac leakage, disconnect the "collector" (yellow) test lead. If the current drops below 10 mA, the loading is caused by ac leakage. Sencore

flyback transformer. Leave the "collector" (yellow) lead open. A current reading greater than 5 mA indicates a dc leakage or short in the horizontal output stage.

Note: Some televisions use a small dc current for horizontal yoke centering, which might be seen during this test. Current in excess of 5 mA indicates a dc short or leakage path in the horizontal output stage components.

Isolating ac loading problems

If the current reading is between 80 mA to 200 mA during the horizontal output load test, but drops to normal when the "collector" (yellow) lead is removed, then the excessive load is the result of a severe ac load in the output stage. The extra current demand might be caused by shorted turns in the flyback or yoke, by a short or leakage in any of the secondary circuits of the flyback, or by the horizontal output transistor or damper diode breaking down. Some possible ac leakage paths are shown in Fig. 12-26.

The best way to isolate ac loading problems is through a process of elimination: disconnect each possible leakage path while you monitor the current. Begin by removing the horizontal output transistor and damper diode. Repeat the horizontal

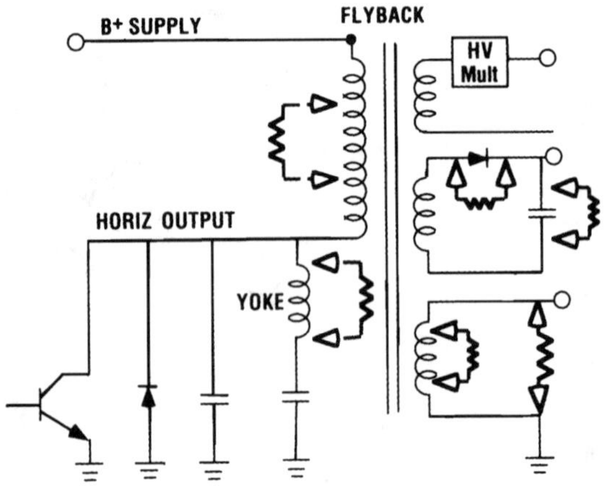

12-26 Possible ac shorts or leakage paths that can load down the B+ power supply. Sencore

output load test and compare the mA reading to the previous reading. If the mA reading decreases, the transistor or damper diode is leaky.

To isolate shorts on the flyback secondaries of the TVA92, use the EXT PPV and DCV input to measure the dc voltages and peak-to-peak flyback pulse amplitudes on the secondary windings of the flyback. Remember that the load test simulates the chassis' normal horizontal output circuit operation at approximately ⅒ of normal. Therefore, the voltages will be approximately ⅒ of the normal values that are shown on the schematic.

Ac or dc voltages that are considerably lower than ⅒ of normal, or that are completely missing, indicate a shorted component or scan-derived circuit associated with that flyback winding. If you suspect a short circuit, open the current path, then repeat the load test and compare the mA readout. If the current decreased substantially, the load on that flyback winding is excessive. Note: Use a voltage multiplier probe such as the Sencore TP212 when measuring the output of voltage multipliers or focus dividers.

If all of the secondary voltages appear normal, the load is likely caused by a shorted turn in the flyback or yoke, or by a component that's breaking down. Use the ringer test to check the flyback and yoke for a shorted turn and check other components for breakdown at their full-rated voltage.

Understanding the μsec readout

The horizontal output load test accurately simulates the normal switching action of the horizontal output stage. During the test, a pulse is produced at the collector of the horizontal output transistor. The shape and width of this flyback pulse reflects the chassis' normal resonant timing. Normal flyback pulse durations (time in μs) vary from about 11.3 μsec to as long as 16 μsec. Any stable microsecond reading in this range indicates normal output stage timing and is accompanied by a "good" indication. If the readout indicates "good," you know that the horizontal output stage is developing normal flyback pulses of the proper duration and that it's not the cause of a dead or shutdown symptom.

A "bad" readout, however, indicates that there's a timing or minor loading problem associated with the horizontal output circuits. You can use the horizontal output load test to further isolate the problem.

Isolating timing problems

There are three common types of problems in the horizontal output and flyback circuits that will cause the timing to be incorrect. Each of these problems produces a different μsec reading.

A reading of dashes during the horizontal output load test indicates that no flyback pulses are being produced. This is caused by an open in the B+ supply path flyback primary or ground path, or by a defective component in the output stage that results in no flyback primary current.

Double check to be sure that the ring/load test leads are connected to the proper circuit test points. If they're correct, this readout indicates an open circuit path. Use an ohmmeter to isolate the open circuit.

Stable pulse widths that are too long or short are caused by a value change in one of the crucial timing components in the horizontal output stage.

If the pulse width is longer than 16 μsec, check the yoke and yoke series capacitor values. If the time is shorter than 11.3 μsec, check the retrace timing capacitors and use the ringer test to check the flyback and yoke. In some cases, a short on the flyback secondary will effectively decrease the inductance value of the flyback transformer and cause the pulse width to fall below 11.3 μsec. Also check for abnormal loading of the flyback secondaries.

Pulse times that randomly change within or outside of the normal time range indicate a short or leakage in the horizontal output stage or flyback secondaries by higher than normal mA readings.

Pulse times that vary by several μsec or more indicate that the waveshape of the flyback pulse is abnormal. The waveshape might be multiple pulses or have abnormal ringing. This symptom is typically caused by loading problems in the flyback secondary or horizontal output stage. Test horizontal output components and flyback secondary circuits to isolate the problem.

Using the dc bias supply with the output loading test

The horizontal output load test energizes the horizontal output stages with a 15-volt B+ supply. This voltage is approximately ⅒ of the normal B+ voltage used in most modern, full-size color television receivers. Some older chassis, however, operate with a normal B+ voltage that's lower than 100 volts. Many of these chassis will indicate greater than 80 mA when tested with the load test, even though the horizontal output stage is working properly. Note the chassis test setup in Fig. 12- 27.

You can accurately test these chassis by using the TVA92's dc bias supply to provide a lower B+ test voltage. Disconnect the orange (B+/15V) lead ringer/load test lead from the chassis. Then use the TVA92's dc bias supply to apply a dc voltage that's ⅒ of the chassis normal B+ voltage. Monitor the dc bias supply current with the output signal monitor/dvm instead of with the horizontal load test readout. It

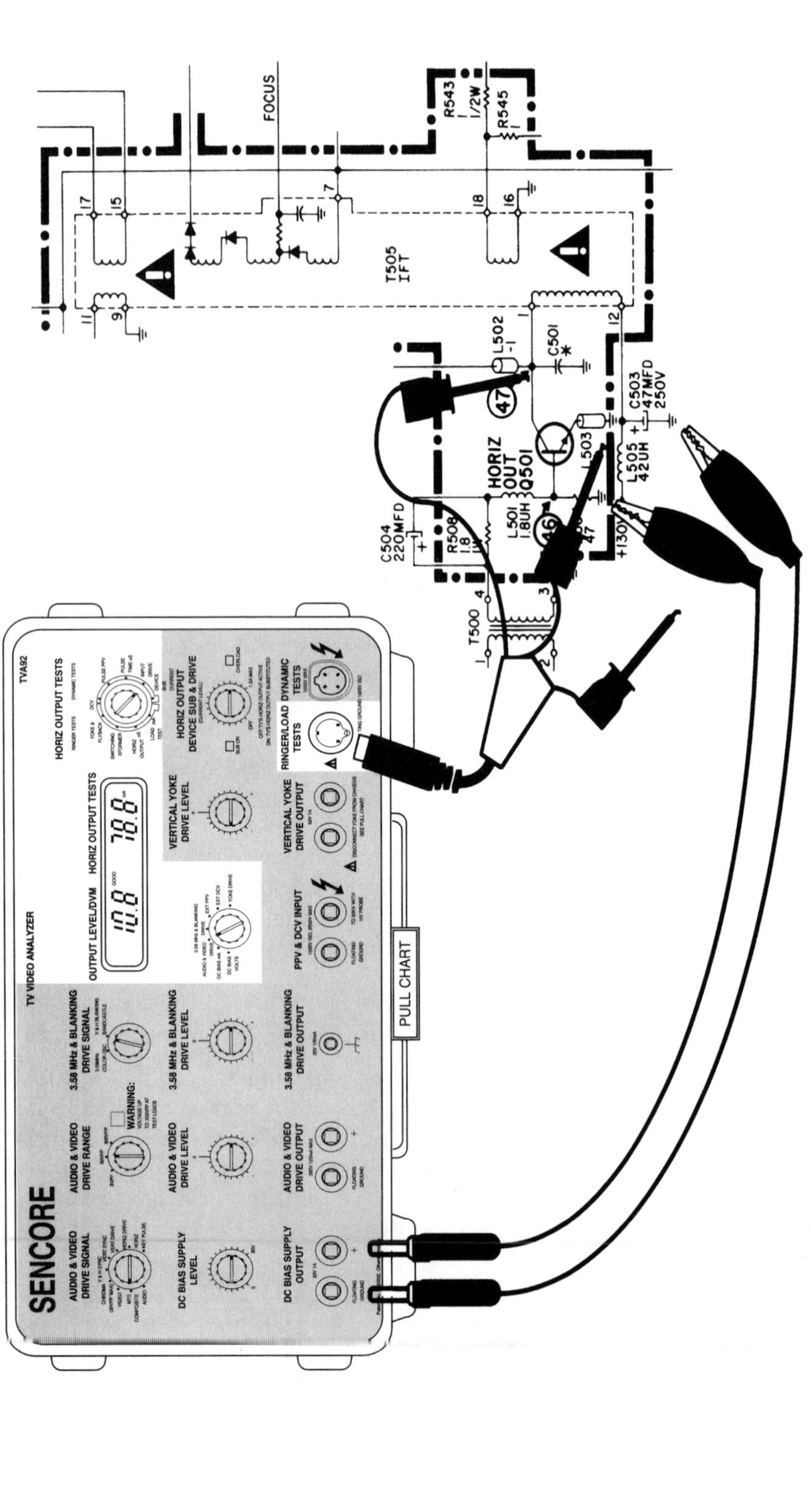

12-27 The dc bias supply can be used with the horizontal output load test to simulate a B+ voltage that is exactly 1/10 of the normal chassis B+ voltage. Sencore should be less than 80 mA. Note: You still use the horizontal load test readout to measure the μsec pulse time.

When testing other chassis, you might want to use the dc bias supply to supply a B+ test voltage that's exactly ⅒ of the normal chassis B+ supply voltage. This will make the voltages in the output and high-voltage circuits very close to ⅒ of normal, compared to the 15 volts supplied by the horizontal load test (unless the normal chassis B+ is 150 volts).

Index

A

ac, isolating loading problems, 359-360
AFT alignment, 33-35
  operations check, 35
alignment
  AFT, 33-35
  bandpass, 82-85
  bar sweep, 102-103
  BW TV, 111-112
  checking without a scope, 103-104
  chroma bar sweep, 72-73
  color circuits, 81-82
  general information, 115-116
  IF, 70-72, 112-114, 247
  Magnavox comb filter, 91-93
  RCA comb filter, 93-95
  SAW filter, 121-123
  spot technique, 115
  synchronous video detectors, 120-121
  trap, 72
  TV, 103
  video heads, 232
  video IF, 100
amplifiers
  bandpass, 73-75, 82-85
  head, 268-276, 288-290, 310
  video, 47-66
analyzers
  NT64 NTSC Pattern Generator, 1, 209-215
  TVA91 TV Video Analyzer, 344-363
  VA62A, 1-27, 152-153
  VC63 VCR Test Accessory, 1, 215-221
  VC93 VCR Analyzer, 278-322
  VG91 Universal Video Generator, 323-344
antennas, troubleshooting using video analyzer, 23-27
audio
  baseband, 311-312
  demodulating composite signal, 343-344
  IC sound systems, 339
  MTS generator, 330
  MTS theory, 340-343
  performance tests, 339-340
  problems in VCRs, 308-313
  stereo separation test, 340
  troubleshooting in VCRs, 338-339
audio detector, 19
automatic color killer (ACK), 77, 249
  adjusting, 85
automatic gain control (AGC), 143-146
automatic phase control (APC), 84

B

bandpass amplifiers, 73-75, 82-85
bandpass coils, setting, 109-111
bar sweep pattern, 100-103
  alignment, 102-103
  comparing to sweep curves, 112
baseband, 311-312
bus operation, computer-autmated, 330

C

capstan
  jitter test, 313, 317-318
  speed test, 313, 316-317
CATV, systems check, 36-37
channel frequency accuracy, 4
channels, 4
  cable frequency allocations, 127-129
  cable-shifted, 124-130
  mono, 342
  professional, 342
  PROG cable, 30-32
  secondary audio program, 342-343
  selecting, 12-14
  standard cable, 29
  standard TV, 28-29
  stereo sub-, 342
  temporary direction shift, 13
  UHF frequency allocations, 130
  VHF frequency allocations, 127-129
chopper circuits, operation, 196-198
chroma, 279 (*see also* color)
  automatic color killer analysis, 77

bandpass alignment, 82-85
bandpass problems, 73-75
circuit alignment/troubleshooting techniques, 67-99
color circuit adjustments, 79
color circuit analysis/alignment, 81-82
comb filters, 85-99
conversion problems, 302-307
demodulator checks, 77-79
detector, 19
down-converted, 298-300
processing circuits, 79-81
recording circuits, 250
substituting 3.58-MHz color oscillator, 302
substituting 30-Hz phase switching signal, 306-307
substituting key pulse signal, 302-306
test patterns, 67-70
troubleshooting, 75-77, 298
VCR playback circuits, 253-258
chroma bar sweep, 3, 8, 67
alignment/checks, 72-73
coils, bandpass, 109-111
circuit adjustments, 79
controller, 42-43
down-converted, 217
frequency response, 211-212
GE's control system, 39
killer, 77, 85, 249
NTSC patterns, 212-215
oscillators, 84-85, 302
record circuits, 250-253
VHS system, 236
color bar pattern, 3, 8
comb filters, 85-86
checking/adjusting, 257-258
Magnavox alignment, 91-93
operation, 86-99, 239-241
RCA alignment, 93-95
RCA operation, 95-97
RCA vs. Magnavox, 90-91
Zenith circuitry, 98
cross patterns, 3, 7
crosshatch, 3, 7-8, 330
crosstalk, 231, 299
color, 239

D

dc
isolating loading problems, 357-359
troubleshooting circuits, 52-54
demodulators, 19, 77-79
detectors
audio, 19
chroma, 19
synchronous video, 118-121, 123-124, 335-336
types in TV receivers, 19
video, 19
digital voltmeter (dvm), 223
dot(s) patterns, 3, 7, 328-329
dot size control, 7-8
double-sideband suppressed-carrier (DSBSC), 343
drive signals, 5-6
drum
jitter test, 313, 318
speed test, 313, 318
dual-trace scope, 222-223

E

Electronic Industries Association (EIA), 209
external modulation, 4

F

filters, SAW, 118-120
flyback pulse path, 163
FM luminance signal, 216-218
cable, 127-129
UHF, 130
VHF, 127-129
frequency counter, 223
frequency interleaving, 239
frequency response, 6
checking in video amplifiers, 51-52
color, 211-212
luminance, 337
stereo, 340
VCRs, 255-257
frequency shift, 30-32
cable, 124-130
direction change, 31
operational tips, 31-32
programmed shift, any channel, 31
programmed shift checks, 30
functional analyzing, 16-19, 278
divide-and-conquer technique, 19-23

G

gain, 106, 245-246
gate, 173, 179
General Electric
color control system, 39
vertical sweep operation, 155
generators (*see* IF generators; RF generators)

H

harmonic related carrier (HRC), 126
head amplifiers, 268-276
playback, 275-276
pre-, 288-290, 310
record, 270-275
headswitchers, 290-292, 310-311
high-voltage shutdown
checking, 206-208
operation, 204-205
horizontal keying pulse, 137-140
horizontal oscillators, 168-169
horizontal output transistor (HOT), 348-357
horizontal sweep systems, 168-208
chopper circuit operation, 196-198
drive signals, 169-172
high-voltage shutdown, 204-205
in-phase driver, 199-200
output SCR gate, 173
output transistor testing, 172-173
output tube grid, 173-174
overview, 192-195
pulse width modulator, 192, 200-202
RCA CTC-108, 207-208
safety tips, 203
SCR B+ regulator circuit tests, 177-179
SCR circuit troubleshooting, 186-191
SCR drive circuit operation, 179-181
SCR functions during, 181-185
split-flyback testing, 175-177
Zenith electronic power system, 204
Zenith system, 191-192
Zenith "Z" chassis, 205-206

I

IF
  circuit testing, 63-64
  response testing, 335
  traps, 16, 103, 106-107
  troubleshooting VCR stages of, 244-247
  video alignment, 100
IF alignment, 70-72, 247
  comments, 113-114
  response curve, 113-114
  tips, 112-113
IF generators, 5, 11, 328
  functions, 13-14
  programmable, 14, 108-109, 247
  trap-setting signals, 16
  troubleshooting, 14-16
in-phase system, driver, 199-200
incremental coherent carrier (ICC), 126
integrated vertical pulse, 137
interlace adder 4, 10, 139-140

J

jitter, 299
  capstan, 313
  drum, 313
  frequency, 267-268

K

keying pulse path, 163

L

line of resolution, 337
luminance, 86-89, 216-218, 234, 253, 279
  frequency response testing, 337
  troubleshooting, 280

M

Magnavox
  comb filter alignment, 91-93
  T-809 vertical module checks, 153-154
  vs. RCA comb filters, 90-91
magnetic tape, 224-231
MATV, systems check, 36-37
modulation
  external, 4
  sound, 5
  video, 4
modulators
  pulse width, 192, 200-202
  RF, 294-295
MTS
  audio generator, 330, 342-343
  signal, 342
  theory, 340-342
multiburst bar sweep pattern, 4, 9-10, 68-69, 102
  testing IF circuits, 63-64

N

NT64 NTSC Pattern Generator, 1, 209-215
  color frequency response, 211-212
  color patterns, 212-215
  vectorscope and, 210-211

O

oscillators
  color, 84-85, 302
  horizontal, 168-169

P

phase error, 238
phase jitter, 238
power supplies, switched-mode, 356-357
pulse width modulator (PWM), 192, 200-202
  service caution statement, 202
pulses
  flyback, 163
  horizontal keying, 137-140
  integrated vertical, 137
  keying, 163, 302-306

R

raster pattern, 328
RCA
  comb filter alignment, 93-95
  comb filter operation, 95-97
  CTC-108 shutdown circuit checks, 207-208
  troubleshooting sandcastle circuits, 159
  vs. Magnavox comb filters, 90-91
retrace SCR, 181-185
RF attenuators, 12
RF generators, 11, 326
RF modulators, 294-295
RF-IF attenuators, 32
RF-IF generators, 11, 28
rotary transformers, 283-285, 310

S

safety, TV receivers and, 203
sandcastle circuits, 163-167
  detailed troubleshooting, 163-165
  flyback pulse path, 163
  generator signals, 160-161
  isolating picture problems, 161
  isolating problems, 165
  keying pulse path, 163, 302-306
  locating retrace line problems, 163
  no picture problems, 159-160, 165-167
  signals, 163
  troubleshooting RCA, 159
SAW filters, 118-120
  alignment, 121-123
SCR
  B+ regulator, 177-179
  drive circuits, operation, 179-181
  functions during sweep operation, 181-185
  gate, 173, 179
  retrace, 181-185
  trace, 181-185
  troubleshooting, 186-191
secondary audio program (SAP), 342-343
servo error, 238
servomechanism, 237
servos, 268, 280, 312-315
  locked test, 313-315
  troubleshooting, 258-260, 313
  VCR, 258
signals
  demodulating composite audio, 343-344
  double-sideband suppressed-carrier, 343
  drive, 5-6
  FM luminance, 216-218
  frequency relationship, 59-62
  generating MTS, 342-343
  generator, 160-161
  horizontal sweep drive, 169-172
  IF trap-setting, 16
  key pulse, 168, 302-306
  MTS, 342
  phase switching, 306-307
  play mode circuit, 266
  record mode circuit, 265-266
  sandcastle circuit, 163
  substituting, 19-23
  tracing, 16
  VCR, 215-216
  VCR crosstalk, 201
  VCR test, 264
  vertical interval reference, 28-46

vertical interval test, 38
vertical sweep drive test, 151
video IF injection, 104-105
VIRS, 28-46
VIRS test, 37-40
silicon-controlled rectifier (*see* SCR)
sound (*see* audio)
sound IF generators, 5
sound modulation, 5
split-flyback testing, 175-177
staircase pattern, 4, 10, 48, 68-70
stereo frequency response, 340
surface acoustic wave filters (*see* SAW filters)
sweep systems
horizontal, 168-208
vertical, 147-163
switched-mode power supply (SMPS), 356-357
synchronous circuits
analyzing defects, 137
applications, 136
closed-loop system testing, 142-143
horizontal keying pulse, 137-140
integrated vertical pulse, 137
testing, 131-135
vertical countdown, 140-142
synchronous video detectors, 118-120, 335-336
aligning, 120-121
analysis, 123-124

T

10-bar staircase pattern, 4, 10, 48, 68
timing, 361-363
tint controller, 43
trace SCR, 181-185
transformers, rotary, 283-285, 310
transistor-diode tester, 223
transistors, horizontal output, 348-357
traps
3.58-MHz video amplifier, 52
alignment, 72
IF, 16, 103, 106-107
notes regarding, 107-109
setting, 106-107
setting with programmable generator, 108
troubleshooting
audio, 338-339
baseband audio, 311-312
chroma, 298, 302-307
confirming TV symptoms, 23
divide-and-conquer technique, 19-23
horizontal sweep systems, 168-208
IF generator and, 14-16
luminance, 280
no picture problems, 159-160
sandcastle circuits, 163-167
SCR circuits, 186-191
servos, 258-260, 313
tuners, 28-46
TV sync/AGC systems, 131-146
VCR audio problems, 308-313
VCR IF stages, 244-247
VCR record circuits, 247-250
VCR VHS-C format, 297
vertical sweep systems, 147-163
video amplifiers, 47-66
video IF/video detector systems, 100-130
Zenith electronic sweep system, 191-192
tuners, 334-335
AFT alignment, 33-35
channels, 28-32
checking overall operation, 32
color controller, 42-43
MATV/CATV systems check, 36-37
servicing TV/VCR VIRS signals, 28-46
substituting, 35-36, 243
tint controller, 43
VCR, 242-243
VHF electronic, 43-45
TVA92 TV Video Analyzer, 344-363
μsec readout, 360-361
connecting to VG91, 347-348
features, 346
horizontal output transistor, 348-357
isolating ac problems, 359-360
isolating dc loading problems, 357-359
isolating timing problems, 361-363
mA readout, 357
sub current comparison, 354-355
sub current readings, 352-354
TVs
aligning BW, 111-112
alignment, 103
confirming symptoms, 23
detectors found in, 19
isolating picture problems, 161
no picture problems, 159-160, 165-167
poor video problems, 59
safety tips when repairing, 203
servicing, 1-207
troubleshooting synchronous/AGC systems, 131-146
tuner servicing VIRS signals, 28-46
video processing BW, 56-59

U

UHF, frequency allocations, 130

V

VA62A video analyzer, 1-27
AFT alignment, 33-35
antenna problems, 23-27
chroma test patterns, 67-70
confirming TV symptoms, 23
drive signals, 5-6
features, 5
functional analyzing, 16-19
IF generator, 5
isolated outputs, 6
operation, 6-7
overview, 1-3
parts, 1
ringing test, 5
signal substitution, 19-23
sound IF generators, 5
video patterns, 3-7
VC63 VCR Test Accessory, 1, 215-221, 260-263
down-converted color, 217
FM luminance signal, 216-218
operation, 260-263
overview, 260
where to use, 218-221
VC93 VCR Analyzer, 278-322
VCRs
analyzing color record circuits, 250-253
audio problems, 308-313
baseband audio, 311-312
capstan jitter test, 313, 317-318
capstan speed test, 313, 316-317
caution statement, 279
chroma, 279, 298-307
chroma playback circuits, 253-258
chroma recording circuits, 250
coax switch/splitter checks, 243-244

VCRs (*cont.*)
- color crosstalk elimination, 239
- color-under system, 238-239
- comb filters, 239-241, 257-258
- control system, 237
- crosstalk, 299
- drum jitter test, 313, 318
- drum speed test, 313, 318
- FM luminance signal, 216-218
- format characteristics, 322
- four-head, 226, 285-286
- frequency response, 255-257
- guard band elimination, 231-232
- head alignment, 232
- head drum assembly, 228
- head preamplifiers, 288-290, 310
- heads, 268-270, 280-283, 300-301, 309-311
- headswitchers, 290-292, 310-311
- helical scan system, 227
- hi-8, 296
- IF stages, 244-247
- jitter-frequency component, 267-268
- luminance, 279
- luminance playback circuits, 253
- magnetic recording fundamentals, 224-231
- overview, 279-280
- play mode signal circuit, 266
- playback circuits, 261-262
- playback head amplifiers, 275-276
- playback head sub, 262
- playback system, 233-235
- record circuits, 247-250
- record head amplifiers, 270-275
- record mode signal circuit, 265-266
- recording circuits, 262-263
- recording system, 233-234
- RF modulator, 294-295
- rotary transformers, 283-285, 310
- servicing, 222-277
- servos, 258, 268, 280, 312-315
- servos locked test, 313-315
- signal crosstalk, 231
- signals, 215-216
- special-effects heads, 287
- speed sensor system, 231
- super-beta, 296
- super-VHS, 295
- system control, 280
- test equipment/tools, 333-334
- test signals, 264
- test tapes, 241-244
- tuner servicing VIRS signals, 28-46
- tuners, 242-243
- VCR/TV switch, 243-244
- VHS color system, 236
- VHS-C format, 297
- video head motor, 231
- video stages, 293-294
- work tapes, 319-321

vectorscope, 210

vertical countdown circuitry, 140-142

vertical interval reference signal (*see* VIRS)

vertical interval test signal (*see* VITS)

vertical sweep systems, 147-163
- analysis, 150-151
- circuits, 155-159
- drive test signal, 151
- GE operation, 155
- module checks/Magnavox T-809, 153-154
- testing stages, 151-152
- VA62A control settings, 152-153
- Zenith, 147-150

VG91 Universal Video Generator, 323-344
- accessories, 326
- applications, 332
- audio systems, 338-340
- commands, 332
- computer-automated bus operation, 330
- controls, 325-326
- crosshatch, 330
- demodulating composite audio signal, 343-344
- error codes, 332
- features, 324-325
- IF generator, 328
- IF response testing, 335
- listener mode, 330-332
- luminance frequency response testing, 337
- MTS audio generator, 330
- MTS generator, 340-343
- operation, 326-328
- operation with VA62A Video Analyzer, 347-348
- overview, 323-324
- performance testing, 333-335
- RF generator, 326
- synchronous detector testing, 335-336
- talker mode, 332
- tuner systems, 334-335
- TV/video functional blocks, 332
- video patterns, 328-329
- video stages, 336-337
- window circle pattern, 329-330

VHF
- electronic tuner, 43-45
- frequency allocations, 127-129

video amplifiers, 47-66
- 3.58-MHz trap, 52
- bar/staircase patterns, 48
- circuitry, 54-56
- frequency response check, 51-52
- isolating problems, 66
- locating faulty video stages, 48-51
- overview, 47-48
- picture patterns, 64-65
- signal frequency relationship, 59-62
- testing IF circuits, 63-64
- testing response by looking at CRT, 65
- troubleshooting dc circuits, 52-54

video analyzers (*see* analyzers)

video circuitry, 54-56

video detectors, 19, 120-121, 123-124, 335-336

video IF, 100-130
- alignment, 100
- signal injection, 104-105
- troubleshooting, 116-118

video modulation, 4

video patterns, 3-7

video processing, BW TV, 56-59

video tape recorders (*see* VCRs)

VIR adder switch, 4, 10

VIRS, 28-46
- GE's color control system, 39
- sensor, 40-43
- system analysis, 39-40
- test signal, 37-40

VITS, 38

W

window circle pattern, 329-330

Z

Zenith
- comb filter circuitry, 98
- electronic power system, 204
- electronic sweep system, 191-192